# Kinematical Theory of Spinning Particles

## Fundamental Theories of Physics

*An International Book Series on The Fundamental Theories of Physics: Their Clarification, Development and Application*



Volume 116

# Kinematical Theory of Spinning Particles

## Classical and Quantum Mechanical Formalism of Elementary Particles

*by*

Martin Rivas

*The University of the Basque Country,*
*Bilbao, Spain*

KLUWER ACADEMIC PUBLISHERS

DORDRECHT / BOSTON / LONDON

A C.I.P. Catalogue record for this book is available from the Library of Congress.

ISBN 0-7923-6824-X

---

Published by Kluwer Academic Publishers,
P.O. Box 17, 3300 AA Dordrecht, The Netherlands.

Sold and distributed in North, Central and South America
by Kluwer Academic Publishers,
101 Philip Drive, Norwell, MA 02061, U.S.A.

In all other countries, sold and distributed
by Kluwer Academic Publishers,
P.O. Box 322, 3300 AH Dordrecht, The Netherlands.

*Printed on acid-free paper*

**To Merche**

# Contents

# List of Figures

*If a spinning particle*
*is not quite a point*
*particle, nor a solid*
*three dimensional top,*
*what can it be?*
**A. O. Barut**

*Eppure si muove ...*
**(the electron)**

# ACKNOWLEDGMENTS

If in any endeavour one has such lovely and unconditional supporters as my wife Merche and daughters Miren and Edurne, the goal will be achieved. Thanks to them for inspiring love and courage.

The analysis of dynamical systems, when an analytical solution does not exist, is in general a difficult task. In those cases, we always have the possibility of making an alternative numerical interpretation. But numerical analysis requires quite often a personal skill in computing or the collaboration of a computer department. The numerical computer program *Dynamics Solver*, developed by Juan M. Aguirregabiria has proven to be a handy tool, even for an inexperienced theoretical physicist. Extensive use of this package has been done for the analysis of many of the examples contained in this book. I enthusiastically recommend its use, even for teaching purposes. I am very much indebted to him by his help and for the many years of sharing life and duties.

Thanks to Aníbal Hernández for pointing out, even in these difficult times, several references related to the zitterbewegung.

I would like to mention the special collaboration of Sanjay Jhingan for a critical reading of several chapters of the manuscript.

Professor Asim O. Barut's quotation in the previous page is taken from a document kindly supplied by professor Erasmo Recami.

Thanks to the editorial board of Kluwer Academic, in particular to professor Alwyn Van der Merwe for his stimulating suggestions.

I acknowledge a grant from the Spanish Ministerio de Educación y Cultura during the final steps of the completion of the manuscript.

# Introduction

After the success of Dirac's equation the electron spin was considered for years as a strict relativistic and quantum mechanical property, without a classical counterpart. This is what one basically reads in some excellent textbooks, mainly written by the mid $20^{\text{th}}$ century. The recently re-edited book *'The Story of Spin'* by Tomonaga, who reviews the main discoveries during a period of 40 years, is not an exception. But, nevertheless, one often reads in other textbooks and research works statements which mention that the spin is neither relativistic nor a quantum mechanical property of the electron, and that a classical interpretation is also possible, giving some answers to this subject. The literature about it, not so extensive as the quantum one, is important.

One of the challenges while writing this book was to give an answer to Professor Barut's quotation, in the preliminary pages. Is it possible to give a comprehensive description, let us say at an undergraduate level, of an elementary particle in the form of what one usually thinks should be the description of a rotating small object? The partial qualitative answer is contained in the other preceding quotation, that, according to some not well confirmed Legend, was pronounced *sotto voce* by Galileo Galilei at the end of his Inquisition trial. In that case this statement was related to Earth's motion. But the electron also moves and the classical description of this motion is what we have been looking for.

The present book is an attempt to produce a classical, and also a quantum mechanical, description of spin and the related properties inherent to it such as the so called *zitterbewegung* and the associated intrinsic dipole structure of elementary particles. The relationship between the classical and the quantum mechanical description of spin will show how both formalisms are able to complement each other, thus producing, for instance, a kinematical explanation of the gyromagnetic ratio of leptons and charged $W^{\pm}$ bosons.

But at the same time, this book also presents a new formalism, based upon group theory, to describe elementary particles from a classical and quantum point of view. In this way the structure of an elementary particle is basically related to the kinematical group of space-time transformations that implements the Special Relativity Principle. It is within the kinematical group where we have to look for the independent variables to describe an elementary object. This relationship has been sought for years and has produced a lot of literature, but is so intimate that it has not been unveiled yet, unfortunately.

The book is organized as follows. The first chapter contains the basic definitions and general Lagrangian formalism for dealing with classical elementary particles, and also a precise mathematical definition of elementary particle is given. Some group theoretical background is necessary to properly understand the intimate connection between the concept of elementarity and the kinematical group of symmetries. This is why we have entitled the book as a *kinematical theory* of elementary particles. Nevertheless, those readers with low group theoretical baggage, will be acquainted with it after the analysis of the first models. A careful reading of the last section of this chapter shows how the formalism works with the simplest kinematical groups, preparing the ground for further theoretical analysis for larger symmetry groups. This is what is done in Chapters 2 and 3, which are devoted to the analysis of non-relativistic and relativistic classical particles. There we consider as the basic symmetry group, the Galilei and Poincaré group, respectively. Different models are explored thus showing how the spin arises when compared with the point particle case.

Chapter 4 takes the challenge of quantizing the previously developed models. This task is accomplished by means of Feynman's path integral approach, because the classical formalism has a well-defined Lagrangian, written in terms of the end-point variables of the variational approach. It is shown how the usual one-particle wave equations can be obtained, by using the methods of standard quantum mechanics, after the choice of the appropriate complete commuting set of operators. These operators are obtained from the algebra produced by the set of generators of the kinematical group in the corresponding irreducible representation. It is fairly simple to get Dirac's equation, once we identify the different classical observables with their quantum equivalent. One of the salient features is the determination of the kind of classical systems that, according to the classical spin structure, quantize either with integer or half integer values.

I have included in Chapter 5 several alternative models of classical spinning particles, taken from the literature, by several authors. It is a collection of models, which in general cannot be found in textbooks, although some of them were published as research articles even before Dirac's theory of the electron. This review is far from being complete but it comprehends most of the models quoted in the literature of this subject. They are discussed in connection with those models obtained from our formalism to show the scope of the different approaches. My apologies for not being as exhaustive as desirable.

The last chapter is devoted to some features, either classical or quantum mechanical, that can be explained because of the spin structure of the particles. It is only a sample to show how, by applying the formalism to some particular problems in which spin plays a role, we can obtain an alternative interpretation that gives a new perspective to old matters.

The subject of the book is already at a seminal level and now needs a deeper improvement. For some readers the contents of the following pages will be considered as a pure academic exercise but, even in this case, it opens new fields of research. If after reading these chapters you have a new view and conceptual ideas concerning particle physics, I will take for granted the time and effort I enjoyed in producing this manuscript. Your criticism will always be welcome.

Martín Rivas

Bilbao, June 2000

# Chapter 1

# GENERAL FORMALISM

This chapter is devoted to general considerations about kinematics and dynamics and the differences between Newtonian dynamics and variational formalism when the mechanical system is a spinning particle. These considerations lead us in a quite natural way to work in a Lagrangian formalism in which Lagrangians depend on higher order derivatives. The advantage is that we shall work in a classical formalism closer to the quantum one, as far as kinematics and dynamics are concerned. We shall develop the main items such as Euler-Lagrange equations, Noether's theorem and canonical formalism in explicit form. The concept of the action of Lie groups on manifolds will be introduced to be used for describing symmetry principles in subsequent chapters for both relativistic and nonrelativistic systems. In particular we shall express the variational problem not only in terms of the independent degrees of freedom, but also as a function of the end point variables of the corresponding action integral. We shall call these variables the kinematical variables of the system. The formalism in terms of kinematical variables proves to be the natural link between classical and quantum mechanics when considered under Feynman's quantization method. It is in terms of the kinematical variables that a group theoretical definition of a classical elementary particle will be stated.

## 1. INTRODUCTION

Historically quantum mechanics has been derived from classical mechanics either as a wave mechanics or by means of a canonical commutation relation formalism or using the recipes of the so-called 'correspondence principle'. Today we know that none of these formalisms are necessary to quantize a system.

Axiomatic quantum mechanics and the algebraic approach show that it is possible to construct a quantum mechanical formalism without any reference to a previous classical model. But nevertheless, to produce a detailed quantum mechanical analysis of a concrete experiment we need to work within a particular and specific representation of the $C^*$-algebra of observables, *i.e.*, we need to define properly the Hilbert space of pure states $\mathcal{H}$, to obtain the mathematical representation of the fundamental observables either as matrix, integral or differential operators. In general $\mathcal{H}$ is a space of complex squared integrable functions $\mathbb{L}^2(X)$, defined on a manifold $X$, in which some measure $d\mu(x)$ is introduced to work out the corresponding Hermitian scalar product. But all infinite-dimensional separable Hilbert spaces are isomorphic and it turns out that the selection of the manifold $X$ and the measure $d\mu(x)$ is important to show the detailed mathematical structure of the different observables we want to work with.

For instance, on the Hilbert space $\mathbb{L}^2(\mathbb{R})$ it is difficult to define the angular momentum observable $\boldsymbol{J}$. We need at least a Hilbert space of the form $\mathbb{L}^2(\mathbb{R}^3)$, although isomorphic to the previous one, to obtain a non-trivial representation of the angular momentum operator $\boldsymbol{J} = -i\hbar\boldsymbol{r} \times \nabla$. This implies that the support of the wave-functions must be a three-dimensional manifold $\mathbb{R}^3$, or that the basic object we are describing is at least a point moving in three-dimensional space. It is nonsense from the physical point of view to define the angular momentum of a point moving in a one-dimensional universe. But this election of the manifold $\mathbb{R}^3$ is dictated by the classical awareness we have about the object we want to describe. Classical mechanics, if properly developed, can help to display the quantum mechanical machinery.

Newtonian mechanics is based upon the hypothesis that the basic object of matter is a point, with the property of having mass $m$ and spin zero. Point dynamics is described by Newton's second law. It supplies in general differential equations for the position of the point and is expressed in terms of the external forces. Larger bodies and material systems are built from these massive and spinless points, and taking into account the constraints and interactions among the different parts of the system, we are able to derive the dynamics of the essential and independent degrees of freedom of any compound system of spinless particles.

Now, let us assume that we have a time-machine and we jump back to Newton's times, we meet him and say: 'Sir, according to the knowledge we have in the future where we come from, elementary objects of nature have spin, in addition to mass $m$, as a separate intrinsic property. What can we do to describe matter?' Probably he will add some extra degrees of freedom to the massive point particle and some plausible dynamical equation for the spin evolution in terms of external torques. The compound systems of spinning particles will become more complex mainly due to the additional degrees of freedom and perhaps because of the entanglement of the new variables with the old ones. Then, coming back to the beginning of our century, when facing the early steps in the dawn of quantum mechanics, it is not difficult to think that the quantization of this more complex classical background will produce a different quantum scenario in which for instance, the spin description would inherit some of the peculiarities of the additional classical variables Newton would have used.

With this preamble, what we want to emphasize is that if we are able to obtain a classical description of spin, then, based upon this picture, we shall produce a different quantization of the model. Going further, if we succeed in describing spin at the classical level we can accept the challenge to describe more and more intrinsic properties from the classical viewpoint. For instance if we want to describe hadronic matter, in addition to spin we have to describe isospin, hypercharge and many other internal properties. For this challenge, we have not only to enlarge the classical degrees of freedom, we also need to establish properly the basic group of kinematical symmetries and also to delve deeper into a plausible geometrical interpretation of the new variables the formalism provides. Probably we shall need more fundamental principles for the new variables that can go beyond our conception of space-time. Whether or not the new variables get an easy interpretation as internal or space-time variables, it is clear that the formalism must be based on invariance principles.

But the classical goal is not important in itself. Nature, at the scale of elementary interactions, behaves according to the laws of quantum mechanics. The finer the classical analysis of basic objects of matter, the richer will be their quantum mechanical description. The quantum mechanical picture when expressed in terms of invariance principles will show the relationship between the classical variables and symmetry group parameters of the manifolds involved. This is our main motivation for the classical analysis of spinning particles: to finally obtain a thorough quantum scheme.

## 1.1 KINEMATICS AND DYNAMICS

When facing the project of getting a classical description of matter we have the recent history of Physics on our back. And, although we have a huge classical luggage, a glance at the successful way quantum mechanics describes both kinematics and dynamics may help us to devise the formalism.

By kinematics we understand the basic statements that define the physical objects we go to work with. In quantum mechanics, the necessary condition for a particle to be considered elementary is based upon group theoretical arguments, related to very general symmetry statements. It is related to the irreducibility of the representation of what is called the kinematical group of space-time transformations and to the so-called internal symmetries group. It is usually called to work '*à la Wigner*'. [1] Intrinsic attributes are then interpreted in terms of the group invariants. We shall also try to derive the basic kinematical ingredients of the classical formalism by group theoretical methods.

Quantum dynamics, in the form of either S-matrix theory, scattering formalism, Wightman's functional method or Feynman's path integral approach, finally describes the probability amplitudes for the whole process in terms of the end point kinematical variables that characterize the initial and final states of the system. The details concerning the intermediate flight of the particles involved, are not explicit in the final form of the result. They are all removed, enhancing the role, as far as the theoretical analysis is concerned, of the initial and final data. Basically it is an input-output formalism.

Therefore the aim of the classical approach we propose, similar to the quantum case, is to first establish group theoretical statements for defining the kinematics of elementary spinning objects, *i.e.*, what are the necessary basic degrees of freedom for an elementary system and second, to express the dynamics in terms of end point variables. We shall start with analyzing the second goal.

## 2. VARIATIONAL VERSUS NEWTONIAN FORMALISM

In a broad sense we understand by Newtonian dynamics a formalism for describing the evolution of classical systems that states a system of differential equations with boundary conditions at a single initial instant of time. This uniquely determines the complete evolution of the system, provided some mathematical regularities of the differential equations are required. In this sense it is a deterministic theory. Just put the system at an initial time $t_1$ in a certain configuration and the evolution follows in a unique way. Roughly speaking it is a *local* formalism in the sense that what the dynamical equations establish is a relationship between different physical magnitudes and external fields at the same space-time point, and this completely characterizes the evolution.

On the other hand, a variational formalism is a global one. For every plausible path to be followed by the system between two fixed end points, although arbitrary, a magnitude is defined. This magnitude, called the action, becomes a real function over the kind of paths joining the end points and physicists call this path dependent function an **action functional**. It is usually written as an integral along the plausible paths of an auxiliary function $L$, called the Lagrangian, which is an explicit function of the different variables and their derivatives up to some order. Dynamics is stated under the condition that the path followed by the system is one for which the action functional is stationary. This leads to a necessary condition to be fulfilled by the Lagrangian and its derivatives, *i.e.*, Euler-Lagrange dynamical equations.

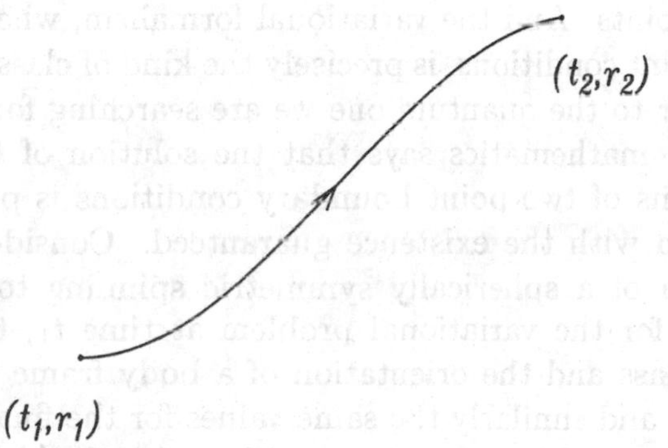


*Figure 1.1.* Evolution in $\{t, \boldsymbol{r}\}$ space.

For instance, in Figure 1.1, we represent a possible path to be followed by a point particle between the initial state expressed in terms of the variables time and position, $t_1$ and $\boldsymbol{r}_1$ respectively and its final state at point $t_2, \boldsymbol{r}_2$ of the evolution space. The action functional is written in terms of the Lagrangian $L$, which is an explicit function of $t$ and $\boldsymbol{r}$ and the first derivative $d\boldsymbol{r}/dt \equiv \dot{\boldsymbol{r}}$. Therefore the variational formalism states that the path followed by the system, $\boldsymbol{r}(t)$, is the one that produces a minimum value for the integral

$$\mathcal{A}[\boldsymbol{r}] = \int_{t_1}^{t_2} L(t, \boldsymbol{r}, \dot{\boldsymbol{r}})\, dt, \tag{1.1}$$

for the class of paths joining $t_1, \boldsymbol{r}_1$ with $t_2, \boldsymbol{r}_2$. The necessary condition for the path $\boldsymbol{r}(t)$ is that it will be at least of class $C^2$ *i.e.*, of continuous second order derivatives and that the system of ordinary differential equations obtained from the Lagrangian

$$\frac{\partial L}{\partial r_i} - \frac{d}{dt}\left(\frac{\partial L}{\partial \dot{r}_i}\right) = 0, \quad i = 1, 2, 3, \tag{1.2}$$

must be satisfied.

Thus the variational method might be interpreted as a mere intermediate trick to obtain in a peculiar way the differential equations of the dynamics of the system from the knowledge of the Lagrangian, properly chosen to achieve this goal. Once the dynamical equations are obtained we can forget about the previous action functional and go onwards as in Newtonian mechanics. But the particular solution we are looking for in the preliminary variational statements is the one that goes through the fixed end points. And the variational formalism, when expressed in terms of end point conditions, is precisely the kind of classical dynamical formalism closer to the quantum one we are searching for.

Nevertheless, mathematics says that the solution of the variational problem in terms of two-point boundary conditions is perhaps neither unique nor even with the existence guaranteed. Consider for instance the free motion of a spherically symmetric spinning top. Let us fix as initial state for the variational problem at time $t_1$, the position of the center of mass and the orientation of a body frame parallel to the laboratory axes and similarly the same values for the final state at time $t_2$. The dynamics of this system, compatible with the above conditions, corresponds to a body rotating with a certain angular velocity around a fixed center of mass. The variational solution implies a center of mass at rest but an infinity of solutions for the rotational motion corresponding to a finite, but arbitrary, number of complete turns of the body around some arbitrary axis. The variational problem has solution with fixed end points, but this solution is not unique. We have a classical indeterminacy in the possible paths followed by the system, at least in the description of the evolution of the orientation.

From the Newtonian viewpoint we need to fix at time $t_1$ the center of mass position and velocity and the initial orientation and angular velocity of the body; the result is a unique trajectory. This contrasts with the many possible trajectories of the variational approach. When comparing both formalisms, this means that it is not possible to express in a unique way our boundary conditions for the variational problem in terms of the single time boundary conditions of Newtonian dynamics.

For point particles this is not usually the case and basically both formalisms are equivalent. But when we have at hand more degrees of freedom, as in the above example in which compact variables that describe orientation are involved, they are completely unlike. Our spinning top is a spinning object and the description of spin is part of our goal; as we shall see later, compact variables of the kind of orientation variables are among the variables we shall need to describe the spin structure of classical elementary particles. Quantization of these systems will lead to the existence of differential operators acting on variables defined in a compact domain. Theorems on representations of compact Lie groups will play a dominant role in the determination of the quantum mechanical spin structure of elementary particles.

Then we have at our disposal a classical dynamical formalism expressed in terms of end point conditions that in a broad sense agrees with Newtonian dynamics for spinless particles, but when particles have spin this produces classical solutions that no longer are equivalent to the Newtonian ones and even suggests a possible classical indeterminacy compatible with the variational statements. This classical indeterminacy cannot be understood as the corresponding quantum uncertainty, because it can be removed by the knowledge of additional information like total energy or linear or angular momentum. It is an indeterminacy related to the non-uniqueness of the solution of the boundary value problem of the variational formalism in general, and in this particular case when acting on variables defined on compact manifolds.

For spinless particles, the matching of both formalisms requires the Lagrangian to be chosen as a function of the first order derivatives of the independent degrees of freedom, because Newton's equations are second order differential equations. But, what about systems involving spinning particles? At this moment of the exposition, if it is not clear what kind of variables are necessary to describe spin at the classical level, and even the agreement of the variational approach with the Newtonian formalism is doubtful because the basic objects they deal with are different, are we able to restrict Lagrangians for spinning particles to dependence on only first order derivatives? This mathematical constraint has to be justified on physical grounds so that we shall not assume this statement any longer.

Then our proposal is to analyze in detail a generalized Lagrangian formalism, in particular under symmetry principles, enhancing the role of the end point variables in order to establish a dynamical formalism quite close to the quantum mechanical one. The variables we need to describe the initial and final data for classical elementary particles, *i.e.*, the classical equivalent to the free asymptotic states of the scattering theory, will be defined by pure kinematical arguments and they will be related to what are called homogeneous spaces of the kinematical group. In this way, they are intimately related to the kinematical group of symmetries, and by assuming as a basic statement a special relativity principle, they will be related to the corresponding space-time transformation group.

Then in the next sections we shall develop the basic features of a generalized Lagrangian formalism and analyze some group theoretical aspects of Lie groups of transformations, to finally express Noether's theorem in terms of the end point variables of the variational formalism. These variables, which will be called kinematical variables, will play a dominant role in the present formalism. They will define very accurately the degrees of freedom of an elementary spinning particle, in the classical and quantum mechanical formalisms.

## 3. GENERALIZED LAGRANGIAN FORMALISM

The Lagrangian formalism of generalized systems depending on higher order derivatives was already worked out by Ostrogradsky. [2] We shall outline it briefly here, mainly to analyze the generalized Lagrangians not only in terms of the independent degrees of freedom but also as functions of what we shall call kinematical variables of the system, *i.e.*, of the end point variables of the variational formulation.

Let us consider a mechanical system of $n$ degrees of freedom, characterized by a Lagrangian that depends on time $t$ and on the $n$ essential coordinates $q_i(t)$, that represent the $n$ independent degrees of freedom, and their derivatives up to a finite order $k$. Because we can have time derivatives of arbitrary order we use a superindex enclosed in brackets to represent the corresponding $k$-th derivative, *i.e.*, $q_i^{(k)}(t) = d^k q_i(t)/dt^k$. The action functional is defined by:

$$\mathcal{A}[q] = \int_{t_1}^{t_2} L(t, q_i(t), q_i^{(1)}(t), \ldots, q_i^{(k)}(t))dt, \qquad (1.3)$$

where $i = 1, \ldots, n$. Using a more compact notation we define $q_i^{(0)} \equiv q_i$, and therefore we shall write

$$L(t, q_i(t), q_i^{(1)}(t), \ldots, q_i^{(k)}(t)) \equiv L(t, q_i^{(s)}(t)),$$

for $s = 0, \ldots, k$.

The trajectory followed by the mechanical system is that path which passing through the fixed end-points $q_i^{(s)}(t_1)$ and $q_i^{(s)}(t_2)$, $i = 1, ..., n$, $s = 0, 1, ..., k-1$, makes extremal the action functional (1.3). Note that we need to fix as boundary values of the variational principle some particular values of time $t$, the $n$ degrees of freedom $q_i$ and their derivatives up to order $k-1$, *i.e.*, one order less than the highest derivative of each variable $q_i$ in the Lagrangian, at both end points of the problem. In other words we can say that the Lagrangian of any arbitrary generalized system is in general an explicit function of the variables we keep fixed as end points of the variational formulation and also of their next order derivative.

Once the action functional (1.3) is defined for some particular path $q_i(t)$, to analyze its variation let us produce an infinitesimal modification of the functions $q_i(t)$, $q_i(t) \to q_i(t) + \delta q_i(t)$ while leaving fixed the end-points of the variational problem, *i.e.*, such that at $t_1$ and $t_2$ the modification of the generalized coordinates and their derivatives up to order $k-1$ vanish, and thus $\delta q_i^{(s)}(t_1) = \delta q_i^{(s)}(t_2) = 0$, for $i = 1, \ldots, n$ and $s = 0, 1, \ldots, k-1$. Then, the variation of the derivatives of the $q_i(t)$ is given by $q_i^{(s)}(t) \to q_i^{(s)}(t) + \delta q_i^{(s)}(t) = q_i^{(s)}(t) + d^s \delta q_i(t)/dt^s$, since the modification of the $s$-th derivative function is just the $s$-th derivative of the modification of the corresponding function. This produces a variation in the action functional $\delta\mathcal{A} = \mathcal{A}[q + \delta q] - \mathcal{A}[q]$, given by:

$$\delta\mathcal{A} = \int_{t_1}^{t_2} L(t, q_i^{(s)}(t) + \delta q_i^{(s)}(t))dt - \int_{t_1}^{t_2} L(t, q_i^{(s)}(t))dt$$

$$= \int_{t_1}^{t_2} dt \sum_{i=1}^{n} \left[ \frac{\partial L}{\partial q_i}\delta q_i + \frac{\partial L}{\partial q_i^{(1)}}\delta q_i^{(1)} + \cdots + \frac{\partial L}{\partial q_i^{(k)}}\delta q_i^{(k)} \right], \qquad (1.4)$$

after expanding to lowest order the first integral. The term

$$\frac{\partial L}{\partial q_i^{(1)}}\delta q_i^{(1)} = \frac{\partial L}{\partial q_i^{(1)}}\frac{d}{dt}\delta q_i = \frac{d}{dt}\left(\frac{\partial L}{\partial q_i^{(1)}}\delta q_i\right) - \frac{d}{dt}\left(\frac{\partial L}{\partial q_i^{(1)}}\right)\delta q_i,$$

and by partial integration of this expression between $t_1$ and $t_2$, it gives:

$$\int_{t_1}^{t_2} \frac{\partial L}{\partial q_i^{(1)}}\delta q_i^{(1)} dt = \frac{\partial L}{\partial q_i^{(1)}}\delta q_i(t_2) - \frac{\partial L}{\partial q_i^{(1)}}\delta q_i(t_1) - \int_{t_1}^{t_2} \frac{d}{dt}\left(\frac{\partial L}{\partial q_i^{(1)}}\right)\delta q_i dt$$

$$= -\int_{t_1}^{t_2} \frac{d}{dt}\left(\frac{\partial L}{\partial q_i^{(1)}}\right)\delta q_i \, dt,$$

because the variations of $\delta q_i$ in $t_1$ and $t_2$ vanish. Similarly for the next term:

$$\frac{\partial L}{\partial q_i^{(2)}}\delta q_i^{(2)} = \frac{\partial L}{\partial q_i^{(2)}}\frac{d}{dt}\delta q_i^{(1)} = \frac{d}{dt}\left(\frac{\partial L}{\partial q_i^{(2)}}\delta q_i^{(1)}\right) - \frac{d}{dt}\left(\frac{\partial L}{\partial q_i^{(2)}}\right)\delta q_i^{(1)},$$

$$\int_{t_1}^{t_2} \frac{\partial L}{\partial q_i^{(2)}} \delta q_i^{(2)} dt = -\int_{t_1}^{t_2} \frac{d}{dt}\left(\frac{\partial L}{\partial q_i^{(2)}}\right) \delta q_i^{(1)} dt = \int_{t_1}^{t_2} \frac{d^2}{dt^2}\left(\frac{\partial L}{\partial q_i^{(2)}}\right) \delta q_i \, dt,$$

because $\delta q_i$ and $\delta q_i^{(1)}$ vanish at $t_1$ and $t_2$, and finally for the last term

$$\int_{t_1}^{t_2} \frac{\partial L}{\partial q_i^{(k)}} \delta q_i^{(k)} dt = (-1)^k \int_{t_1}^{t_2} \frac{d^k}{dt^k}\left(\frac{\partial L}{\partial q_i^{(k)}}\right) \delta q_i \, dt,$$

so that each term of (1.4) is written only in terms of the variations $\delta q_i$. Remark that to reach these final expressions, it has been necessary to assume the vanishing of all $\delta q_i^{(s)}$, for $s = 0, \ldots, k-1$, at times $t_1$ and $t_2$. By collecting all terms we get

$$\delta \mathcal{A} = \int_{t_1}^{t_2} dt \sum_{i=1}^{n} \left[ \frac{\partial L}{\partial q_i} - \frac{d}{dt}\left(\frac{\partial L}{\partial q_i^{(1)}}\right) + \cdots + (-1)^k \frac{d^k}{dt^k}\left(\frac{\partial L}{\partial q_i^{(k)}}\right) \right] \delta q_i.$$

If the action functional is extremal along the path $q_i(t)$, its variation must vanish, $\delta \mathcal{A} = 0$. The variations $\delta q_i$ are arbitrary and therefore all terms between squared brackets cancel out. We obtain a system of $n$ differential equations, the Euler-Lagrange equations

$$\frac{\partial L}{\partial q_i} - \frac{d}{dt}\left(\frac{\partial L}{\partial q_i^{(1)}}\right) + \cdots + (-1)^k \frac{d^k}{dt^k}\left(\frac{\partial L}{\partial q_i^{(k)}}\right) = 0, \quad i = 1, \ldots, n, \quad (1.5)$$

which can be written in condensed form as:

$$\sum_{s=0}^{k} (-1)^s \frac{d^s}{dt^s}\left(\frac{\partial L}{\partial q_i^{(s)}}\right) = 0, \qquad i = 1, \ldots, n. \qquad (1.6)$$

## 4. KINEMATICAL VARIABLES

In general, the system (1.6) is a system of $n$ ordinary differential equations of order $2k$, and thus existence and uniqueness theorems guarantee only the existence of a solution of this system for the $2kn$ boundary conditions $q_i^{(s)}(t_1)$, $i = 1, \ldots, n$ and $s = 0, 1, \ldots, 2k-1$, at the initial instant $t_1$. However the variational problem has been stated by the requirement that the solution goes through the two fixed endpoints, a condition that does not guarantee either the existence or the uniqueness of the solution. Nevertheless, let us assume that with the fixed endpoint conditions of the variational problem, $q_i^{(s)}(t_1)$ and $q_i^{(s)}(t_2)$, $i = 1, \ldots, n$ and $s = 0, 1, \ldots, k-1$, at times $t_1$ and $t_2$, respectively, there exists a solution of (1.6) perhaps non-unique. This implies that the $2kn$ boundary conditions at time $t_1$ required by the existence and uniqueness theorems, can be expressed perhaps in a non-uniform way, as functions of the $kn$ conditions at each of the two endpoints. From now on, we shall consider systems in which this condition is satisfied. It turns out that a particular solution passing through these points will be expressed as a function of time with some explicit dependence of the end point values

$$\widetilde{q}_i(t) \equiv q_i(t; q_j^{(r)}(t_1), q_l^{(r)}(t_2)), \tag{1.7}$$

$i, j, l = 1, \ldots, n,\ r = 0, 1, \ldots k-1$, in terms of these boundary end point conditions.

**Definition:** The **Action Function** of the system along a classical path is the value of the action functional (1.3) when we introduce in the integrand a particular solution (1.7) passing through those endpoints:

$$\int_{t_1}^{t_2} L(t, \widetilde{q}_i(t))\, dt = A\left(t_1, q_i^{(r)}(t_1); t_2, q_i^{(r)}(t_2)\right). \tag{1.8}$$

Once the time integration is performed, we see that it will be an explicit function of the $kn+1$ variables at the initial instant, $q_j^{(r)}(t_1)$, $r = 0, \ldots, k-1$ including the time $t_1$, and of the corresponding $kn+1$ variables at final time $t_2$. We write it as

$$A\left(t_1, q_i^{(r)}(t_1); t_2, q_i^{(r)}(t_2)\right) \equiv A(x_1, x_2).$$

We thus arrive at the following

**Definition:** The **kinematical variables** of the system are the time $t$ and the $n$ degrees of freedom $q_i$ and their time derivatives up to order $k-1$. The manifold $X$ they span is the **kinematical space** of the system.

The kinematical space for ordinary Lagrangians is just the configuration space spanned by variables $q_i$ enlarged with the time variable $t$. It is usually called the **enlarged configuration space.** But for generalized Lagrangians it also includes higher order derivatives up to one order less than the highest derivative. Thus, the action function of a system becomes a function of the values the kinematical variables take at the end points of the trajectory, $x_1$ and $x_2$. From now on we shall consider systems for which the action function is defined and is a continuous and differentiable function of the kinematical variables at the end points of its possible evolution. This function clearly has the property $A(x,x)=0$.

The constancy of speed of light in special relativity brings space and time variables on the same footing. So, the next step is to remove the time observable as the evolution parameter of the variational formalism and express the evolution as a function of some arbitrary parameter to be chosen properly. Then, let us assume that the trajectory of the system can be expressed in parametric form, in terms of some arbitrary evolution parameter $\tau$, $\{t(\tau), q_i(\tau)\}$. The functional (1.3) can be rewritten in terms of the kinematical variables and their derivatives and becomes:

$$\mathcal{A}[t,q] = \int_{\tau_1}^{\tau_2} L\left(t(\tau), q_i(\tau), \frac{\dot{q}_i(\tau)}{\dot{t}(\tau)}, \ldots, \frac{\dot{q}_i^{(k-1)}(\tau)}{\dot{t}(\tau)}\right) \dot{t}(\tau) d\tau$$

$$= \int_{\tau_1}^{\tau_2} \widehat{L}\left(x(\tau), \dot{x}(\tau)\right) d\tau, \qquad (1.9)$$

where the dot means derivative with respect to the evolution variable $\tau$ that without loss of generality can be taken dimensionless. Therefore $\widehat{L} \equiv L(t(\tau), \dot{q}_i^{(s)}/\dot{t}(\tau))\, \dot{t}(\tau)$ has dimensions of action.

It seems that (1.9) represents the variational problem of a Lagrangian system depending only on first order derivatives and of $kn+1$ degrees of freedom. However the kinematical variables, considered as generalized coordinates, are not all independent. There exist among them the following $(k-1)n$ differential constraints

$$q_i^{(s)}(\tau) = \dot{q}_i^{(s-1)}(\tau)/\dot{t}(\tau), \quad i=1,\ldots,n, \quad s=1,\ldots,k-1. \qquad (1.10)$$

We can also see that the integrand $\widehat{L}$ is a homogeneous function of first degree as a function of the derivatives of the kinematical variables. In fact, each time derivative function $q_i^{(s)}(t)$ has been replaced by the quotient $\dot{q}_i^{(s-1)}(\tau)/\dot{t}(\tau)$ of two derivatives with respect to $\tau$. Even the highest order $k$-th derivative function $q_i^{(k)} = \dot{q}_i^{(k-1)}/\dot{t}$ is expressed in terms of the derivatives of the kinematical variables $q_i^{(k-1)}$ and $t$. Thus the function $L$ is a homogeneous function of zero degree in the derivatives of the kinematical variables. Finally, the last term $\dot{t}(\tau)$, gives to $\widehat{L}$ the character of homogeneous function of first degree. Then, Euler's theorem on homogeneous functions gives rise to the additional relation:

$$\widehat{L}(x(\tau), \dot{x}(\tau)) = \sum_j \frac{\partial \widehat{L}}{\partial \dot{x}^j} \dot{x}^j = \sum_j F_j(x, \dot{x}) \dot{x}^j. \qquad (1.11)$$

With the above $(k-1)n$ differentiable constraints among the kinematical variables (1.10) and condition (1.11), it reduces to $n$ the number of essential degrees of freedom of the system (1.9).

This possibility of expressing the Lagrangian as a homogeneous function of first degree of the derivatives was already considered in 1933 by Dirac [3] on aesthetical grounds.

Function $\widehat{L}$ is not an explicit function of the evolution parameter $\tau$ and thus we can see that the variational problem (1.9), is invariant with respect to any arbitrary change of evolution parameter $\tau$. [4]

**In fact, if we change the evolution parameter $\tau = \tau(\theta)$, then the derivative $\dot{t}(\tau) = (dt/d\theta)(d\theta/d\tau)$ and $\dot{q}_i^{(s)}(\tau) = (dq_i^{(s)}(\theta)/d\theta)(d\theta/d\tau)$ such that the quotients**

$$\frac{\dot{q}_i^{(s)}(\tau)}{\dot{t}(\tau)} = \frac{(dq_i^{(s)}(\theta)/d\theta)\,\dot{\theta}(\tau)}{(dt(\theta)/d\theta)\,\dot{\theta}(\tau)} \equiv \frac{\dot{q}_i^{(s)}(\theta)}{\dot{t}(\theta)},$$

**where once again this last dot means derivation with respect to $\theta$. It turns out that (1.9) can be written as:**

$$A[t,q] = \int_{\tau_1}^{\tau_2} L(t(\theta), q_i(\theta), \ldots, \dot{q}_i^{(k-1)}(\theta)/\dot{t}(\theta))\frac{dt(\theta)}{d\theta}d\theta$$

$$= \int_{\theta_1}^{\theta_2} \widehat{L}(x(\theta), \dot{x}(\theta))d\theta. \qquad (1.12)$$

The formalism thus stated has the advantage that it is independent of the evolution parameter, and if we want to come back to a time evolution description, we just use the time as evolution parameter and make the replacement $\tau = t$, and therefore $\dot{t} = 1$. From now on we shall consider those systems for which the evolution can be described in a parametric form, and we shall delete the symbol $\widehat{}$ over the Lagrangian, which is understood as written in terms of the kinematical variables and their first order derivatives.

If we know the action function of the system $A(x_1, x_2)$, as a function of the kinematical variables at the end points we can proceed conversely and recover the Lagrangian $L(x, \dot{x})$ by the limiting process:

$$L(x, \dot{x}) = \lim_{y \to x} \frac{\partial A(x,y)}{\partial y^j}\dot{x}^j, \qquad (1.13)$$

where the usual addition convention on repeated or dummy index $j$, extended to the whole set of kinematical variables, has been assumed.

**If in (1.9) we consider two very close points $x_1 \equiv x$ and $x_2 \equiv x + dx$, we have that the action function $A(x, x+dx) = A(x, x+\dot{x}d\tau) = L(x, \dot{x})d\tau$ and making a Taylor expansion of the function $A$ with the condition $A(x,x) = 0$ we get (1.13).**

The function of the kinematical variables and their derivatives (1.13) together with the homogeneity condition (1.11) and the $(k-1)n$ constraints among the kinematical variables (1.10) reduce the problem to that of a system with $n$ degrees of freedom but whose Lagrangian is a function of the derivatives up to order $k$ of the essential coordinates $q_i$.

The formulation in terms of kinematical variables leads to the dynamical equations (1.6) although the system looks like a system of a greater number of variables. Let us first consider an example such that $k=1$, *i.e.*, it is an ordinary first order Lagrangian. There are no constraints among the kinematical variables, and thus a system of $n$ degrees of freedom has exactly $n+1$ kinematical variables, the time $t \equiv x^0$ and the $n$ degrees of freedom $q_i \equiv x^i$. The Lagrangian (1.11) in terms of the kinematical variables produces a variational problem with $n+1$ equations:

$$\frac{\partial L}{\partial x^i} - \frac{d}{d\tau}\left(\frac{\partial L}{\partial \dot{x}^i}\right) = 0, \qquad i = 0, 1, 2, \ldots, n. \tag{1.14}$$

However, not all of equations (1.14) are independent, because if every left-hand side of each equation (1.14) is multiplied by the corresponding $\dot{x}^i$ and added all together, we get:

$$\left[\frac{\partial L}{\partial x^i} - \frac{d}{d\tau}\left(\frac{\partial L}{\partial \dot{x}^i}\right)\right]\dot{x}^i = \frac{\partial L}{\partial x^i}\dot{x}^i - \frac{d}{d\tau}\left(\frac{\partial L}{\partial \dot{x}^i}\dot{x}^i\right) + \frac{\partial L}{\partial \dot{x}^i}\ddot{x}^i. \tag{1.15}$$

Because of the homogeneity of $L = (\partial L/\partial \dot{x}^i)\dot{x}^i$ it happens that the term:

$$\frac{d}{d\tau}\left(\frac{\partial L}{\partial \dot{x}^i}\dot{x}^i\right) \equiv \frac{dL}{d\tau} = \frac{\partial L}{\partial x^i}\dot{x}^i + \frac{\partial L}{\partial \dot{x}^i}\ddot{x}^i, \tag{1.16}$$

and thus (1.15) vanishes identically. Now, if we assume for instance that the time variable $x^0$ is a monotonic function of parameter $\tau$ such that $\dot{x}^0(\tau) \neq 0, \forall \tau \in (\tau_1, \tau_2)$, then we can express the term

$$\frac{\partial L}{\partial x^0} - \frac{d}{d\tau}\left(\frac{\partial L}{\partial \dot{x}^0}\right) = -\sum_{i=1}^{n}\left[\frac{\partial L}{\partial x^i} - \frac{d}{d\tau}\left(\frac{\partial L}{\partial \dot{x}^i}\right)\right]\dot{x}^i/\dot{x}^0, \tag{1.17}$$

and the dynamical equation of (1.14) corresponding to $i=0$ is a function of the others and therefore only $n$ dynamical equations are functionally independent.

If the Lagrangian depends on higher order derivatives there will be constraints among the kinematical variables, and the variational problem must be solved with the method of Lagrange multipliers. Let us consider for simplicity another example of only one degree of freedom

$q$, but a Lagrangian that depends up to the second derivative of $q$, *i.e.*, $L(t, q, q^{(1)}, q^{(2)})$. The only dynamical equation is:

$$\frac{\partial L}{\partial q} - \frac{d}{dt}\left(\frac{\partial L}{\partial q^{(1)}}\right) + \frac{d^2}{dt^2}\left(\frac{\partial L}{\partial q^{(2)}}\right) = 0. \tag{1.18}$$

The kinematical variables of the system are $x^0 = t, x^1 = q, x^2 = q^{(1)} \equiv dq/dt$, and thus $q^{(2)} \equiv dq^{(1)}/dt = \dot{x}^2/\dot{x}^0$. The Lagrangian in a parametric description in terms of these variables is expressed as $\widehat{L} = \dot{x}^0 L(x^0, x^1, x^2, \dot{x}^2/\dot{x}^0) \equiv \widehat{L}(x, \dot{x})$. It is a homogeneous function of first degree in the variables $\dot{x}^i$, with the constraint $x^2 = \dot{x}^1/\dot{x}^0$, or $\dot{x}^1 - x^2\dot{x}^0 = 0$, and we see that $\widehat{L}$ is independent of $\dot{x}^1$ by construction. The variational problem must be solved from the modified Lagrangian $G(x, \dot{x}) = \widehat{L}(x, \dot{x}) + \lambda(\tau)(\dot{x}^1 - x^2\dot{x}^0)$, which is still a homogeneous function of first degree in the derivatives of the kinematical variables and where $\lambda(\tau)$ is a function of $\tau$ to be determined, called a Lagrange multiplier.

Dynamical equations in the parametric description are now the three equations:

$$\frac{\partial G}{\partial x^i} - \frac{d}{d\tau}\left(\frac{\partial G}{\partial \dot{x}^i}\right) = 0, \qquad i = 0, 1, 2, \tag{1.19}$$

and because of the homogeneity of $G$ in terms of the $\dot{x}^i$ the equation for $i = 0$ can be expressed as a function of the other two, similar to the previous example, but now we have the additional unknown $\lambda(\tau)$. The two independent dynamical equations are:

$$\frac{\partial \widehat{L}}{\partial x^1} - \frac{d}{d\tau}\left(\frac{\partial \widehat{L}}{\partial \dot{x}^1} + \lambda(\tau)\right) = 0, \tag{1.20}$$

$$\frac{\partial \widehat{L}}{\partial x^2} - \lambda(\tau)\dot{x}^0 - \frac{d}{d\tau}\left(\frac{\partial \widehat{L}}{\partial \dot{x}^2}\right) = 0, \tag{1.21}$$

but since by construction $\widehat{L}$ is not an explicit function of $\dot{x}^1$, then (1.20) is reduced to the equation $\partial\widehat{L}/\partial x^1 - \dot{\lambda}(\tau) = 0$, and by replacing $\lambda(\tau)$ from (1.21) in (1.20), and recovering the generalized coordinate $q$ and its derivatives, we get(1.18).

In the general case we obtain $kn+1$ dynamical equations, one for each kinematical variable. However, one of these equations can be expressed in terms of the others because of the homogeneity of the Lagrangian; now we have in addition $(k-1)n$ new variables $\lambda_i$, the Lagrange multipliers that can be eliminated between the remaining equations. We thus finally obtain the $n$ independent equations that satisfy the $n$ variables $q_i$ associated to the $n$ degrees of freedom.

The action function plays an important role since in a broad sense it characterizes the dynamics in a global way. Its knowledge determines through (1.13) the Lagrangian and by (1.6) the dynamical equations satisfied by our system.

## 4.1 EXAMPLES

The action function of a nonrelativistic Galilei point particle of mass $m$ is given by:

$$A(t_1, \boldsymbol{r}_1, t_2, \boldsymbol{r}_2) = \frac{m}{2} \frac{(\boldsymbol{r}_2 - \boldsymbol{r}_1)^2}{(t_2 - t_1)}, \tag{1.22}$$

and thus it gives rise to the Lagrangian:

$$\begin{aligned} \widehat{L} &= \lim_{x_2 \to x} \left[ \frac{\partial A(x, x_2)}{\partial t_2} \dot{t} + \frac{\partial A(x, x_2)}{\partial r_{2i}} \dot{r}_i \right] \\ &= \lim_{x_2 \to x} \left[ -\frac{m}{2} \frac{(\boldsymbol{r}_2 - \boldsymbol{r})^2}{(t_2 - t)^2} \dot{t} + m \frac{(\boldsymbol{r}_2 - \boldsymbol{r})}{(t_2 - t)} \cdot \dot{\boldsymbol{r}} \right], \end{aligned}$$

but

$$\lim_{x_2 \to x} \frac{(\boldsymbol{r}_2 - \boldsymbol{r})}{(t_2 - t)} = \frac{d\boldsymbol{r}}{dt} = \frac{\dot{\boldsymbol{r}}}{\dot{t}}.$$

Then $\widehat{L} = m\dot{\boldsymbol{r}}^2/2\dot{t}$, homogeneous of first degree in terms of the derivatives of the kinematical variables, such that in a time evolution description $\tau = t$ and thus $\dot{t} = 1$; taking into account that there are no constraints among the kinematical variables we get:

$$L = \frac{m}{2} \left( \frac{d\boldsymbol{r}}{dt} \right)^2. \tag{1.23}$$

The action function of a relativistic point particle of mass $m$ is given by:

$$A(t_1, \boldsymbol{r}_1, t_2, \boldsymbol{r}_2) = -mc\sqrt{c^2(t_2 - t_1)^2 - (\boldsymbol{r}_2 - \boldsymbol{r}_1)^2}, \tag{1.24}$$

and similarly

$$\widehat{L} = \lim_{x_2 \to x} \left[ \frac{\partial A(x, x_2)}{\partial t_2} \dot{t} + \frac{\partial A(x, x_2)}{\partial r_{2i}} \dot{r}_i \right] = -mc\sqrt{c^2 \dot{t}^2 - \dot{\boldsymbol{r}}^2},$$

also homogeneous of first degree in the derivatives of the kinematical variables and in a time evolution description we arrive at:

$$L = -mc\sqrt{c^2 - \left( \frac{d\boldsymbol{r}}{dt} \right)^2}. \tag{1.25}$$

Conversely, we can recover the action functions (1.22) and (1.24) after integration of the corresponding Lagrangians (1.23) and (1.25), respectively, along the classical free path joining the end points $(t_1, \boldsymbol{r}_1)$ and $(t_2, \boldsymbol{r}_2)$. This is the normal way we are used to. In general we shall obtain prescriptions for guessing some plausible Lagrangians and afterwards finding the corresponding action functions, but we have proceeded in this example in the reverse way to enhance the role of the action function and the homogeneity in both cases of Lagrangian $\widehat{L}$.

These two Lagrangians for relativistic and nonrelativistic point particles will be obtained in the following chapters as a result of the application of the formalism we propose to Lagrangian systems whose kinematical space is the four-dimensional manifold spanned by variables time $t$ and position $\boldsymbol{r}$. The formalism determines them uniquely with no other equivalence or possibility.

## 5. CANONICAL FORMALISM

In ordinary Lagrangian systems that depend only on first order derivatives of the independent degrees of freedom, the canonical approach associates to every generalized coordinate $q_i$ a dynamical variable $p^i$, called its conjugate momentum that is defined by

$$p^i = \frac{\partial L}{\partial \dot{q}_i}.$$

As a generalization of this, in Lagrangian systems with higher order derivatives, a generalized canonical formalism can be obtained by defining various canonical conjugate momenta (up to a total of $k$ of them) associated to each of the independent degrees of freedom $q_i$: [5]

$$p^i_{(s)} = \sum_{r=0}^{k-s} (-1)^r \frac{d^r}{dt^r} \left( \frac{\partial L}{\partial q_i^{(r+s)}} \right), \quad s = 1, \ldots, k, \quad i = 1, \ldots, n, \quad (1.26)$$

and it is said that $p^i_{(s)}$ is the conjugate momentum of order $s$ of the variable $q_i$. It can be checked from their definition that they satisfy the property:

$$p^i_{(s)} - \frac{\partial L}{\partial q_i^{(s)}} = - \frac{dp^i_{(s+1)}}{dt}. \quad (1.27)$$

The generalized Hamiltonian is similarly defined as:

$$H = p^i_{(s)} q_i^{(s)} - L, \quad (1.28)$$

where addition on repeated indexes $i = 1, \ldots, n$ and $s = 1, \ldots, k$ is assumed, and for this reason we have written the corresponding up and down indexes of the momenta in the opposite position from the corresponding indexes of the variables $q_i^{(s)}$. We see that $\partial H/\partial p^i_{(s)} = q_i^{(s)}$ and taking into account (1.27) and doing the derivative with respect to the variable $q_i^{(s)}$, we get $\partial H/\partial q_i^{(s)} = -dp^i_{(s+1)}/dt$. Finally, $\partial H/\partial t = -\partial L/\partial t$ and it turns out that in the canonical formulation the generalized Hamilton's equations are:

$$-\frac{\partial H}{\partial q_i^{(s)}} = \frac{dp^i_{(s+1)}}{dt}, \tag{1.29}$$

$$\frac{\partial H}{\partial p^i_{(s+1)}} = \frac{dq_i^{(s)}}{dt}, \tag{1.30}$$

$$\frac{\partial H}{\partial t} = -\frac{\partial L}{\partial t}, \tag{1.31}$$

where there are $kn$ pairs of canonical conjugate variables, $(q_i^{(s)}, p^i_{(s+1)})$, for $s = 0, \ldots, k-1$, $i = 1, \ldots, n$, the generalized coordinates being now the $n$ degrees of freedom $q_i$ and their derivatives up to order $k-1$, *i.e.*, the kinematical variables with the time excluded.

The Poisson bracket of any two functions $A(q,p)$ and $B(q,p)$ is defined as usual in terms of the corresponding pairs of canonical conjugate variables:

$$\{A(q,p), B(q,p)\} = \sum_{i=1}^{n}\sum_{s=0}^{k-1}\left(\frac{\partial A}{\partial q_i^{(s)}}\frac{\partial B}{\partial p^i_{(s+1)}} - \frac{\partial A}{\partial p^i_{(s+1)}}\frac{\partial B}{\partial q_i^{(s)}}\right), \tag{1.32}$$

that satisfies the antisymmetry $\{A, B\} = -\{B, A\}$, and the distributive properties

$$\{A, B + C\} = \{A, B\} + \{A, C\}, \qquad \{A, BC\} = \{A, B\}C + B\{A, C\}$$

and for any three dynamical variables $A$, $B$ and $C$, the Jacobi identities:

$$\{A, \{B, C\}\} + \{B, \{C, A\}\} + \{C, \{A, B\}\} = 0.$$

Then, Hamilton's equations (1.29) and (1.30) can also be written as

$$\{p^i_{(s)}, H\} = \frac{dp^i_{(s)}}{dt}, \quad \{q_i^{(s)}, H\} = \frac{dq_i^{(s)}}{dt},$$

and in general, for any time dependent dynamical variable $A(q,p,t)$, its time derivative, using (1.29,1.30), will be

$$\frac{dA}{dt} = \frac{\partial A}{\partial q_i^{(s)}}\frac{dq_i^{(s)}}{dt} + \frac{\partial A}{\partial p^i_{(s)}}\frac{dp^i_{(s)}}{dt} + \frac{\partial A}{\partial t}$$

$$= \frac{\partial A}{\partial q_i^{(s)}}\frac{\partial H}{\partial p^i_{(s+1)}} - \frac{\partial A}{\partial p^i_{(s+1)}}\frac{\partial H}{\partial q_i^{(s)}} + \frac{\partial A}{\partial t}$$
$$= \{A, H\} + \frac{\partial A}{\partial t}.$$

# 6. LIE GROUPS OF TRANSFORMATIONS

Let us introduce the notation and general features of the action of Lie groups on continuous manifolds to analyze the transformation properties of the different magnitudes we can work with in either classical or quantum mechanics. We shall use these features all throughout this book.

Let us consider the transformation of an $n$-dimensional manifold $X$, $x' = gx$ given by $n$ continuous and differentiable functions depending on a set $g \in G$ of $r$ continuous parameters of the form

$$x'^i = f^i(x^j; g^\sigma), \quad \forall x \in X, \quad \forall g \in G, \quad i, j = 1, \ldots, n, \quad \sigma = 1, \ldots, r.$$

This transformation is said to be the action of a Lie group of transformations if it fulfils the two conditions:
(i) $G$ is a Lie group, *i.e.*, there exists a group composition law $c = \phi(a, b) \in G$, $\forall a, b \in G$, in terms of $r$ continuous and differentiable functions $\phi^\sigma$.
(ii) The transformation equations satisfy

$$x'' = f(x'; b) = f(f(x; a); b) = f(x; c) = f(x; \phi(a, b)).$$

The group parametrization can be chosen such that the coordinates that characterize the neutral element $e$ of the group are $e \equiv (0, \ldots, 0)$, so that an infinitesimal element of the group is the one with infinitesimal coordinates $\delta g^\sigma, \sigma = 1, \ldots, r$.

Under the action of an infinitesimal element $\delta g$ of the group $G$, the change in the coordinates $x^i$ of a point $x \in X$ is given by

$$x^i + dx^i = f^i(x; \delta g) = x^i + \left.\frac{\partial f^i(x; g)}{\partial g^\sigma}\right|_{g=e} \delta g^\sigma,$$

after a Taylor expansion up to first order in the group parameters and with $x^i = f^i(x; 0)$. There are $nr$ auxiliary functions of the group that are defined as

$$u^i_\sigma(x) = \left.\frac{\partial f^i(x; g)}{\partial g^\sigma}\right|_{g=e}, \tag{1.33}$$

and therefore to first order in the group parameters, $dx^i = u^i_\sigma(x)\delta g^\sigma$.

The group action on the manifold $X$ can be extended to the action on the set $\mathcal{F}(X)$ of continuous and differentiable functions defined on $X$ by means of:

$$g : h(x) \to h'(x) \equiv h(gx). \tag{1.34}$$

If the group element is infinitesimal, then

$$h'(x) = h(x^i + dx^i) = h(x^i + u^i_\sigma(x)\delta g^\sigma) = h(x) + \frac{\partial h(x)}{\partial x^i}\, u^i_\sigma(x)\delta g^\sigma,$$

after a Taylor expansion to first order in the infinitesimal group parameters. The infinitesimal transformation on $\mathcal{F}(X)$ can be represented by the action of a differential operator in the form

$$h'(x) = \left(\mathbb{I} + \delta g^\sigma\, u^i_\sigma(x)\frac{\partial}{\partial x^i}\right) h(x) = (\mathbb{I} + \delta g^\sigma X_\sigma)\, h(x) = U(\delta g)h(x),$$

where $\mathbb{I}$ is the identity operator and the linear differential operators

$$X_\sigma = u^i_\sigma(x)\frac{\partial}{\partial x^i}. \tag{1.35}$$

In particular, when acting with the operator $U(\delta g) \equiv (\mathbb{I} + \delta g^\sigma X_\sigma)$ on the coordinate $x^j$ we get $x^j + dx^j = x^j + u^j_\sigma(x)\delta g^\sigma$.

The operators $X_\sigma$ are called the **generators** of the infinitesimal transformations. They are $r$ linearly independent operators that span an $r$-dimensional real vector space such that its commutator $[X_\sigma, X_\lambda]$ also belongs to the same vector space, *i.e.*,

$$[X_\sigma, X_\lambda] = c^\alpha_{\sigma\lambda}\, X_\alpha, \quad \alpha, \sigma, \lambda = 1, \ldots, r. \tag{1.36}$$

The coefficients $c^\alpha_{\sigma\lambda}$ are a set of real constant numbers, called the **structure constants** of the group, and the vector space spanned by the generators is named the **Lie algebra** $\mathcal{L}(G)$, associated to the Lie group $G$. The structure constants are antisymmetric in their lower indexes $c^\alpha_{\sigma\lambda} = -c^\alpha_{\lambda\sigma}$, and satisfy Jacobi's indentitites:

$$c^\alpha_{\sigma\lambda} c^\beta_{\mu\alpha} + c^\alpha_{\lambda\mu} c^\beta_{\sigma\alpha} + c^\alpha_{\mu\sigma} c^\beta_{\lambda\alpha} = 0, \qquad \forall \sigma, \lambda, \mu, \beta = 1, \ldots, r.$$

Equations (1.36) are the commutation relations that characterize the structure of the Lie algebra of the group.

If a finite group transformation of parameters $g^\sigma$ can be done in $n$ smaller steps of parameters $g^\sigma/n$, with $n$ sufficiently large, then a finite transformation $U(g)h(x)$ can be obtained as

$$U(g)h(x) \equiv \lim_{n\to\infty} \left(\mathbb{I} + \frac{g^\sigma}{n} X_\sigma\right)^n h(x) = \exp(g^\sigma X_\sigma)\, h(x).$$

This defines the exponential mapping and in this case the group parameters $g^\sigma$ are called **normal** or **canonical** parameters. In the normal parameterization the composition law of one-parameter subgroups reduces to the addition of the corresponding parameters of the involved group elements.

Consider $\mathcal{F}(X)$ a Hilbert space of states of a quantum system; (1.34) can be interpreted as the transformed wave function under the group element $g$. Then if the operator $U(g)$ is unitary it is usually written in the explicit form

$$U(g) = \exp\left(\frac{i}{\hbar} g^\sigma \widetilde{X}_\sigma\right),$$

in terms of the imaginary unit $i$ and Planck's constant $\hbar$, such that in this case the new $\widetilde{X}_\sigma$ above are self-adjoint operators and therefore represent certain observables of the system. The physical dimensions of these observables depend on the dimensions of the group parameters $g^\sigma$, since the argument of the exponential function is dimensionless and because of the introduction of Planck's constant this implies that $g^\sigma \widetilde{X}_\sigma$ has dimensions of action. These observables, taking into account (1.35), are represented in a unitary representation by the differential operators

$$\widetilde{X}_\sigma = \frac{\hbar}{i} u^i_\sigma(x) \frac{\partial}{\partial x^i}. \qquad (1.37)$$

However, (1.34) is not the most general form of transformation of the wave function of a quantum system, as we shall see in Chapter 4, but once we know the way it transforms we shall be able to obtain the explicit expression of the group generators by a similar procedure as the one developed so far. In general the wave function transforms under continuous groups with what is called a projective unitary representation of the group, which involves in general some additional phase factors.

Let us consider the following simple examples. The action of the translation group $\{\mathbb{R}, +\}$ acting on the real line is given by $x' = f(x; g) \equiv x + a$. The neutral element is $a = 0$ and the composition law is the addition of parameters $a'' = a' + a$. Because we have only one variable and one group parameter we obtain one auxiliary function and according to (1.33) it is

$$u(x) = \left.\frac{\partial(x+a)}{\partial a}\right|_{a=0} = 1.$$

Therefore the group has a generator with general expression given by (1.35) and is written as $P = \partial/\partial x$. In a quantum mechanical description, when acting on functions $\phi(x)$, according to (1.37) it becomes $P = -i\hbar\partial/\partial x$, with dimensions of action divided by length, *i.e.*, of linear momentum.

Another example is a rotation of the plane of angle $\alpha$ around the origin. The transformation equations are

$$x' = x\cos\alpha - y\sin\alpha, \quad y' = x\sin\alpha + y\cos\alpha.$$

It is the one-parameter group $SO(2)$ acting on a two-dimensional manifold $\mathbb{R}^2$. The neutral element corresponds to $\alpha = 0$ and the group composition law is again $\alpha'' = \alpha' + \alpha$. The two auxiliary functions are

$$u^1(x,y) = \left.\frac{\partial(x\cos\alpha - y\sin\alpha)}{\partial\alpha}\right|_{\alpha=0} = -y,$$

$$u^2(x,y) = \left.\frac{\partial(x\sin\alpha + y\cos\alpha)}{\partial\alpha}\right|_{\alpha=0} = x,$$

so that according to (1.37) the infinitesimal generator of the group is

$$J = -y\frac{\partial}{\partial x} + x\frac{\partial}{\partial y}. \tag{1.38}$$

In the quantum case its representation is equation (1.38) multiplied by $\hbar/i$, and therefore $J$ has dimensions of action or angular momentum.

Finally let us consider the action of the Euclidean group on two-dimensional space. It consists of rotations and translations of the plane. It is a three-parameter group and its action on a point of $\mathbb{R}^2$ is given by

$$x' = x\cos\alpha - y\sin\alpha + a, \quad y' = x\sin\alpha + y\cos\alpha + b.$$

An arbitrary group element is characterized by the three parameters $g \equiv (\alpha, a, b)$. The neutral element is $e \equiv (0,0,0)$. The composition of the group $g'' = g'g$ is

$$\alpha'' = \alpha' + \alpha, \quad a'' = a\cos\alpha' - b\sin\alpha' + a', \quad b'' = a\sin\alpha' + b\cos\alpha' + b',$$

which are analytic functions of the group parameters. The three linearly independent infinitesimal generators associated to the three one-parameter subgroups of parameters $\alpha$, $a$ and $b$ are respectively

$$J = -y\frac{\partial}{\partial x} + x\frac{\partial}{\partial y}, \quad P_x = \frac{\partial}{\partial x}, \quad P_y = \frac{\partial}{\partial y},$$

and the commutation relations that define its Lie algebra are

$$[P_x, P_y] = 0, \quad [J, P_x] = -P_y, \quad [J, P_y] = P_x.$$

In all these examples the different subgroups are given in the canonical parametrization, so that the exponential mapping works. For instance, a rotation of finite value $\alpha$ will be expressed as $\exp(\alpha J)$ and its action on the $x$ coordinate will be:

$$\exp(\alpha J)\,x = \left(1 + \frac{1}{1!}\alpha J + \frac{1}{2!}(\alpha J)(\alpha J) + \frac{1}{3!}(\alpha J)(\alpha J)(\alpha J) + \ldots\right) x,$$

and taking into account the expression (1.38) for $J$ when acting on the $x$ coordinate we get

$$= x - \frac{\alpha}{1!}y - \frac{\alpha^2}{2!}x + \frac{\alpha^3}{3!}y + \ldots = x\cos\alpha - y\sin\alpha,$$

as it should be. We can similarly obtain its action on the $y$ coordinate.

## 6.1 CASIMIR OPERATORS

When we have a representation of a Lie group either by linear operators or by matrices acting on a linear space, we can define there what are called the Casimir operators. They are operators $C$ that can be expressed as functions of the generators $X_\sigma$ of the Lie algebra with the property that they commute with all of them, *i.e.*, they satisfy $[C, X_\sigma] = 0, \quad \forall \sigma = 1, \ldots, r$. In general they are not expressed as real linear combinations of the $X_\sigma$ and therefore they do not belong to the Lie algebra of the group. They belong to what is called the **group algebra**, *i.e.*, the associative, but in general non-commutative algebra, spanned by the real or complex linear combinations of products of the $X_\sigma$, in the corresponding group representation.

In those representations where the $X_\sigma$ are represented by self-adjoint operators as in a quantum formalism, the Casimir operators may be also self-adjoint and will represent those observables that remain invariant under the group transformations. In particular, when we consider later the kinematical groups that relate the space-time measurements between inertial observers, the Casimir operators of these groups will represent the intrinsic properties of the system. They are those properties of the physical system whose measured values are independent of the inertial observers.

For semisimple groups, *i.e.*, for groups that do not have Abelian invariant subgroups like the rotation group $SO(3)$, the unitary groups $SU(n)$ and many others, it is shown that the Casimir operators are real homogeneous polynomials of the generators $X_\sigma$, but this is no longer the case for general Lie groups. Nevertheless, for most of the interesting Lie groups in physics, like Galilei, Poincaré, De Sitter, $SL(4,\mathbb{R})$, the inhomogeneous $ISL(4,\mathbb{R})$ and Conformal $SU(2,2)$ groups, the Casimir operators can be taken as real polynomial functions of the generators.

In the examples shown in the previous section, the generators of $\{\mathbb{R},+\}$ and $SO(2)$, $P$ and $J$ respectively are Casimir operators of those groups because they are one-parameter groups. For the Euclidean group in two dimensions, it can be checked that the operator $P_x^2 + P_y^2$, which is a polynomial function of the generators, commutes with all of them and therefore it is a Casimir operator for this group.

## 6.2 EXPONENTS OF A GROUP

The concept of exponent of a continuous group $G$ was developed by Bargmann in his work on the projective unitary representations of continuous groups. [6]

Wigner's theorem about the symmetries of a physical system is well known in Quantum Mechanics. [7]

It states that if $\mathcal{H}$ is a Hilbert space that characterizes the pure quantum states of a system, and the system has a symmetry $S$, then there exists a unitary or antiunitary operator $U(S)$, defined up to a phase, that implements that symmetry on $\mathcal{H}$, *i.e.*, if $\phi$ and $\psi \in \mathcal{H}$ are two possible vector states of the system and $|<\phi|\psi>|^2$ is the transition probability between them and $U(S)\phi$ and $U(S)\psi$ represent the transformed states under the operation $S$, then

$$|<U(S)\phi|U(S)\psi>|^2 = |<\phi|\psi>|^2.$$

If the system has a whole group of symmetry operations $G$, then to each element $g \in G$ there is associated an operator $U(g)$ unitary or antiunitary, but if $G$ is a continuous group, in that case $U(g)$ is necessarily unitary. This can be seen by the fact that the product of two antiunitary operators is a unitary one.

Because there is an ambiguity in the election of the phase of the unitary operator $U(g)$, it implies that in general $U(g_1)U(g_2) \neq U(g_1g_2)$ and therefore the transformation of the wave function is not given by an expression of the form (1.34), but it also involves in general a phase factor. However in the case of continuous groups we can properly choose the corresponding phases of all elements in such a way that

$$U(g_1)U(g_2) = \omega(g_1, g_2)U(g_1g_2), \qquad (1.39)$$

where $\omega(g_1, g_2) = \exp\{i\xi(g_1, g_2)\}$ is a phase that is a continuous function of its arguments. The real continuous function on $G \times G$, $\xi(g_1, g_2)$ is called an **exponent** of $G$. The operators $U(g)$ do not reproduce the composition law of the group $G$ and (1.39) represents what Bargmann calls a **projective representation** of the group.

If we use the associative property of the group law, we get

$$\begin{aligned}(U(g_1)U(g_2))\,U(g_3) &= \omega(g_1, g_2)U(g_1g_2)U(g_3)\\ &= \omega(g_1, g_2)\omega(g_1g_2, g_3)U(g_1g_2g_3),\end{aligned}$$

and also

$$\begin{aligned}U(g_1)\,(U(g_2)U(g_3)) &= U(g_1)\omega(g_2, g_3)U(g_2g_3)\\ &= \omega(g_1, g_2g_3)\omega(g_2, g_3)U(g_1g_2g_3).\end{aligned}$$

Therefore

$$\omega(g_1, g_2)\omega(g_1g_2, g_3) = \omega(g_1, g_2g_3)\omega(g_2, g_3), \qquad (1.40)$$

which in terms of the exponents becomes:

$$\xi(g_1, g_2) + \xi(g_1g_2, g_3) = \xi(g_1, g_2g_3) + \xi(g_2, g_3). \qquad (1.41)$$

Because of the continuity of the exponents,

$$\xi(g,e) = \xi(e,g) = 0, \quad \forall g \in G, \tag{1.42}$$

where $e$ is the neutral element of the group.

Any continuous function on $G$, $\phi(g)$, with the condition $\phi(e) = 0$, can generate a trivial exponent by

$$\xi(g,g') = \phi(gg') - \phi(g) - \phi(g'),$$

that satisfies (1.41) and (1.42). All trivial exponents are equivalent to zero exponents, and in a unitary representation (1.39) can be compensated into the phases of the factors, thus transforming the projective representation (1.39) into a true unitary one.

Given a continuous group, the existence or not of non-trivial exponents is an intrinsic group property related to the existence or not of central extensions of the group. [8]

## 6.3 HOMOGENEOUS SPACE OF A GROUP

A manifold $X$ is called a homogeneous space of a group $G$, if $\forall\, x_1, x_2 \in X$ there exists at least one element $g \in G$ such that $x_2 = gx_1$. In that case it is said that $G$ acts on $X$ in a transitive way. The term homogeneous reminds us that the local properties of the manifold at a point $x$ are translated to any other point of the manifold by means of the group action, and therefore all points of $X$ share the same local properties.

The **orbit** of a point $x$ is the set of points of the form $gx$, $\forall g \in G$, such that if $X$ is a homogeneous space of $G$, then the whole $X$ is the orbit of any of its points.

Given a point $x_0 \in X$, the **stabilizer group** (little group) of $x_0$ is the subgroup $H_{x_0}$ of $G$, that leaves invariant the point $x_0$, *i.e.*, $\forall h \in H_{x_0}$, $hx_0 = x_0$.

If $H$ is a subgroup of $G$, then every element $g \in G$ can be written as $g = g'h$, where $h \in H$, and $g'$ is an element of $G/H$, the set of left cosets generated by the subgroup $H$. If $X$ is a homogeneous space of $G$, it can be generated by the action of $G$ on an arbitrary point $x_0 \in X$. Then $\forall x \in X$, $x = gx_0 = g'hx_0 = g'x_0$, and thus the homogeneous space $X$ is isomorphic to the manifold $G/H_{x_0}$.

The homogeneous spaces of a group can be constructed as quotient manifolds of the group by all its possible continuous subgroups. Conversely, it can also be shown that if $X$ a homogeneous space of a group $G$, then there exists a subgroup $H$ of $G$ such that $X$ is isomorphic to $G/H$. Therefore, the largest homogeneous space of a group is the group itself.

In the previous example of the Euclidean group in two dimensions, $\mathcal{E}_2$, the group manifold is spanned by the variables $\{(\alpha, a, b)\}$ with domains $\alpha \in [0, 2\pi]$, $a, b \in \mathbb{R}$. Since the set of rotations $\{(\alpha, 0, 0), \forall\alpha \in [0, 2\pi]\}$, forms a subgroup $SO(2)$ of $\mathcal{E}_2$, then the quotient manifold $X \equiv \mathcal{E}_2/SO(2)$ is the set of classes of elements of the form $\{(\beta, x, y)\}$ with $\beta$ arbitrary and $x$ and $y$ fixed. Each class can be characterized by the pair $(x, y) \in \mathbb{R}^2$, and therefore the manifold $X$ is isomorphic to the Euclidean plane $\mathbb{R}^2$. Given two arbitrary points in this plane $(x_1, y_1)$ and $(x_2, y_2)$, they correspond respectively to the sets of group elements $(\beta, x_1, y_1)$ and $(\beta', x_2, y_2)$ of $\mathcal{E}_2$, and therefore $(\beta, x_1, y_1) = (\alpha, a, b) * (\beta', x_2, y_2)$ leads to $(\alpha, a, b) = (\beta, x_1, y_1) * (\beta', x_2, y_2)^{-1}$. Thus, there exists at least a group element that links both points of the manifold $\mathbb{R}^2$. It is in fact a homogeneous space of $\mathcal{E}_2$.

## 7. GENERALIZED NOETHER'S THEOREM

Noether's analysis for generalized Lagrangian systems also states the following

**Theorem:** To every one-parameter group of continuous transformations that transform the action function of the system in the form $A(\delta g x_1, \delta g x_2) = A(x_1, x_2) + B(x_2)\delta g - B(x_1)\delta g$, leaving dynamical equations invariant, there is associated a classical observable $N$, which is a constant of the motion.

Let us assume the existence of an $r$-parameter continuous group of transformations $G$, of the enlarged configuration space $(t, q_i)$, that can be extended as a transformation group to the whole kinematical space $X$. Let $\delta g$ be an infinitesimal element of $G$ with coordinates $\delta g^\alpha$, $\alpha = 1, \ldots, r$ and its action on these variables be given by:

$$t \to t' = t + \delta t = t + M_\alpha(t, q)\delta g^\alpha, \qquad (1.43)$$

$$q_i(t) \to q_i'(t') = q_i(t) + \delta q_i(t) = q_i(t) + M_{i\alpha}^{(0)}(t, q)\delta g^\alpha, \qquad (1.44)$$

and its extension on the remaining kinematical variables by

$$q'^{(1)}_i(t') = q_i^{(1)}(t) + \delta q_i^{(1)}(t) = q_i^{(1)}(t) + M_{i\alpha}^{(1)}(t, q, q^{(1)})\delta g^\alpha, \qquad (1.45)$$

and in general

$$q'^{(s)}_i(t') = q_i^{(s)}(t) + \delta q_i^{(s)}(t) = q_i^{(s)}(t) + M_{i\alpha}^{(s)}(t, q, \ldots, q^{(s)})\delta g^\alpha, \qquad (1.46)$$

for $s = 0, 1, \ldots, k-1$ and where $M_\alpha$ and $M_{i\alpha}^{(0)}$ are functions only of $q_i$ and $t$ while the functions $M_{i\alpha}^{(s)}$ with $s \geq 1$, obtained in terms of the derivatives of the previous ones, will be functions of the time $t$ and of the variables $q_i$ and their time derivatives up to order $s$.

For instance,

$$q'^{(1)}_i(t') \equiv \frac{dq'_i(t')}{dt'} = \frac{d(q_i(t) + M^{(0)}_{i\alpha}\delta g^\alpha)}{dt}\frac{dt}{dt'},$$

but up to first order in $\delta g$

$$\frac{dt}{dt'} = 1 - M_\alpha(t,q)\delta g^\alpha,$$

and thus

$$q'^{(1)}_i(t') = q^{(1)}_i(t) + \left(\frac{dM^{(0)}_{i\alpha}(t,q)}{dt} - q^{(1)}_i M_\alpha(t,q)\right)\delta g^\alpha,$$

and comparing with (1.45) we get

$$M^{(1)}_{i\alpha}(t,q,q^{(1)}) = \frac{dM^{(0)}_{i\alpha}(t,q)}{dt} - q^{(1)}_i M_\alpha(t,q),$$

where the total time derivative

$$\frac{dM^{(0)}_{i\alpha}(t,q)}{dt} = \frac{\partial M^{(0)}_{i\alpha}(t,q)}{\partial t} + \sum_j \frac{\partial M^{(0)}_{i\alpha}(t,q)}{\partial q_j} q^{(1)}_j.$$

The remaining $M^{(s)}_{i\alpha}$ for $s > 1$, are obtained in the same way from the previous $M^{(s-1)}_{i\alpha}$.

Under $\delta g$ the change of the action functional of the system is:

$$\begin{aligned}\delta\mathcal{A}[q] &= \int_{t'_1}^{t'_2} L(t', q'^{(s)}_i(t'))dt' - \int_{t_1}^{t_2} L(t, q^{(s)}_i(t))dt \\ &= \int_{t'_1}^{t'_2} L(t+\delta t, q^{(s)}_i(t) + \delta q^{(s)}_i(t))dt' - \int_{t_1}^{t_2} L(t, q^{(s)}_i(t))dt.\end{aligned}$$

By replacing in the first integral the integration range $(t'_1, t'_2)$ by $(t_1, t_2)$ having in mind the Jacobian of $t'$ in terms of $t$, this implies that the differential $dt' = (1 + d(\delta t)/dt)dt$, and thus:

$$\begin{aligned}\delta\mathcal{A}[q] &= \int_{t_1}^{t_2} L(t+\delta t, q^{(s)}_i + \delta q^{(s)}_i)\left(1 + \frac{d(\delta t)}{dt}\right)dt - \int_{t_1}^{t_2} L(t, q^{(s)}_i)dt \\ &= \int_{t_1}^{t_2}\left(L\frac{d(\delta t)}{dt} + \frac{\partial L}{\partial t}\delta t + \frac{\partial L}{\partial q^{(s)}_i}\delta q^{(s)}_i(t)\right)dt,\end{aligned}$$

keeping only first order terms in the Taylor expansion of the first Lagrangian $L(t+\delta t, q^{(s)} + \delta q^{(s)})$.

Now, in the total variation of $\delta q_i^{(s)}(t) = q'^{(s)}_i(t') - q_i^{(s)}(t)$ is contained a variation in the form of the function $q_i^{(s)}(t)$ and a variation in its argument $t$, that is also affected by the transformation of the group, *i.e.*,

$$\begin{aligned}\delta q_i^{(s)} = q'^{(s)}_i(t+\delta t) - q_i^{(s)}(t) &= q'^{(s)}_i(t) - q_i^{(s)}(t) + (dq_i^{(s)}(t)/dt)\delta t\\ &= \bar{\delta} q_i^{(s)}(t) + q_i^{(s+1)}(t)\delta t,\end{aligned}$$

where $\bar{\delta} q_i^{(s)}(t)$ is the variation in form of the function $q_i^{(s)}(t)$ at the instant of time $t$. Taking into account that for the variation in form

$$\bar{\delta} q_i^{(s)}(t) = d^s(\bar{\delta} q_i(t))/dt^s = d(\bar{\delta} q_i^{(s-1)}(t))/dt,$$

it follows that

$$\begin{aligned}\delta\mathcal{A}[q] &= \int_{t_1}^{t_2}\left(L\frac{d(\delta t)}{dt} + \frac{\partial L}{\partial t}\delta t + \frac{\partial L}{\partial q_i^{(s)}}\bar{\delta} q_i^{(s)}(t) + \frac{\partial L}{\partial q_i^{(s)}}\frac{dq_i^{(s)}}{dt}\delta t\right) dt\\ &= \int_{t_1}^{t_2}\left(\frac{d(L\delta t)}{dt} + \frac{\partial L}{\partial q_i^{(s)}}\bar{\delta} q_i^{(s)}(t)\right) dt. \qquad (1.47)\end{aligned}$$

Making the replacements

$$\begin{aligned}\frac{\partial L}{\partial q_i}\bar{\delta} q_i &= \frac{\partial L}{\partial q_i}\bar{\delta} q_i,\\ \frac{\partial L}{\partial q_i^{(1)}}\bar{\delta} q_i^{(1)} &= \frac{\partial L}{\partial q_i^{(1)}}\frac{d(\bar{\delta} q_i)}{dt} = \frac{d}{dt}\left(\frac{\partial L}{\partial q_i^{(1)}}\bar{\delta} q_i\right) - \frac{d}{dt}\left(\frac{\partial L}{\partial q_i^{(1)}}\right)\bar{\delta} q_i,\\ \frac{\partial L}{\partial q_i^{(2)}}\bar{\delta} q_i^{(2)} &= \frac{d}{dt}\left(\frac{\partial L}{\partial q_i^{(2)}}\bar{\delta} q_i^{(1)}\right) - \frac{d}{dt}\left(\frac{\partial L}{\partial q_i^{(2)}}\right)\bar{\delta} q_i^{(1)}\\ &= \frac{d}{dt}\left(\frac{\partial L}{\partial q_i^{(2)}}\bar{\delta} q_i^{(1)}\right) - \frac{d}{dt}\left(\frac{d}{dt}\left(\frac{\partial L}{\partial q_i^{(2)}}\right)\bar{\delta} q_i\right) + \frac{d^2}{dt^2}\left(\frac{\partial L}{\partial q_i^{(2)}}\right)\bar{\delta} q_i,\end{aligned}$$

$$\frac{\partial L}{\partial q_i^{(k)}}\bar{\delta} q_i^{(k)} = \frac{d}{dt}\left(\frac{\partial L}{\partial q_i^{(k)}}\bar{\delta} q_i^{(k-1)}\right) - \frac{d}{dt}\left(\frac{d}{dt}\left(\frac{\partial L}{\partial q_i^{(k)}}\right)\bar{\delta} q_i^{(k-2)}\right) + \cdots,$$

and collecting terms we get

$$\delta\mathcal{A}[q] = \int_{t_1}^{t_2} dt\left\{\frac{d(L\delta t)}{dt}\right.$$

$$+\bar{\delta}q_i\left[\frac{\partial L}{\partial q_i}-\frac{d}{dt}\left(\frac{\partial L}{\partial q_i^{(1)}}\right)+\cdots+(-1)^k\frac{d^k}{dt^k}\left(\frac{\partial L}{\partial q_i^{(k)}}\right)\right]$$
$$+\frac{d}{dt}\left(\bar{\delta}q_i\left[\frac{\partial L}{\partial q_i^{(1)}}-\frac{d}{dt}\left(\frac{\partial L}{\partial q_i^{(2)}}\right)+\cdots+(-1)^{k-1}\frac{d^{k-1}}{dt^{k-1}}\left(\frac{\partial L}{\partial q_i^{(k)}}\right)\right]\right)$$
$$+\frac{d}{dt}\left(\bar{\delta}q_i^{(1)}\left[\frac{\partial L}{\partial q_i^{(2)}}-\frac{d}{dt}\left(\frac{\partial L}{\partial q_i^{(3)}}\right)+\cdots+(-1)^{k-2}\frac{d^{k-2}}{dt^{k-2}}\left(\frac{\partial L}{\partial q_i^{(k)}}\right)\right]\right)$$
$$+\cdots+\frac{d}{dt}\left(\bar{\delta}q_i^{(k-1)}\left[\frac{\partial L}{\partial q_i^{(k)}}\right]\right)\Bigg\}.$$

The terms between squared brackets are precisely the conjugate momenta of order $s$, $p_{(s)}^i$, except the first one, which is the left-hand side of (1.5) and vanishes identically if the functions $q_i$ satisfy the dynamical equations. Now if we introduce in the integrand the variables $q_i$ that satisfy Euler-Lagrange equations, the variation of the action functional (1.47) is transformed into the variation of the action function along the classical trajectory, and therefore, the variation of the action function can be written as,

$$\delta A(x_1,x_2)=\int_{t_1}^{t_2}\frac{d}{dt}\left\{L\delta t+\left(\bar{\delta}q_i p_{(1)}^i+\bar{\delta}q_i^{(1)}p_{(2)}^i+\cdots+\bar{\delta}q_i^{(k-1)}p_{(k)}^i\right)\right\}dt, \tag{1.48}$$

with $p_{(s)}^i$ given in (1.26). If we replace in (1.48) the form variation $\bar{\delta}q_i^{(s)}=\delta q_i^{(s)}-q_i^{(s+1)}$, then

$$\delta A(x_1,x_2)=\int_{t_1}^{t_2}\frac{d}{dt}\left\{L\delta t+\delta q_i^{(s)}p_{(s+1)}^i-q_i^{(s+1)}p_{(s+1)}^i\delta t\right\}dt \tag{1.49}$$

with the usual addition convention. By substitution of the variations $\delta t$ and $\delta q_i^{(s)}$ in terms of the infinitesimal element of the group $\delta g^\alpha$, (1.44-1.46), we get:

$$\delta A(x_1,x_2)=\int_{t_1}^{t_2}\frac{d}{dt}\left\{\left(L-p_{(s)}^i q_i^{(s)}\right)M_\alpha+p_{(u+1)}^i M_{i\alpha}^{(u)}\right\}\delta g^\alpha dt, \tag{1.50}$$

with the following range for repeated indexes for the addition convention, $i=1,\ldots,n$, $s=1,\ldots,k$, $u=0,1,\ldots,k-1$ and $\alpha=1,\ldots,r$.

In the above integral we are using the solution of dynamical equations, and therefore the variation of the action function is

$$\delta A(x_1,x_2)=A(\delta gx_1,\delta gx_2)-A(x_1,x_2).$$

If it happens to be of first order in the group parameters in the form

$$\delta A(x_1,x_2) = B_\alpha(x_2)\delta g^\alpha - B_\alpha(x_1)\delta g^\alpha, \tag{1.51}$$

then equating to (1.50) we can perform the trivial time integral. By considering that parameters $\delta g^\alpha$ are arbitrary, rearranging terms depending on $t_1$ and $t_2$ on the left- and right-hand side, respectively, we get several observables that take the same values at the two arbitrary times $t_1$ and $t_2$. They are thus constants of the motion and represent the time conserved physical quantities,

$$N_\alpha = B_\alpha(x) - \left(L - p^i_{(s)}q_i^{(s)}\right) M_\alpha - p^i_{(s+1)}M^{(s)}_{i\alpha}, \qquad \alpha = 1,\ldots,r. \tag{1.52}$$

These are the $r$ Noether constants of the motion related to the infinitesimal transformations (1.51) of the action function under the corresponding $r$-parameter group.

To express the different magnitudes in terms of the kinematical variables, let us define the variables $x^j$ according to the rule: $x^0 = t$, $x^i = q_i$, $x^{n+i} = q_i^{(1)}, \ldots, x^{(k-1)n+i} = q_i^{(k-1)}$. Since $L = \widehat{L}/\dot{x}^0$, and $q_i^{(s)} = \dot{q}_i^{(s-1)}/\dot{x}^0$, the derivatives in the definition of the canonical momenta can be written as:

$$\frac{\partial L}{\partial q_i^{(s)}} = \frac{\partial(\widehat{L}/\dot{x}^0)}{\partial\left(\dot{x}^{(s-1)n+i}/\dot{x}^0\right)} = \frac{\partial \widehat{L}}{\partial \dot{x}^{(s-1)n+i}} = F_{(s-1)n+i}, \tag{1.53}$$

in terms of the functions $F_i$ of the expansion (1.11) of the Lagrangian. The different conjugate momenta appear in the form:

$$p^i_{(s)} = \sum_{r=0}^{k-s}(-1)^r \frac{d^r}{dt^r}F_{(r+s-1)n+i}, \tag{1.54}$$

in terms of the functions $F_i$ and their time derivatives. Therefore the Noether constants of the motion are written as

$$N_\alpha = B_\alpha(x) - \left(F_j\frac{\dot{x}^j}{\dot{x}^0} - p^i_{(s)}\frac{\dot{x}^{(s-1)n+i}}{\dot{x}^0}\right) M_\alpha - p^i_{(s+1)}M^{(s)}_{i\alpha}. \tag{1.55}$$

We see that the Noether constants of the motion $N_\alpha$ are finally expressed in terms of the functions $F_i$ and their time derivatives, of the functions $M^{(s)}_{i\alpha}$ that represent the way the different kinematical variables transform under infinitesimal transformations, and of the functions $B_\alpha$ which, as we shall see below, are related to the exponents of the group

$G$. Functions $F_i$ and their time derivatives are homogeneous functions of zero degree in terms of the derivatives of the kinematical variables $\dot{x}^i$. Functions $B_\alpha(x)$ and $M_{i\alpha}^{(s)}(x)$ depend only on the kinematical variables. Consequently, Noether constants of the motion are also homogeneous functions of zero degree in terms of the derivatives of kinematical variables and thus invariant under arbitrary changes of evolution parameter.

In the previous exposition we have assumed for simplicity that the dependence of the Lagrangian on the derivatives of the coordinates $q_i$ is up to the same order $k$ for each variable $q_i$. However, if this dependence is of a different order $k_i$ for every variable $q_i$, we have just to replace in the dynamical equations (1.6) the variable $k$ by the corresponding $k_i$, *i.e.*,

$$\sum_{s=0}^{k_i}(-1)^s \frac{d^s}{dt^s}\left(\frac{\partial L}{\partial q_i^{(s)}}\right) = 0 \qquad i = 1, \ldots, n, \tag{1.56}$$

and the definition of the conjugate momenta (1.54) by

$$p_{(s)}^i = \sum_{r=0}^{k_i - s}(-1)^r \frac{d^r}{dt^r} F_{(r+s-1)n+i}, \tag{1.57}$$

without any further change, and the kinematical space is not of dimension $kn+1$, but rather of dimension $k_1 + k_2 + \cdots + k_n + 1$.

## 8. LAGRANGIAN GAUGE FUNCTIONS

In the variational formulation of classical mechanics

$$\mathcal{A}[q] = \int_{t_1}^{t_2} L(t, q_i^{(s)}(t))dt \equiv \int_{\tau_1}^{\tau_2} L(x, \dot{x})d\tau, \tag{1.58}$$

$\mathcal{A}[q]$ is a path functional, *i.e.*, it takes in general different values for the different paths joining the fixed end points $x_1$ and $x_2$. Then it is necessary that $Ld\tau$ be a non-exact differential. Otherwise, if $Ldt = d\lambda$, then $\mathcal{A}[q] = \lambda_2 - \lambda_1$ and the functional does not distinguish between the different paths and the action function of the system from $x_1$ to $x_2$, $A(x_1, x_2) = \lambda(x_2) - \lambda(x_1)$, is expressed in terms of the potential function $\lambda(x)$, and is thus, path independent.

If $\lambda(x)$ is a real function defined on the kinematical space $X$ of a Lagrangian system with action function $A(x_1, x_2)$, then the function $A'(x_1, x_2) = A(x_1, x_2) + \lambda(x_2) - \lambda(x_1)$ is another action function equivalent to $A(x_1, x_2)$. In fact it gives rise by (1.13) to the Lagrangian $L'$ that differs from $L$ in a total $\tau$-derivative. [9]

Using (1.13), we have

$$L'(x, \dot{x}) = L(x, \dot{x}) + \frac{d\lambda}{d\tau}, \tag{1.59}$$

and therefore $L$ and $L'$ produce the same dynamical equations and $A(x_1, x_2)$ and $A'(x_1, x_2)$ are termed as equivalent action functions.

Let $G$ be a transformation group of the enlarged configuration space $(t, q_i)$, that can be extended to a transformation group of the kinematical space $X$. Let $g \in G$ be an arbitrary element of $G$ and $x' = gx$, the transform of $x$. Consider a mechanical system characterized by the action function $A(x_1, x_2)$ that under the transformation $g$ is changed into $A(x'_1, x'_2)$. If $G$ is a symmetry group of the system, *i.e.*, the dynamical equations in terms of the variables $x'$ are the same as those in terms of the variables $x$, this implies that $A(x'_1, x'_2)$ and $A(x_1, x_2)$ are necessarily equivalent action functions, and thus they will be related by:

$$A(gx_1, gx_2) = A(x_1, x_2) + \alpha(g; x_2) - \alpha(g; x_1). \tag{1.60}$$

The function $\alpha$ will be in general a continuous function of $g$ and $x$. This real function $\alpha(g; x)$ defined on $G \times X$ is called a **gauge function** of the group $G$ for the kinematical space $X$. Because of the continuity of the group it satisfies $\alpha(e; x) = 0$, $e$ being the neutral element of $G$. If the transformation $g$ is infinitesimal, let us represent it by the coordinates $\delta g^\sigma$, then $\alpha(\delta g; x) = \delta g^\sigma B_\sigma(x)$ to first order in the group parameters. The transformation of the action function takes the form

$$A(\delta g x_1, \delta g x_2) = A(x_1, x_2) + \delta g^\sigma B_\sigma(x_2) - \delta g^\sigma B_\sigma(x_1),$$

*i.e.*, in the form required by Noether's theorem to obtain the corresponding conserved quantities. In general, $B_\sigma$ functions for gauge-variant Lagrangians are obtained by

$$B_\sigma(x) = \left.\frac{\partial \alpha(g; x)}{\partial g^\sigma}\right|_{g=0}. \tag{1.61}$$

Because of the associative property of the group law, any gauge function satisfies the identity

$$\alpha(g'; gx) + \alpha(g; x) - \alpha(g'g; x) = \xi(g', g), \tag{1.62}$$

where the function $\xi$, defined on $G \times G$, is independent of $x$ and is an exponent of the group $G$.

This can be seen by the mentioned associative property of the group law. From (1.60) we get:

$$A(g'gx_1, g'gx_2) = A(x_1, x_2) + \alpha(g'g; x_2) - \alpha(g'g; x_1), \tag{1.63}$$

and also

$$A(g'gx_1, g'gx_2) = A(gx_1, gx_2) + \alpha(g'; gx_2) - \alpha(g'; gx_1)$$

$$= A(x_1, x_2) + \alpha(g; x_2) - \alpha(g; x_1) + \alpha(g'; gx_2) - \alpha(g'; gx_1),$$

**and therefore by identification of this with the above (1.63), when collecting terms with the same $x$ argument we get**

$$\alpha(g'; gx_2) + \alpha(g; x_2) - \alpha(g'g; x_2) = \alpha(g'; gx_1) + \alpha(g; x_1) - \alpha(g'g; x_1),$$

**and since $x_1$ and $x_2$ are two arbitrary points of $X$, this expression is (1.62) and defines a function $\xi(g', g)$, independent of $x$.**

If we substitute this function $\xi(g', g)$ into (1.41) we see that it is satisfied identically. For $g' = g = e$, it reduces to $\xi(e, e) = \alpha(e; x) = 0$, and thus $\xi$ is an exponent of $G$.

It is shown by Levy-Leblond, [10] that if $X$ is a homogeneous space of $G$, *i.e.*, if there exists a subgroup $H$ of $G$ such that $X = G/H$, then, the exponent $\xi$ is equivalent to zero on the subgroup $H$, and gauge functions for homogeneous spaces become:

$$\alpha(g; x) = \xi(g, h_x), \tag{1.64}$$

where $h_x$ is any group element of the coset space represented by $x \in G/H$.

For the Poincaré group $\mathcal{P}$ all its exponents are equivalent to zero and thus the gauge functions when $X$ is a homogeneous space of $\mathcal{P}$ are identically zero. Lagrangians of relativistic systems whose kinematical spaces are homogeneous spaces of $\mathcal{P}$ can be taken strictly invariant.

However, the Galilei group $\mathcal{G}$ has nontrivial exponents, that are characterized by a parameter $m$ that is interpreted as the total mass of the system, and thus Galilei Lagrangians for massive systems are not in general invariant under $\mathcal{G}$. In the quantum formalism, the Hilbert space of states of a massive nonrelativistic system carries a projective unitary representation of the Galilei group instead of a true unitary representation. [11]

# 9. RELATIVITY PRINCIPLE. KINEMATICAL GROUPS

A **Special Relativity Principle** postulates the existence of a class of equivalent observers, called inertial observers, for which the laws that govern the different mechanical processes must have the same form. This idea stated first for mechanical systems is usually extended to the analysis of all physical phenomena, thus giving rise to a kind of universality of the physical laws when considered their invariance for a wide class of observers, as large as possible.

Therefore any theoretical frame, devised to study some physical system, must contain the different interpretations provided by the different inertial observers. Then, when measuring some particular observable

$A$ by observers $O$ and $O'$, the formalism has to describe on theoretical grounds not only the values of their respective measurements but also the way both observers relate them, in terms of their relative situation or motion.

In this way a relativity principle states that there are no privileged observers for analyzing general physical phenomena, but rather that once a single particular observer is given, there exists a complete class of them and therefore the theoretical problem is translated to properly define this class, *i.e.*, how many they are and how they are related.

A second part of the relativity principle is that the way two inertial observers relate their respective measurements of any physical magnitude depends only on how they relate their measurements of arbitrary space-time events. It turns out that a sample of measurements of the instant and the three spatial coordinates of different space-time events are enough to establish the relationship between measurements of any other observable.

The composition of different space-time transformations produces new space-time transformations. Because the composition of applications is associative and given a transformation there always exists its inverse, we have a complete group of space-time transformations associated to any relativity principle.

This set of space-time transformations that defines the class of inertial observers related to some specific special relativity principle is called a **kinematical group.**

We know that the Poincaré $\mathcal{P}$ and Galilei $\mathcal{G}$ groups are two possible space-time transformation groups that define respectively the relativity principle for relativistic and nonrelativistic mechanical processes. These groups have only two independent Casimir invariants and therefore the intrinsic features of elementary particles these groups allow description are only mass and spin. To describe more and more intrinsic attributes the basic kinematical group has to be enlarged.

One of the goals of theoretical physics is to search for the appropriate kinematical group, an endeavour that is still open. Nevertheless, since the scope of this book is at least to achieve a description of spin, either relativistic or nonrelativistic, we shall deal with these two groups in the next chapters.

## 10. ELEMENTARY SYSTEMS

In Newtonian mechanics the simplest geometrical object is a point of mass $m$. Starting with massive points we can construct arbitrary systems of any mass and shape, and thus any distribution of matter. The massive point can be considered as the elementary particle of Newtonian mechanics. In the modern view of particle physics it corresponds to a spinless particle. We know that there exist spinning objects like electrons, muons, photons, neutrinos, quarks and perhaps many others, that can be considered as elementary particles in the sense that they cannot be considered as compound systems of other objects. Even more, we do not find in Nature any spinless elementary particles. It is clear that the Newtonian point does not give account of the spin structure of particles and the existence of spin is a fundamental intrinsic attribute of an elementary particle, which is lacking in Newtonian mechanics, but it has to be accounted for.

In quantum mechanics, Wigner's work [12] on the representations of the inhomogeneous Lorentz group provides a very precise mathematical definition of the concept of elementary particle. An **elementary particle** is a quantum mechanical system whose Hilbert space of pure states is the representation space of a projective unitary irreducible representation of the Poincaré group. Irreducible representations of the Poincaré group are characterized by two invariant parameters $m$ and $S$, the mass and the spin of the system, respectively. By finding the different irreducible representations, we can obtain the quantum description of massless and massive particles of any spin.

The very important expression of the above mathematical definition, with physical consequences, lies in the term **irreducible**. Mathematically it means that the Hilbert space is an invariant vector space under the group action and that it has no other invariant subspaces. But it also means that there are no other states for a single particle than those that can be obtained by just taking any arbitrary vector state, form all its possible images in the different inertial frames and finally produce the closure of all finite linear combinations of these vectors.

We see that starting from a single state and by a simple change of inertial observer, we obtain the state of the particle described in this new frame. Take the orthogonal part of this vector to the previous one and normalize it. Repeat this operation with another kinematical transformation acting on the same first state, followed by the corresponding orthonormalization procedure, as many times as necessary to finally obtain a complete orthonormal basis of the whole Hilbert space of states. All states in this basis are characterized by the physical parameters that define the first state and a countable collection of group transformations of the kinematical group $G$. And this can be done starting from any arbitrary state.

This idea allows us to define a concept of physical equivalence among states of any arbitrary quantum mechanical system in the following way: Two states are said to be physically equivalent if they can produce by the above method an orthonormal basis of the same Hilbert subspace, or in an equivalent way, if they belong to the same invariant subspace under the group action. It is easy to see that this is an equivalence relation. But if the representation is irreducible, all states are equivalent as basic pieces of physical information for describing the elementary system. There is one and only one single piece of basic physical information to describe an elementary object. That is what the term elementary might mean.

But this definition of elementary particle is a pure group theoretical one. The only quantum mechanical ingredient is that the group operates on a Hilbert space. Then one question arises. Can we translate this quantum mechanical definition into the classical domain and obtain an equivalent group theoretical definition for a classical elementary particle?

Following with the above idea, in classical mechanics we have no vector space structure to describe the states of a system. What we have are manifolds of points where each point represents either the configuration state, the kinematical state or the phase state of the system depending on which manifold we work. But the idea that any point that represents the state of an elementary particle is physically equivalent to any other, is in fact the very mathematical concept of homogeneity of the manifold under the corresponding group action. In this way, the irreducibility assumption of the quantum mechanical definition is translated into the realm of classical mechanics in the concept of homogeneity of the corresponding manifold under the Poincaré group or any other kinematical group we consider as the symmetry group of the theory. But, what manifold? Configuration space? Phase space? The answer as has been shown in previous works, [13] is that the appropriate manifold is the **kinematical space.**

In the Lagrangian approach of classical mechanics, the kinematical space $X$ is the manifold where the dynamics is developed as an input-output formalism. When quantizing the system we will obtain the natural link between the classical and quantum formalisms through Feynman's path integral approach, as will be shown later. This manifold is the natural space on which to define the Hilbert space structure of the quantized system. In a formal way we can say that each point $x \in X$ that represents the kinematical state of a system is spread out and is transformed through Feynman's quantization into the particle wave function $\psi(x)$ defined around $x$. This wave function is in general a squared integrable complex function defined on $X$. This will be done in Chapter 4.

We can also analyze the elementarity condition from a different point of view. Let us consider an inertial observer $O$ that is measuring a certain observable $A(\tau)$ of an arbitrary system at an instant $\tau$. This observable takes the value $A'(\tau)$ for a different inertial observer $O'$. It can be expressed in terms of $A(\tau)$ in the form $A'(\tau) = f(A(\tau), g)$, where $g$ is the kinematical transformation between both observers. At instant $\tau + d\tau$, the corresponding measured values of that observable will have changed but $A'(\tau + d\tau) = f(A(\tau + d\tau), g)$ with the same $g$ as before, and assuming that the evolution parameter $\tau$ is group invariant.

But if the system is elementary, we take as an assumption that the modifications of the observables produced by the dynamics can be compensated by a change of inertial reference frame. Then, given an observer $O$, it is always possible to find at instant $\tau + d\tau$ another inertial observer $O'$ that measures the value of an essential observable $A'(\tau + d\tau)$ with the same value as $O$ does at instant $\tau$, *i.e.*, $A'(\tau + d\tau) \equiv A(\tau)$. If the system is not elementary, this will not be possible in general because the external interaction might change its internal structure, and thus it will not be possible to compensate the modification of the observable by a simple change of inertial observer. Think about a non-relativistic description of an atom that goes into some excited state. The new internal energy, which is Galilei invariant, cannot be transformed into the old one by a simple change of reference frame.

But the essential observables are the kinematical variables. From the dynamical point of view we can take as initial and final points any $x_1$ and $x_2 \in X$, compatible with the causality requirements. This means that any $x$ can be considered as the initial point of the variational formalism. In this way, at any instant $\tau$ if the system is elementary, we can find an infinitesimal kinematical transformation $\delta g(\tau)$ such that

$$x'(\tau + d\tau) = f(x(\tau + d\tau), \delta g(\tau)) \equiv x(\tau),$$

or by taking the inverse of this transformation,

$$x(\tau + d\tau) = f^{-1}(x(\tau), \delta g(\tau)).$$

This equation represents the dynamical evolution equation in $X$ space. Knowledge of the initial state $x_1$ and the function $\delta g(\tau)$ completely determines the evolution of the system. In general, $\delta g(\tau)$ will depend on the instant $\tau$, because the change of the observables depends on the external interaction. But if the system is elementary and the motion is free, all $\delta g(\tau)$ have necessarily to be the same, and thus $\tau$ independent. We cannot distinguish in a free motion one instant from any other. Then, starting from $x_1$ we shall arrive at $x_2$ by the continuous action of the same infinitesimal group element $\delta g$, and the free particle motion is the action of the one-parameter group generated by $\delta g$ on the initial state. Therefore, there should exist a finite group element $g \in G$ such that $x_2 = gx_1$. We thus arrive at the:

**Definition: A classical elementary particle** is a Lagrangian system whose kinematical space $X$ is a **homogeneous space** of the kinematical group $G$.

Usually the Lagrangian of any classical Newtonian system is restricted to depend only on the first order derivative of each of the coordinates $q_i$ that represent the independent degrees of freedom, or equivalently, that the $q_i$ satisfy second order differential equations. But at this stage, if we do not know what are the basic variables we need to describe our elementary system, how can we state that they necessarily satisfy second order differential equations? If some of the degrees of freedom, say $q_1$, $q_2$ and $q_3$, represent the center of mass position of the system, Newtonian mechanics implies that in this particular case $L$ will depend on the first order derivatives of these three variables. But what about other degrees of freedom? It is this condition on the kinematical space to be considered as a homogeneous space of $G$, as the mathematical statement of elementarity, that will restrict the dependence of the Lagrangian on these higher order derivatives. It is this definition of elementary particle with the proper election of the kinematical group, which will supply information about the structure of the Lagrangian.

The Galilei and Poincaré groups are ten-parameter Lie groups and therefore the largest homogeneous space we can find for these groups is a ten-dimensional manifold. The variables that define the different homogeneous spaces will share the same domains and dimensions as the corresponding variables we use to parameterize the group. Both groups, as we shall see later, are parameterized in terms of the following variables $(b, \boldsymbol{a}, \boldsymbol{v}, \boldsymbol{\alpha})$ with domains and dimensions respectively like $b \in \mathbb{R}$ that represents the time parameter of the time translation and $\boldsymbol{a} \in \mathbb{R}^3$, the three spatial coordinates for the space translation. Parameter $\boldsymbol{v} \in \mathbb{R}^3$ are the three components of the relative velocity between the inertial observers, restricted to $v < c$ in the Poincaré case. Finally $\boldsymbol{\alpha} \in SO(3)$ are three dimensionless variables which characterize the relative orientation of the corresponding Cartesian frames and whose compact domain is expressed in terms of a suitable parametrization of the rotation group.

In this way the maximum number of kinematical variables, for a classical elementary particle, is also ten. We represent them by $x \equiv (t, \boldsymbol{r}, \boldsymbol{u}, \boldsymbol{\alpha})$ with the same domains and dimensions as above and interpret them respectively as the time, position, velocity and orientation of the particle.

Because the Lagrangian must also depend on the next order derivatives of the kinematical variables, we arrive at the conclusion that $L$ must also depend on the acceleration and angular velocity of the particle. The particle is a system of six degrees of freedom, three $\boldsymbol{r}$, represent the position of a point and other three $\alpha$, its orientation in space. We can visualize this by assuming a system of three orthogonal unit vectors linked to point $r$ as a body frame. But the Lagrangian will depend up to the second time derivative of $r$, or acceleration of that point, and on the first derivative of $\alpha$, *i.e.*, on the angular velocity. The Galilei and Poincaré groups lead to generalized Lagrangians depending up to second order derivatives of the position.

By this definition it is the kinematical group $G$ that implements the special Relativity Principle that completely determines the structure of the kinematical space where the Lagrangians that represent classical elementary particles have to be defined. [14] Point particles are particular cases of the above definition and their kinematical spaces are just the quotient structures between the group $G$ and subgroup of rotations and boosts, and thus their kinematical variables reduce only to time and position $(t, r)$. Therefore, the larger the kinematical group of space-time transformations, the greater the number of allowed classical variables to describe elementary objects with a more detailed and complex structure. In this way, the proposed formalism can be accommodated to any symmetry group.

## 10.1 ELEMENTARY LAGRANGIAN SYSTEMS

An elementary Lagrangian system will be characterized by the Lagrangian function $L(x, \dot{x})$ where the variables $x \in X$ lie in a homogeneous space $X$ of $G$. $L$ is a homogeneous function of first degree of the derivatives of the kinematical variables, and this allows us to write

$$L(x, \dot{x}) = F_i(x, \dot{x})\, \dot{x}^i. \tag{1.65}$$

Functions $F_i(x, \dot{x})$ are therefore homogeneous functions of zero degree in the variables $\dot{x}^i$ and summation convention on repeated indexes as usual is assumed.

Under $G$, $x$ transforms as $x' = gx$ or more explicitly its coordinates by $x'^i = f^i(g, x)$, and their derivative variables

$$\dot{x}'^i = \frac{\partial x'^i}{\partial x^j} \dot{x}^j, \tag{1.66}$$

transform like the components of a contravariant vector.

The Lagrangian transforms under $G$,

$$L(x'(x), \dot{x}'(x, \dot{x})) = L(x, \dot{x}) + \frac{d\alpha(g; x)}{d\tau}, \tag{1.67}$$

i.e.,

$$F_i(x', \dot{x}')\dot{x}'^i = F_j(x, \dot{x})\dot{x}^j + \frac{\partial \alpha(g;x)}{\partial x^j}\dot{x}^j. \tag{1.68}$$

Taking into account the way the different variables transform, we thus arrive at:

$$F_i(x', \dot{x}') = \frac{\partial x^j}{\partial x'^i}\left[F_j(x, \dot{x}) + \frac{\partial \alpha(g;x)}{\partial x^j}\right]. \tag{1.69}$$

In the case when $\alpha(g;x) = 0$, they transform like the components of a covariant vector over the kinematical space $X$. But in general this will not be the case and $\alpha(g;x)$ contains basic physical information about the system.

We thus find that for a fixed kinematical space $X$, the knowledge of the group action of $G$ on $X$, and the gauge function $\alpha(g;x)$, will give us information about the possible structure of the functions $F_i(x, \dot{x})$, and therefore about the structure of the Lagrangian.

In practice, if we restrict ourselves to the Galilei $\mathcal{G}$ and Poincaré $\mathcal{P}$ groups, we see that $\mathcal{P}$ has gauge functions equivalent to zero and thus Poincaré Lagrangians that describe elementary particles can be taken strictly invariant. In the case of the Galilei group, it has only one class of gauge functions that define the mass of the system, and thus nonrelativistic Lagrangians will be in general not invariant. In the particular case of Galilei invariant Lagrangians, they will describe massless systems, as we shall see in the next chapters.

## 11. THE FORMALISM WITH THE SIMPLEST KINEMATICAL GROUPS

Let us start first with the Newtonian point particle. By definition its kinematical variables for its Lagrangian formalism are time $t$ and position $\boldsymbol{r}$. To establish a relativity principle we have to fix the kinematical group of space-time transformations that defines the class of equivalent or inertial observers. Let us assume that the set of inertial observers are all at rest with their Cartesian frames parallel with respect to each other. Then the kinematical relation between observers is given by the group action

$$t'(\tau) = t(\tau) + b, \quad \boldsymbol{r}'(\tau) = \boldsymbol{r}(\tau) + \boldsymbol{a}, \tag{1.70}$$

at any instant of the evolution parameter $\tau$, so that the kinematical group is the space-time translation group $\{\mathbb{R}^4, +\}$. It is a four-parameter group with four generators $H$ and $\boldsymbol{P}$, where as usual we represent three-vector magnitudes by bold face characters. In the action (1.70) and according to the general representation (1.35) the generators are given by the differential operators

$$H = \partial/\partial t, \quad \boldsymbol{P} = \nabla, \tag{1.71}$$

and satisfy the commutation relations

$$[H, P_i] = [P_i, P_j] = 0. \tag{1.72}$$

This group has four functionally invariant Casimir operators, which are precisely these four generators. The group law $g'' = g'g$, is

$$b'' = b' + b, \quad \boldsymbol{a}'' = \boldsymbol{a}' + \boldsymbol{a}. \tag{1.73}$$

We see that the kinematical space of our point particle is in fact isomorphic to the whole kinematical group $\{\mathbb{R}^4, +\}$, where we consider the group action from the left, so that our kinematical variables $x \equiv (t, \boldsymbol{r})$ have the same domains and dimensions as the group parameters $(b, \boldsymbol{a})$, respectively. The kinematical space $X$ is clearly the largest homogeneous space of the kinematical group.

The Lagrangian for a point particle will be a function of $(t, \boldsymbol{r}, \dot{t}, \dot{\boldsymbol{r}})$, and according to (1.11), a homogeneous function of first degree in terms of $(\dot{t}, \dot{\boldsymbol{r}})$. Then it can be written as

$$L = T\dot{t} + \boldsymbol{R} \cdot \dot{\boldsymbol{r}}, \tag{1.74}$$

with $T = \partial L/\partial \dot{t}$ and $\boldsymbol{R} = \partial L/\partial \dot{\boldsymbol{r}}$, where the vector function $\boldsymbol{R}$ is a shorthand notation for $R_i = \partial L/\partial \dot{r}_i$. Since this Abelian group has no nontrivial exponents, the Lagrangian will be invariant under this group, and thus independent of $(t, \boldsymbol{r})$. By assuming a group invariant evolution parameter $\tau$, the derivatives of the kinematical variables transform as

$$\dot{t}'(\tau) = \dot{t}(\tau), \quad \dot{\boldsymbol{r}}'(\tau) = \dot{\boldsymbol{r}}(\tau), \tag{1.75}$$

by simply taking the $\tau$-derivative in (1.70). Then the functions $T$ and $\boldsymbol{R}$ transform as $T' = T$ and $\boldsymbol{R}' = \boldsymbol{R}$. Dynamical equations can be any autonomous second order differential equation of the functions $\boldsymbol{r}(t)$, not depending explicitly on the variables $\boldsymbol{r}$ and $t$.

When applying Noether's theorem for this kinematical group we obtain as constants of the motion, the energy $H = -T$ and the linear momentum $\boldsymbol{P} = \boldsymbol{R}$. Possible Lagrangians for this kind of systems are very general and might be any arbitrary function of the components of the velocity $dr_i/dt$. The homogeneity condition in terms of kinematical variables implies that, for instance, expressions of the form

$$a_{ij}\frac{\dot{r}_i \dot{r}_j}{\dot{t}} + b_{ijk}\frac{\dot{r}_i \dot{r}_j \dot{r}_k}{\dot{t}^2} + \cdots,$$

with arbitrary constants $a_{ij}$, $b_{ijk}$, etc., or expressions like

$$\sqrt{a_0\dot{t}^2 + a_i\dot{t}\dot{r}_i + b_{ij}\dot{r}_i\dot{r}_j + c_{ijk}\dot{r}_i\dot{r}_j\dot{r}_k/\dot{t} + \cdots} \quad ,$$

that are homogeneous of first degree in the derivatives, can be taken as possible Lagrangians. Linear terms of the form $a\dot{t}+b_i\dot{r}_i = d(at+b_ir_i)/d\tau$ are also homogeneous functions of first degree in the derivatives, but are not considered because they can be written as a total $\tau$-derivative, and can be withdrawn.

Let us go further and extend the class of inertial observers in such a way as to allow them to set up their Cartesian frames in arbitrary orientations. This amounts physically to assuming that the three-dimensional space is **isotropic**. All directions in three-space, are equivalent. Then, our kinematical transformations are

$$t'(\tau) = t(\tau) + b, \quad \boldsymbol{r}'(\tau) = R(\boldsymbol{\beta})\boldsymbol{r}(\tau) + \boldsymbol{a}, \tag{1.76}$$

where $R(\boldsymbol{\beta})$ represents a rotation matrix written in terms of three parameters $\beta_i$ of a suitable parametrization of the rotation group. This new kinematical group is the Euclidean group in three dimensions including time translations. It is usually called the Aristotle group $G_A$. It has in addition to $H$ and $\boldsymbol{P}$, three new generators $\boldsymbol{J}$, that in the above action (1.76) are given by the operators $\boldsymbol{J} = \boldsymbol{r}\times\nabla$, and the commutation relations of the Lie algebra of this group are

$$[H, P_i] = [H, J_i] = [P_i, P_j] = 0, \; [J_i, J_j] = -\epsilon_{ijk}J_k, \; [J_i, P_j] = -\epsilon_{ijk}P_k. \tag{1.77}$$

It has three functionally independent Casimir operators, $H$, $\boldsymbol{P}^2$ and $\boldsymbol{P}\cdot\boldsymbol{J}$, but it does not have central extensions, and thus no nontrivial exponents. Lagrangians in this case will be also invariant. This additional rotation invariance leads to the conclusion that $L$, which still has the general form (1.74), will be an arbitrary function of $\dot{\boldsymbol{r}}^2$.

When applying Noether's theorem, we have in addition to the energy $H = -\partial L/\partial\dot{t} = -T$ and linear momentum $\boldsymbol{P} = \partial L/\partial\dot{\boldsymbol{r}} = \boldsymbol{R}$, a new observable, related to the invariance under rotations, the angular momentum $\boldsymbol{J} = \boldsymbol{r}\times\boldsymbol{P}$.

The group elements are parameterized in terms of the seven parameters $g \equiv (b, \boldsymbol{a}, \boldsymbol{\beta})$ and the group $G_A$ has the composition law $g'' = g'g$ given by:

$$b'' = b' + b, \quad \boldsymbol{a}'' = R(\boldsymbol{\beta}')\boldsymbol{a} + \boldsymbol{a}', \quad R(\boldsymbol{\beta}'') = R(\boldsymbol{\beta}')R(\boldsymbol{\beta}). \tag{1.78}$$

We clearly see, by comparing (1.78) with (1.76), that the kinematical space $X$ of our point particle is isomorphic to the homogeneous space of the group, $X \simeq G_A/SO(3)$. It corresponds to the coset space of elements of the form $(t, \boldsymbol{r}, \boldsymbol{0})$ when acting on the subgroup $SO(3)$ of elements $(0, \boldsymbol{0}, \boldsymbol{\beta})$. The kinematical variables $(t, \boldsymbol{r})$ span the same manifold and have the same dimensions as the set of group elements of the form $(b, \boldsymbol{a}, \boldsymbol{0})$.

But once we have a larger symmetry group, we can extend our definition of elementary particle to the whole group $G_A$. The physical system might have three new kinematical variables $\boldsymbol{\alpha}$, the angular variables. In a $\tau$-evolution description of the dynamics, with the identification in $g'' = g'g$ of $g'' \equiv x'(\tau)$, $g \equiv x(\tau)$ and $g'$ playing the role of the group element $g$ acting on the left, we get $x' = gx$. Taking into account (1.78), they explicitly transform as:

$$t'(\tau) = t(\tau) + b, \; \boldsymbol{r}'(\tau) = R(\boldsymbol{\beta})\boldsymbol{r}(\tau) + \boldsymbol{a},$$

as in (1.76) and also for the new degrees of freedom

$$R(\boldsymbol{\alpha}'(\tau)) = R(\boldsymbol{\beta})R(\boldsymbol{\alpha}(\tau)). \tag{1.79}$$

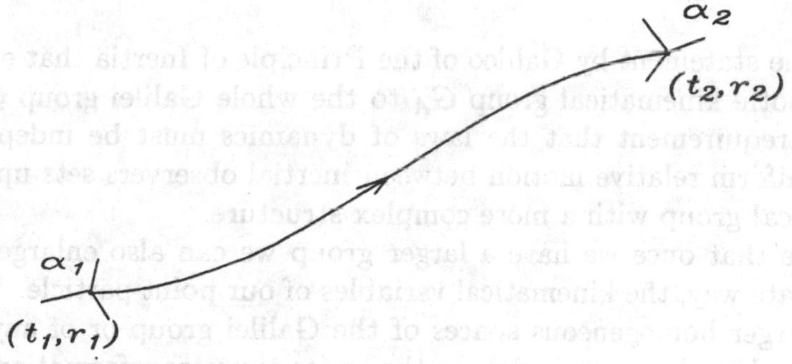


*Figure 1.2.* **Evolution in kinematical $\{t, \boldsymbol{r}, \boldsymbol{\alpha}\}$ space.**

The seven kinematical variables of our elementary particle are now time, position and orientation. Our system can be interpreted as a point with three unit vectors $\boldsymbol{e}_i$ attached to it. This local frame can rotate, and rotation of this frame is described by the evolution of the new variables $\boldsymbol{\alpha}$. We schematically represent them as in Figure 1.2. The components of unit vectors $\boldsymbol{e}_i$ are the matrix elements of the orthogonal rotation matrix $(\boldsymbol{e}_i)_j = R_{ji}(\boldsymbol{\alpha}(\tau))$, when considered by columns. These nine components or matrix elements are expressed in this way in terms of the three essential parameters $\boldsymbol{\alpha}$, the three new degrees of freedom.

Then the Lagrangian for this system will be also a function of $\boldsymbol{\alpha}$ and $\dot{\boldsymbol{\alpha}}$, or equivalently of the angular velocity $\boldsymbol{\omega}$ of the moving frame. The homogeneity condition allows us to write $L$ as

$$L = T\dot{t} + \boldsymbol{R} \cdot \dot{\boldsymbol{r}} + \boldsymbol{W} \cdot \boldsymbol{\omega}, \tag{1.80}$$

where $T$ and $\boldsymbol{R}$, are defined as before (1.74), and $\boldsymbol{W} = \partial L/\partial \boldsymbol{\omega}$. These functions transform under the group $G_A$ as:

$$T' = T, \quad \boldsymbol{R}' = R(\boldsymbol{\beta})\boldsymbol{R}, \quad \boldsymbol{W}' = R(\boldsymbol{\beta})\boldsymbol{W}.$$

As a remark all functions $T$, $\boldsymbol{R}$ and $\boldsymbol{W}$ are translation invariant. Now, total energy is $H = -T$, linear momentum $\boldsymbol{P} = \boldsymbol{R}$, but the angular momentum takes the form

$$\boldsymbol{J} = \boldsymbol{r} \times \boldsymbol{P} + \boldsymbol{W}.$$

The particle, in addition to the angular momentum $\boldsymbol{r} \times \boldsymbol{P}$, has now a translation invariant angular momentum. The particle, a point and a rotating frame like the usual description of a rigid body, has spin $\boldsymbol{W}$. Nevertheless we have seen that while restricting ourselves to the Aristotle kinematical group we do not obtain generalized Lagrangians. All above Lagrangians depend only on the first order derivative of variables $\boldsymbol{r}$ and $\boldsymbol{\alpha}$.

It is the statement by Galileo of the Principle of Inertia that enlarges the Aristotle kinematical group $G_A$ to the whole Galilei group $\mathcal{G}$. The physical requirement that the laws of dynamics must be independent of the uniform relative motion between inertial observers sets up a new kinematical group with a more complex structure.

We see that once we have a larger group we can also enlarge, in an appropriate way, the kinematical variables of our point particle. We can obtain larger homogeneous spaces of the Galilei group or of any other kinematical group we consider, as the space-time transformation group that implements the relativity principle. But this will be guided by physical conditions. If we want to describe electromagnetic interactions, the relativity principle is extended to include invariance of Maxwell's equations. Then this principle requires that the kinematical group must be, at least, the Poincaré group. But the formalism can be further extended to the conformal group or by considering hyperbolic rotations in space, changes of scale in space and time or more general transformations to more and more general groups.

Therefore the variables we need to describe the structure of elementary particles are directly connected with the structure of the kinematical group. This is one of the reasons why we call to the different homogeneous spaces of the corresponding kinematical group the kinematical space of the system, and to the above formalism the Kinematical Theory of Elementary Particles.

When restricted to ten-parameter groups, Bacry and Levy-Leblond [15] found by very general hypothesis different possible kinematical groups. Among them there are the Galilei $\mathcal{G}$ and Poincaré $\mathcal{P}$ groups, but also the De Sitter groups $SO(3,2)$ and $SO(4,1)$, which can be considered to describe the relativity principle in a cosmological background in a static curved space-time universe of constant curvature. They depend on two universal constants $R$, the scale factor of the universe that defines its constant curvature, and $c$ the speed of light that defines the upper bound of the velocity of signals, and also of a signature $k = \pm 1$ that selects either positive or negative curvature. With these two universal scalars, all ten group parameters are dimensionless. When the curvature vanishes, *i.e.*, $R \to \infty$, both De Sitter groups have as a limit the Poincaré group. De Sitter space-time, either of positive or negative curvature, derives into flat Minkowski space-time. When $c \to \infty$, then $\mathcal{P} \to \mathcal{G}$. Each time a universal constant disapears by a limiting process, a group parameter obtains a physical dimension, thus producing what has been called a **dimensional splitting**. [16]

In the next two chapters we shall deal with the Galilei and Poincaré groups respectively. We will analyze different kinds of nonrelativistic and relativistic spinning particles these groups allow us to define, when considered as the kinematical groups of the theory and according to the formalism proposed so far. Extension of the approach to the conformal group is delayed till Chapter 6.

## Notes

1 E.P. Wigner, *Ann. Math.* **40**, 149 (1939).

2 M. Ostrogradsky, *Mem. Acad. St. Petersburg*, **6**(4), 385 (1850).

3 P.A.M. Dirac, *Proc. Cam. Phil. Soc.* **29**, 389 (1933): "a greater elegance is obtained", "a symmetrical treatment suitable for relativity."

4 R. Courant, D. Hilbert, *Methods of Mathematical Physics*, Vol. 1, Interscience, N.Y. (1970); I.M. Gelfand, S.V. Fomin, *Calculus of Variations* Prentice Hall, Englewood Cliffs, N.J. (1963).

5 E.T.Whittaker, *Analytical Dynamics*, Cambridge University Press, Cambridge (1927), p. 265.

6 V.Bargmann, *Ann. Math.* **59**, 1 (1954).

7 E.P. Wigner, *Group theory and its application to the quantum mechanics of atomic spectra*, Acad. Press, NY (1959); V. Bargmann, *J. Math. Phys.* **5**, 862 (1964).

8 see ref.7 and J.M. Levy-Leblond, *Comm. Math. Phys.*, **12**, 64 (1969); A.A. Kirillov, *Élements de la theorie des représentations*, Mir, Moscow (1974).

9 J.M. Levy-Leblond, *Comm. Math. Phys.* **12**, 64 (1969).

10 see ref.10.

11 see ref.9 and also J.M. Levy-Leblond, *Galilei Group and Galilean Invariance*, in E.M. Loebl, *Group Theory and its applications*, Acad. Press, NY (1971), vol. 2, p. 221.

12 see ref.1.

13 M. Rivas, *J. Phys.* A **18**, 1971 (1985); *J. Math. Phys.* **30**, 318 (1989); *J. Math. Phys.* **35**, 3380 (1994).

14 see ref.14.

15 H. Bacry and J.M. Levy-Leblond, *J. Math. Phys.* **9**, 1605 (1968).

16 J.F.Cariñena, M.A. del Olmo and M. Santander, *J. Phys.* A **14**, 1 (1981).

# Chapter 2

# NONRELATIVISTIC ELEMENTARY PARTICLES

In this chapter different kinds of nonrelativistic particles are analyzed, by considering the Galilei group $\mathcal{G}$ as the kinematical group of space-time transformations that implements the relativity principle. To begin with, we shall consider first a detailed analysis of the Galilei group and its Lie algebra in terms of some particular parametrization of the rotation group, to properly describe the orientation variables when considered as kinematical variables of an elementary particle. In this way we shall be able to obtain a mathematical and physical description of the variables of the different homogeneous spaces of $\mathcal{G}$.

One important feature of the spinning particles, either relativistic or nonrelativistic, is the separation between the center of mass and center of charge. Thus when talking about the position of the particle we have to distinguish between the position of the center of charge and the position of the center of mass. This comes about from the analysis of the conserved kinematical momentum $\boldsymbol{K}$ and linear momentum $\boldsymbol{P}$, which is not lying along the velocity of the charge. Linear momentum is related as usual with the velocity of the center of mass. The free particle can be interpreted as a point charge with a constant magnetic moment and an oscillating electric dipole moment. This dipole structure is produced by the zitterbewegung, which is interpreted as the motion of the point charge around the center of mass.

The spin is related to the zitterbewegung and also to the rotation of the particle, and will be expressed in terms of the kinematical variables as any other observable. It will be unnecessary to postulate torque-like dynamical equations for the spin variables because its dynamics is a consequence of the evolution of the kinematical variables, and we shall obtain torque equations of the usual form that will suggest the interpretation of the dipole structure of the particle.

## 1. GALILEI GROUP

The Galilei group is a group of space-time transformations characterized by ten parameters $g \equiv (b, \boldsymbol{a}, \boldsymbol{v}, \boldsymbol{\alpha})$. The action of $g$ on a space-time point $x \equiv (t, \boldsymbol{r})$ is given by $x' = gx$, and is considered in the form

$$x' = \exp(bH)\exp(\boldsymbol{a}\cdot\boldsymbol{P})\exp(\boldsymbol{v}\cdot\boldsymbol{K})\exp(\boldsymbol{\alpha}\cdot\boldsymbol{J})\,x$$

as the action of a rotation followed by a pure Galilei transformation and finally a space and time translation. In this way all parameters that define each one-parameter subgroup are normal, because the exponential mapping works. Explicitly

$$t' = t + b, \tag{2.1}$$
$$\boldsymbol{r}' = R(\boldsymbol{\alpha})\boldsymbol{r} + \boldsymbol{v}t + \boldsymbol{a}, \tag{2.2}$$

and the composition law of the group $g'' = g'g$ is:

$$b'' = b' + b, \tag{2.3}$$
$$\boldsymbol{a}'' = R(\boldsymbol{\alpha}')\boldsymbol{a} + \boldsymbol{v}'b + \boldsymbol{a}', \tag{2.4}$$
$$\boldsymbol{v}'' = R(\boldsymbol{\alpha}')\boldsymbol{v} + \boldsymbol{v}', \tag{2.5}$$
$$R(\boldsymbol{\alpha}'') = R(\boldsymbol{\alpha}')R(\boldsymbol{\alpha}). \tag{2.6}$$

For rotations we shall alternatively use two different parametrizations. One is the normal or canonical parametrization in terms of a three vector $\boldsymbol{\alpha} = \alpha\boldsymbol{n}$, where $\boldsymbol{n}$ is a unit vector along the rotation axis, and $\alpha \in [0, \pi]$ is the clockwise rotation angle in radians, when looking along $\boldsymbol{n}$. Another, in terms of a three vector $\boldsymbol{\mu} = \boldsymbol{n}\tan(\alpha/2)$, which is more suitable to represent algebraically the composition of rotations.

The rotation matrix $R(\boldsymbol{\alpha}) = \exp(\boldsymbol{\alpha}\cdot\boldsymbol{J})$ is expressed in terms of the normal parameters $\alpha_i$ and in terms of the antisymmetric matrix generators $J_i$ which have the usual matrix representation

$$J_1 = \begin{pmatrix} 0 & 0 & 0 \\ 0 & 0 & -1 \\ 0 & 1 & 0 \end{pmatrix}, \quad J_2 = \begin{pmatrix} 0 & 0 & 1 \\ 0 & 0 & 0 \\ -1 & 0 & 0 \end{pmatrix}, \quad J_3 = \begin{pmatrix} 0 & -1 & 0 \\ 1 & 0 & 0 \\ 0 & 0 & 0 \end{pmatrix},$$

and satisfy the commutation relations $[J_i, J_k] = \epsilon_{ikl}J_l$, such that if we write the normal parameters $\boldsymbol{\alpha} = \alpha\boldsymbol{n}$ in terms of the rotation angle $\alpha$ and the unit vector $\boldsymbol{n}$ along the rotation axis, it is written as

$$R(\boldsymbol{\alpha})_{ij} = \delta_{ij}\cos\alpha + n_i n_j(1 - \cos\alpha) - \epsilon_{ijk}n_k\sin\alpha, \quad i, j, k = 1, 2, 3. \tag{2.7}$$

In the parametrization $\boldsymbol{\mu} = \boldsymbol{n}\tan(\alpha/2)$, the rotation matrix is

$$R(\boldsymbol{\mu})_{ij} = \frac{1}{1+\mu^2}\left((1-\mu^2)\delta_{ij} + 2\mu_i\mu_j - 2\epsilon_{ijk}\mu_k\right), \quad i,j,k = 1,2,3. \tag{2.8}$$

In terms of these variables, $R(\boldsymbol{\mu}'') = R(\boldsymbol{\mu}')R(\boldsymbol{\mu})$ is equivalent to

$$\boldsymbol{\mu}'' = \frac{\boldsymbol{\mu}' + \boldsymbol{\mu} + \boldsymbol{\mu}' \times \boldsymbol{\mu}}{1 - \boldsymbol{\mu}' \cdot \boldsymbol{\mu}}. \tag{2.9}$$

This can be seen in a simple manner by using the homomorphism between the rotation group and the group $SU(2)$. The matrix generators of $SU(2)$ are $\boldsymbol{J} = -i\boldsymbol{\sigma}/2$ in terms of Pauli matrices $\boldsymbol{\sigma}$. In the normal parametrization the rotation matrix $\exp(\boldsymbol{\alpha}\cdot\boldsymbol{J}) = \exp(-i\boldsymbol{\alpha}\cdot\boldsymbol{\sigma}/2)$ is written in the form

$$R(\boldsymbol{\alpha}) = \cos(\alpha/2)\mathbb{I} - i(\boldsymbol{n}\cdot\boldsymbol{\sigma})\sin(\alpha/2).$$

By defining $\boldsymbol{\mu} = \boldsymbol{n}\tan(\alpha/2)$, this rotation matrix is expressed as

$$R(\boldsymbol{\mu}) = \frac{1}{\sqrt{1+\mu^2}}(\mathbb{I} - i\boldsymbol{\mu}\cdot\boldsymbol{\sigma}), \tag{2.10}$$

where $\mathbb{I}$ is the $2\times 2$ unit matrix and in this form we can get the composition law (2.9). [1]

If the rotation is of value $\pi$, then eqs. (2.7) or (2.8) lead to

$$R(\boldsymbol{n},\pi)_{ij} = -\delta_{ij} + 2n_in_j.$$

Even if the two rotations $R(\boldsymbol{\mu})$ and $R(\boldsymbol{\mu}')$ involved in (2.9) are of value $\pi$, although $\tan(\pi/2) = \infty$, this expression is defined and gives:

$$\boldsymbol{n}''\tan(\alpha''/2) = \frac{\boldsymbol{n}\times\boldsymbol{n}'}{\boldsymbol{n}\cdot\boldsymbol{n}'}.$$

The absolute value of this relation leads to $\tan(\alpha''/2) = \tan\theta$, *i.e.*, $\alpha'' = 2\theta$, where $\theta$ is the angle between the two unit vectors $\boldsymbol{n}$ and $\boldsymbol{n}'$. We obtain the known result that every rotation of value $\alpha$ around an axis $\boldsymbol{n}$ can be obtained as the composition of two rotations of value $\pi$ around two axes orthogonal to $\boldsymbol{n}$ and separated by an angle $\alpha/2$.

For the orientation variables we shall use throughout the book the early Greek variables $\boldsymbol{\alpha}, \boldsymbol{\beta}, \ldots$ whenever we consider the normal parametrization, while for the $\tan(\alpha/2)$ parametrization we will express rotations in terms of the intermediate Greek variables $\boldsymbol{\mu}, \boldsymbol{\nu}, \boldsymbol{\rho}, \ldots$.

The neutral element of the Galilei group is $(0, \boldsymbol{0}, \boldsymbol{0}, \boldsymbol{0})$ and the inverse of every element is

$$(b, \boldsymbol{a}, \boldsymbol{v}, \boldsymbol{\alpha})^{-1} = (-b, -R(-\boldsymbol{\alpha})(\boldsymbol{a} - b\boldsymbol{v}), -R(-\boldsymbol{\alpha})\boldsymbol{v}, -\boldsymbol{\alpha}).$$

The generators of the group in the realization (2.1, 2.2) are the differential operators

$$H = \partial/\partial t, \quad P_i = \partial/\partial x^i, \quad K_i = t\partial/\partial x^i, \quad J_k = \varepsilon_{kli} x^l \partial/\partial x^i \qquad (2.11)$$

and the commutation rules of the Galilei Lie algebra are

$$[\boldsymbol{J}, \boldsymbol{J}] = -\boldsymbol{J}, \quad [\boldsymbol{J}, \boldsymbol{P}] = -\boldsymbol{P}, \quad [\boldsymbol{J}, \boldsymbol{K}] = -\boldsymbol{K}, \quad [\boldsymbol{J}, H] = 0, \qquad (2.12)$$

$$[H, \boldsymbol{P}] = 0,\ [H, \boldsymbol{K}] = \boldsymbol{P},\ [\boldsymbol{P}, \boldsymbol{P}] = 0,\ [\boldsymbol{K}, \boldsymbol{K}] = 0,\ [\boldsymbol{K}, \boldsymbol{P}] = 0. \qquad (2.13)$$

All throughout this book, except when explicitly stated, we shall use the following shorthand notation for commutators of scalar and 3-vector operators, that as usual, are represented by bold face characters:

$$\begin{aligned}
[\boldsymbol{A}, \boldsymbol{B}] &= \boldsymbol{C}, &\Longrightarrow\quad & [A_i, B_j] = \epsilon_{ijk} C_k,\\
[\boldsymbol{A}, \boldsymbol{B}] &= C, &\Longrightarrow\quad & [A_i, B_j] = \delta_{ij} C,\\
[\boldsymbol{A}, B] &= \boldsymbol{C}, &\Longrightarrow\quad & [A_i, B] = C_i,\\
[B, \boldsymbol{A}] &= \boldsymbol{C}, &\Longrightarrow\quad & [B, A_i] = C_i,
\end{aligned}$$

where $\delta_{ij} = \delta_{ji}$ is Kronecker's delta and $\epsilon_{ijk}$ is the completely antisymmetric symbol, so that Latin indexes match on both sides of commutators.

The group action (2.1)-(2.2) represents the relationship between the coordinates $(t, \boldsymbol{r})$ of a space-time event as measured by the inertial observer $O$ and the corresponding coordinates $(t', \boldsymbol{r}')$ of the same space-time event as measured by another inertial observer $O'$. The ten group parameters have the following meaning. If we consider the event $(0, \boldsymbol{0})$ measured by $O$, for instance the flashing of a light beam from its origin at time $t = 0$, it takes the values $(b, \boldsymbol{a})$ in $O'$, where $b$ is the time parameter that represents the time translation and $\boldsymbol{a}$ is the space translation. The parameter $\boldsymbol{v}$ of dimensions of velocity represents the velocity of the origin of the Cartesian frame of $O$ as measured by $O'$, and finally the parameters $\boldsymbol{\alpha}$, or $R(\boldsymbol{\alpha})$, represent the orientation of the Cartesian frame of $O$ as measured by $O'$. In a certain sense the ten parameters $(b, \boldsymbol{a}, \boldsymbol{v}, \boldsymbol{\alpha})$ with dimensions respectively of time, position, velocity and orientation describe the relative motion of the Cartesian frame of $O$ by $O'$.

The Galilei group has non-trivial exponents given by [2]

$$\xi(g, g') = m\left(\frac{1}{2}\boldsymbol{v}^2 b' + \boldsymbol{v}\cdot R(\boldsymbol{\alpha})\boldsymbol{a}'\right). \qquad (2.14)$$

They are characterized by the non-vanishing parameter $m$.

The central extension of the Galilei group [3] is an 11-parameter group with an additional generator $I$ which commutes with the other ten,

$$[I, H] = [I, \boldsymbol{P}] = [I, \boldsymbol{K}] = [I, \boldsymbol{J}] = 0, \tag{2.15}$$

and the remaining commutation relations are the same as above (2.12, 2.13), except the last one which appears as

$$[K_i, P_j] = -m\delta_{ij} I, \quad \text{or} \quad [\boldsymbol{K}, \boldsymbol{P}] = -mI, \tag{2.16}$$

using our shorthand notation, in terms of a non-vanishing parameter $m$. If we define the following polynomial operators on the group algebra

$$\boldsymbol{W} = I\boldsymbol{J} - \frac{1}{m}\boldsymbol{K} \times \boldsymbol{P}, \quad U = IH - \frac{1}{2m}\boldsymbol{P}^2, \tag{2.17}$$

$U$ commutes with all generators of the extended Galilei group and $\boldsymbol{W}$ satisfies the commutation relations:

$$[\boldsymbol{W}, \boldsymbol{W}] = -I\boldsymbol{W}, \quad [\boldsymbol{J}, \boldsymbol{W}] = -\boldsymbol{W}, \quad [\boldsymbol{W}, \boldsymbol{P}] = [\boldsymbol{W}, \boldsymbol{K}] = [\boldsymbol{W}, H] = 0,$$

so that $\boldsymbol{W}^2$ also commutes with all generators. It turns out that the extended Galilei group has three functionally independent Casimir operators which, in those representations in which the operator $I$ becomes the unit operator, for instance in irreducible representations, are interpreted as the mass, $M = mI$, the internal energy $H_0 = H - \boldsymbol{P}^2/2m$, and the absolute value of spin

$$S^2 = \left(\boldsymbol{J} - \frac{1}{m}\boldsymbol{K} \times \boldsymbol{P}\right)^2. \tag{2.18}$$

The spin operator $\boldsymbol{S}$ in those representations in which $I = \mathbb{I}$, satisfy the commutation relations:

$$[\boldsymbol{S}, \boldsymbol{S}] = -\boldsymbol{S}, \quad [\boldsymbol{J}, \boldsymbol{S}] = -\boldsymbol{S}, \quad [\boldsymbol{S}, \boldsymbol{P}] = [\boldsymbol{S}, H] = [\boldsymbol{S}, \boldsymbol{K}] = 0,$$

*i.e.*, it is an angular momentum operator, transforms like a vector under rotations and is invariant under space and time translations and under Galilei boosts, respectively. It reduces to the total angular momentum operator $\boldsymbol{J}$ in those frames in which $\boldsymbol{P} = \boldsymbol{K} = 0$.

## 2. NONRELATIVISTIC POINT PARTICLE

Let us consider a mechanical system whose kinematical space is the manifold $X = \mathcal{G}/\mathcal{H}$, where $\mathcal{H}$ is the six-dimensional subgroup of the homogeneous Galilei transformations of elements of the form $(0, 0, \boldsymbol{v}, \boldsymbol{\mu})$.

Then $X$ is a four-dimensional manifold spanned by the variables $(t, \boldsymbol{r}) \equiv x$, with domains $t \in \mathbb{R}$, $\boldsymbol{r} \in \mathbb{R}^3$, similar to the group parameters $b$ and $\boldsymbol{a}$ respectively. We assume that they are functions of some evolution parameter $\tau$ and at any instant $\tau$ of the evolution two different inertial observers relate their measurements by:

$$t'(\tau) = t(\tau) + b, \tag{2.19}$$

$$\boldsymbol{r}'(\tau) = R(\boldsymbol{\mu})\boldsymbol{r}(\tau) + \boldsymbol{v}t(\tau) + \boldsymbol{a}. \tag{2.20}$$

Because of the way they transform, we can interpret them respectively as the time and position of the system. If we assume that the evolution parameter $\tau$ is group invariant, by taking the $\tau$–derivative of both sides of the above expressions, it turns out that the derivatives of the kinematical variables at any instant $\tau$ transform as:

$$\dot{t}'(\tau) = \dot{t}(\tau), \tag{2.21}$$

$$\dot{\boldsymbol{r}}'(\tau) = R(\boldsymbol{\mu})\dot{\boldsymbol{r}}(\tau) + \boldsymbol{v}\dot{t}(\tau). \tag{2.22}$$

There are no constraints among these variables. It is only the homogeneity of the Lagrangian in terms of their derivatives (1.11) which reduces to three the number of independent degrees of freedom. This homogeneity leads to the general form:

$$L = T\dot{t} + \boldsymbol{R}\cdot\dot{\boldsymbol{r}}, \tag{2.23}$$

where $T = \partial L/\partial\dot{t}$ and $R_i = \partial L/\partial\dot{r}_i$ are still some unknown functions of the kinematical variables and their derivatives, which are homogeneous of zero degree in terms of the derivatives.

Associated to this manifold $X$, the gauge function for this system is

$$\alpha(g; x) = \xi(g, x) = m\left(v^2 t/2 + \boldsymbol{v}\cdot R(\boldsymbol{\mu})\boldsymbol{r}\right), \tag{2.24}$$

where the parameter $m$ is interpreted as the mass of the system and $\xi(g, g')$ is the exponent of $\mathcal{G}$, so that the transformation of the Lagrangian under the Galilei group is

$$L(x', \dot{x}') = L(x, \dot{x}) + m\left(v^2\dot{t}/2 + \boldsymbol{v}\cdot R(\boldsymbol{\mu})\dot{\boldsymbol{r}}\right). \tag{2.25}$$

Then

$$T' = \frac{\partial L'}{\partial\dot{t}'} = \left(\frac{\partial L}{\partial\dot{t}} + \frac{1}{2}mv^2\right)\frac{\partial\dot{t}}{\partial\dot{t}'} + \left(\frac{\partial L}{\partial\dot{r}_i} + mv_j R(\boldsymbol{\mu})_{ji}\right)\frac{\partial\dot{r}_i}{\partial\dot{t}'}, \tag{2.26}$$

but from (2.21) and (2.22) we get $\partial\dot{t}/\partial\dot{t}' = 1$ and $\partial\dot{r}_i\partial\dot{t}' = -R^{-1}(\boldsymbol{\mu})_{ik}v_k$, respectively, and thus

$$T' = T - \frac{1}{2}mv^2 - \boldsymbol{v}\cdot R(\boldsymbol{\mu})\boldsymbol{R}. \tag{2.27}$$

Similarly

$$R' = R(\mu)R + mv. \tag{2.28}$$

The conjugate momenta of the independent degrees of freedom $q_i = r_i$, are $p_i = \partial L/\partial \dot{r}_i$, and consequently Noether's theorem leads to the following constants of the motion:
a) Under time translations the gauge function (2.24) vanishes, $\delta t = \delta b$, $M = 1$, while $\delta r_i = 0$ and the constant reduces to the following expression $\boldsymbol{R} \cdot d\boldsymbol{r}/dt - L/\dot{t} = -T$.
b) Under space translations also $\alpha(g; x) \equiv 0$, $\delta t = 0$, $M = 0$, while $\delta r_i = \delta a_i$, $M_{ij} = \delta_{ij}$ and the conserved observable is $\boldsymbol{R}$.
c) Under pure Galilei transformations $\delta t = \delta b$ and $M = 0$, while $\delta r_i = t\delta v_i$ and $M_{ij} = t\delta_{ij}$, but now the gauge function to first order in the velocity parameters is $\alpha(\delta v; x) = m\boldsymbol{r} \cdot \delta \boldsymbol{v}$, and we get $m\boldsymbol{r} - \boldsymbol{R}t$.
d) Under rotations $\alpha(g; x) \equiv 0$, $\delta t = 0$ and $M = 0$, while $\delta r_i = -\varepsilon_{ijk} r_j n_k \delta\alpha$ and $M_{ik} = -\varepsilon_{ijk} r_j$ the conserved quantity is $\boldsymbol{r} \times \boldsymbol{R}$.

Collecting all terms we can give them the following names:

$$\text{Energy} \quad H = -T, \tag{2.29}$$

$$\text{linear momentum} \quad \boldsymbol{P} = \boldsymbol{R} = \boldsymbol{p}, \tag{2.30}$$

$$\text{kinematical momentum} \quad \boldsymbol{K} = m\boldsymbol{r} - \boldsymbol{P}t, \tag{2.31}$$

$$\text{angular momentum} \quad \boldsymbol{J} = \boldsymbol{r} \times \boldsymbol{P}. \tag{2.32}$$

We reserve for these observables the same symbols as the corresponding group generators which produce the space-time transformations that leave dynamical equations invariant. Even their names make reference to the corresponding group transformation parameter, except the energy which in this context should be called the 'temporal momentum'. For the kinematical momentum we can find in the literature alternative names like 'Galilei momentum' or 'static momentum'. The first because it is related to invariance of dynamical equations under pure Galilei transformations or boosts, while the second because for systems of point particles it reduces to $\boldsymbol{K} = \sum m_i \boldsymbol{r}_i - t \sum \boldsymbol{P}_i$, such that for the center of mass observer, $\sum \boldsymbol{P}_i = 0$, and therefore it appears as the static momentum of the masses with respect to the origin of the observer's frame. With the habitual definition of the center of mass position it then reduces to $\boldsymbol{K} = m\boldsymbol{q} - t\boldsymbol{P}$, where $m = \sum m_i$ is the total mass of the system, $\boldsymbol{P} = \sum \boldsymbol{P}_i$ the total momentum and $\boldsymbol{q}$ the center of mass position. Being consistent with this notation, we should call it 'Poincaré or Lorentz momentum' in a relativistic approach. Nevertheless we shall use throughout this book the name of kinematical momentum for this observable $\boldsymbol{K}$.

The linear momentum takes the general expression $\boldsymbol{P} = m\dot{\boldsymbol{r}}/\dot{t} = m\boldsymbol{u}$ because taking the $\tau$-derivative in (2.31) of the kinematical momentum, $\dot{\boldsymbol{K}} = 0$, implies $\boldsymbol{P} = m\boldsymbol{u}$, where $\boldsymbol{u}$ is the time derivative of the position of the system, *i.e.*, the velocity of the particle.

The six conditions $\boldsymbol{P} = 0$ and $\boldsymbol{K} = 0$, imply $\boldsymbol{u} = 0$ and $\boldsymbol{r} = 0$, so that the system is at rest and placed at the origin of the reference frame. There is still an arbitrary rotation and a time translation to fix a unique inertial observer. Nevertheless we call this class of observers, for which $\boldsymbol{P} = 0$ and $\boldsymbol{K} = 0$, the center of mass observer. These six conditions will be also used as the definition of the center of mass observer for any other system even in a relativistic approach.

From (2.27) and (2.28) we see that the energy and linear momentum transform as:

$$H' = H + \boldsymbol{v} \cdot R(\boldsymbol{\mu})\boldsymbol{P} + \frac{1}{2}mv^2, \tag{2.33}$$

$$\boldsymbol{P}' = R(\boldsymbol{\mu})\boldsymbol{P} + m\boldsymbol{v}. \tag{2.34}$$

Then, if $H_0$ and $\boldsymbol{P} = 0$ are the energy and linear momentum measured by the center of mass observer, for any arbitrary observer that sees the particle moving with velocity $\boldsymbol{u}$, it follows from (2.33) and (2.34) that

$$H = H_0 + \frac{1}{2}mu^2 = H_0 + \boldsymbol{P}^2/2m, \quad \boldsymbol{P} = m\boldsymbol{u}.$$

The above ten constants of the motion (2.29-2.32) are the generating functions of the corresponding canonical transformations of the system, such that on phase space in terms of the generalized coordinates $\boldsymbol{q} \equiv \boldsymbol{r}$ and their canonical conjugate momenta $\boldsymbol{p}$, they take the form

$$H = \boldsymbol{p}^2/2m, \quad \boldsymbol{P} = \boldsymbol{p}, \quad \boldsymbol{K} = m\boldsymbol{q} - \boldsymbol{p}t, \quad \boldsymbol{J} = \boldsymbol{q} \times \boldsymbol{p}. \tag{2.35}$$

Taking the Poisson bracket of these functions we get the following commutation relations

$$\{\boldsymbol{J}, \boldsymbol{J}\} = \boldsymbol{J}, \quad \{\boldsymbol{J}, \boldsymbol{P}\} = \boldsymbol{P}, \quad \{\boldsymbol{J}, \boldsymbol{K}\} = \boldsymbol{K}, \quad \{\boldsymbol{J}, H\} = 0, \tag{2.36}$$

$$\{H, \boldsymbol{P}\} = 0, \{H, \boldsymbol{K}\} = -\boldsymbol{P}, \{\boldsymbol{P}, \boldsymbol{P}\} = 0, \{\boldsymbol{K}, \boldsymbol{K}\} = 0, \{\boldsymbol{K}, \boldsymbol{P}\} = m, \tag{2.37}$$

where we have used the same short-hand notation as for commutators. When compared with relations (2.12), (2.13) and (2.16) they are, up to a global sign, the same commutation relations of the extended Galilei group, where the eleventh variable is the constant unit function.

The internal energy of the system $H_0 = H - P^2/2m$, which is reduced to the energy measured by the center of mass observer, is a group invariant. In this case it is also a constant of the motion so that it reduces to a scalar of constant magnitude $H_0$, independent of the kinematical variables and their derivatives, in such a way that $H$ takes the form $H = H_0 + mu^2/2$, and therefore the general form of the Lagrangian for this system, by substituting in (2.23) the corresponding expressions for $T$ and $\boldsymbol{R}$, is

$$L = -H_0\dot{t} - \frac{m}{2}\frac{\dot{\boldsymbol{r}}^2}{\dot{t}^2}\dot{t} + m\frac{\dot{\boldsymbol{r}}}{\dot{t}}\cdot\dot{\boldsymbol{r}} = \frac{m}{2}\frac{\dot{\boldsymbol{r}}^2}{\dot{t}} - H_0\dot{t}. \qquad (2.38)$$

Here the last term can be removed because it is a total derivative. This is equivalent to considering that a Galilei point particle has zero internal energy or that this invariant $H_0$ plays no role in the dynamics of this system. This part of the Lagrangian also corresponds to the $-mc^2\dot{t}$ term of the nonrelativistic limit of the point particle relativistic Lagrangian, as we shall see in the next chapter. Therefore we have finally obtained the usual nonrelativistic Lagrangian for the point particle which is uniquely defined.

If we define the spin of the system, as in (2.18), by

$$\boldsymbol{S} \equiv \boldsymbol{J} - \frac{1}{m}\boldsymbol{K}\times\boldsymbol{P} = \boldsymbol{J} - \boldsymbol{r}\times\boldsymbol{P} = 0, \qquad (2.39)$$

it vanishes, so that the point particle is a spinless system.

## 3. GALILEI SPINNING PARTICLES

The most general nonrelativistic particle [4] is the system whose kinematical space $X$ is the whole Galilei group $\mathcal{G}$. Then the kinematical variables are the ten real variables $x(\tau) \equiv (t(\tau), \boldsymbol{r}(\tau), \boldsymbol{u}(\tau), \boldsymbol{\rho}(\tau))$ with domains $t \in \mathbb{R}$, $\boldsymbol{r} \in \mathbb{R}^3$, $\boldsymbol{u} \in \mathbb{R}^3$ and $\boldsymbol{\rho} \in \mathbb{R}^3_c$ similarly as the corresponding group parameters. The relationship between the values $x'(\tau)$ and $x(\tau)$ they take at any instant $\tau$ for two arbitrary inertial observers, is given by:

$$t'(\tau) = t(\tau) + b, \qquad (2.40)$$

$$\boldsymbol{r}'(\tau) = R(\boldsymbol{\mu})\boldsymbol{r}(\tau) + \boldsymbol{v}t(\tau) + \boldsymbol{a}, \qquad (2.41)$$

$$\boldsymbol{u}'(\tau) = R(\boldsymbol{\mu})\boldsymbol{u}(\tau) + \boldsymbol{v}, \qquad (2.42)$$

$$\boldsymbol{\rho}'(\tau) = \frac{\boldsymbol{\mu} + \boldsymbol{\rho}(\tau) + \boldsymbol{\mu}\times\boldsymbol{\rho}(\tau)}{1 - \boldsymbol{\mu}\cdot\boldsymbol{\rho}(\tau)}. \qquad (2.43)$$

In a group theoretical language, this can also be interpreted as $x' = gx$, *i.e.*, as the action of the group element $g \in \mathcal{G}$ acting on the point $x$

considered also as another element of the group, to produce the group element $x'$. In the above transformations we have considered the group action on the left, *i.e.*, $x' = gx$, where in (2.3-2.6) or (2.9), the unprimed group element $g \equiv (b, \boldsymbol{a}, \boldsymbol{v}, \boldsymbol{\mu})$ is replaced by the unprimed kinematical variables $x \equiv (t, \boldsymbol{r}, \boldsymbol{u}, \boldsymbol{\rho})$, the double-primed group element $g'' \equiv (b'', \boldsymbol{a}'', \boldsymbol{v}'', \boldsymbol{\mu}'')$ by the primed kinematical variables $x' \equiv (t', \boldsymbol{r}', \boldsymbol{u}', \boldsymbol{\rho}')$ and the $g' \equiv (b', \boldsymbol{a}', \boldsymbol{v}', \boldsymbol{\mu}')$ group element by $g \equiv (b, \boldsymbol{a}, \boldsymbol{v}, \boldsymbol{\mu})$. The way they transform, allows us to interpret them respectively as the time, position, velocity and orientation of the particle.

Among these kinematical variables there exist the differential constraints $\boldsymbol{u}(\tau) = \dot{\boldsymbol{r}}(\tau)/\dot{t}(\tau)$, that together with the homogeneity condition of the Lagrangian $L$ in terms of the derivatives of the kinematical variables:

$$L(x, \dot{x}) = (\partial L/\partial \dot{x}_i)\dot{x}_i, \tag{2.44}$$

reduce from ten to six the essential degrees of freedom of the system.

These degrees of freedom are the position $\boldsymbol{r}(t)$ and the orientation $\boldsymbol{\rho}(t)$ and the Lagrangian depends on the second derivative of $\boldsymbol{r}(t)$ and the first derivative of $\boldsymbol{\rho}(t)$. Expression (2.44) is explicitly given by:

$$L = T\dot{t} + \boldsymbol{R} \cdot \dot{\boldsymbol{r}} + \boldsymbol{U} \cdot \dot{\boldsymbol{u}} + \boldsymbol{V} \cdot \dot{\boldsymbol{\rho}}, \tag{2.45}$$

where the functions $T = \partial L/\partial \dot{t}$, $R_i = \partial L/\partial \dot{r}^i$, $U_i = \partial L/\partial \dot{u}^i$, $V_i = \partial L/\partial \dot{\rho}^i$ will be in general functions of the ten kinematical variables $(t, \boldsymbol{r}, \boldsymbol{u}, \boldsymbol{\rho})$ and homogeneous functions of zero degree in terms of the derivatives $(\dot{t}, \dot{\boldsymbol{r}}, \dot{\boldsymbol{u}}, \dot{\boldsymbol{\rho}})$. By assuming that the evolution parameter $\tau$ is group invariant, these derivatives transform under $\mathcal{G}$:

$$\begin{aligned}
\dot{t}'(\tau) &= \dot{t}(\tau), && (2.46)\\
\dot{\boldsymbol{r}}'(\tau) &= R(\boldsymbol{\mu})\dot{\boldsymbol{r}}(\tau) + \boldsymbol{v}\dot{t}(\tau), && (2.47)\\
\dot{\boldsymbol{u}}'(\tau) &= R(\boldsymbol{\mu})\dot{\boldsymbol{u}}(\tau), && (2.48)\\
\dot{\boldsymbol{\rho}}'(\tau) &= \frac{(\dot{\boldsymbol{\rho}}(\tau) + \boldsymbol{\mu} \times \dot{\boldsymbol{\rho}}(\tau))(1 - \boldsymbol{\mu} \cdot \boldsymbol{\rho}(\tau))}{(1 - \boldsymbol{\mu} \cdot \boldsymbol{\rho}(\tau))^2} \\
&\quad + \frac{\boldsymbol{\mu} \cdot \dot{\boldsymbol{\rho}}(\tau)(\boldsymbol{\mu} + \boldsymbol{\rho}(\tau) + \boldsymbol{\mu} \times \boldsymbol{\rho}(\tau))}{(1 - \boldsymbol{\mu} \cdot \boldsymbol{\rho}(\tau))^2}. && (2.49)
\end{aligned}$$

Instead of the derivative $\dot{\boldsymbol{\rho}}(\tau)$ that transforms in a complicated way, we can define the angular velocity of the particle $\boldsymbol{\omega}$ as a function of it in the form

$$\boldsymbol{\omega} = \frac{2}{1 + \rho^2}(\dot{\boldsymbol{\rho}} + \boldsymbol{\rho} \times \dot{\boldsymbol{\rho}}). \tag{2.50}$$

It is a linear function of $\dot{\boldsymbol{\rho}}$, with inverse transformation

$$\dot{\boldsymbol{\rho}} = \frac{1}{2}(\boldsymbol{\omega} - \boldsymbol{\rho} \times \boldsymbol{\omega} + \boldsymbol{\rho}(\boldsymbol{\rho} \cdot \boldsymbol{\omega})) \tag{2.51}$$

linear in $\omega$, and under $\mathcal{G}$ it transforms as:

$$\omega'(\tau) = R(\mu)\omega(\tau). \tag{2.52}$$

We interpret the rotation matrix $R(\rho)$ as the rotation that carries the initial frame linked to the body at instant $\tau = 0$ to the frame at instant $\tau$, as in a rigid body. Then, the three columns of matrix $R(\rho)$ represent the Cartesian components of the three unit vectors linked to the body when chosen parallel to the laboratory frame at instant $\tau = 0$.

If $\boldsymbol{k}(\tau)$ is any internal vector of a rigid body with origin at point $\boldsymbol{r}$, then its dynamics is contained in the expression $\boldsymbol{k}(\tau) = R(\rho(\tau))\boldsymbol{k}(0)$. The velocity of point $\boldsymbol{k}$ is

$$\dot{\boldsymbol{k}}(\tau) = \dot{R}(\rho(\tau))\boldsymbol{k}(0) = \dot{R}(\rho(\tau))R^{-1}(\rho(\tau))\boldsymbol{k}(\tau) = \Omega(\tau)\boldsymbol{k}(\tau)$$

where matrix $\Omega = \dot{R}R^{-1} = \dot{R}R^T$ is an antisymmetric matrix. At any instant $\tau$, $R(\rho(\tau))R^T(\rho(\tau)) = \mathbb{I}$, where superscript $T$ means the transposed matrix and $\mathbb{I}$ is the $3 \times 3$ unit matrix. Taking the $\tau$-derivative of this expression, $\dot{R}R^T + R\dot{R}^T = \Omega + \Omega^T = 0$, and thus the three essential components of the antisymmetric matrix $\Omega$ define a three-vector $\boldsymbol{\omega}$

$$\Omega = \begin{pmatrix} 0 & -\omega_z & \omega_y \\ \omega_z & 0 & -\omega_x \\ -\omega_y & \omega_x & 0 \end{pmatrix},$$

such that we can also write $\dot{\boldsymbol{k}}(\tau) = \Omega(\tau)\boldsymbol{k}(\tau) \equiv \boldsymbol{\omega}(\tau) \times \boldsymbol{k}(\tau)$ and $\boldsymbol{\omega}$ is interpreted as the instantaneous angular velocity. The different components of $\boldsymbol{\omega}$, expressed as functions of the variables $\boldsymbol{\rho}$ and $\dot{\boldsymbol{\rho}}$ are given in (2.50).

Expression (2.43) corresponds to $R(\rho'(\tau)) = R(\mu)R(\rho(\tau))$. Therefore

$$\begin{aligned} \Omega' &= \dot{R}(\rho'(\tau))R^T(\rho'(\tau)) = R(\mu)\dot{R}(\rho(\tau))R^T(\rho(\tau))R^T(\mu) \\ &= R(\mu)\Omega R^{-1}(\mu), \end{aligned}$$

and this leads to the equation (2.52) in terms of the essential components $\boldsymbol{\omega}$ of the antisymmetric matrix $\Omega$.

In this way the last part of the Lagrangian $(\partial L/\partial\dot{\rho}^i)\dot{\rho}^i$ can be writen as

$$\boldsymbol{V}\cdot\dot{\boldsymbol{\rho}} \equiv \frac{\partial L}{\partial\dot{\rho}^i}\dot{\rho}^i = \frac{\partial L}{\partial\omega^j}\frac{\partial\omega^j}{\partial\dot{\rho}^i}\dot{\rho}^i = \boldsymbol{W}\cdot\boldsymbol{\omega}, \tag{2.53}$$

due to the linearity of $\boldsymbol{\omega}$ in terms of $\dot{\boldsymbol{\rho}}$ and where $W_i = \partial L/\partial\omega^i$. Thus the most general form of the Lagrangian of a nonrelativistic particle can also be written instead of (2.45) as:

$$L = T\dot{t} + \boldsymbol{R}\cdot\dot{\boldsymbol{r}} + \boldsymbol{U}\cdot\dot{\boldsymbol{u}} + \boldsymbol{W}\cdot\boldsymbol{\omega}. \tag{2.54}$$

Since $X$ is a homogeneous space of $\mathcal{G}$ we have seen that the most general gauge function for the group is just the group exponent:

$$\alpha(g;x) = \xi(g,h_x) = m(v^2 t(\tau)/2 + \boldsymbol{v}\cdot R(\boldsymbol{\mu})\boldsymbol{r}(\tau)), \tag{2.55}$$

similar to (2.24), and this allows us to interpret the parameter $m$ as the mass of the system. Under the action of an arbitrary element of the Galilei group, the Lagrangian $L$ transforms according to:

$$L(gx(\tau), d(gx(\tau))/d\tau) = L(x(\tau), \dot{x}(\tau)) + d\alpha(g;x(\tau))/d\tau. \tag{2.56}$$

This leads through some straightforward calculations, similar to the ones performed in (2.26)-(2.28), to the following form of transformation of the functions:

$$\begin{aligned} T'(\tau) &= T(\tau) - \boldsymbol{v}\cdot R(\boldsymbol{\mu})\boldsymbol{R}(\tau) - mv^2/2, & (2.57)\\ \boldsymbol{R}'(\tau) &= R(\boldsymbol{\mu})\boldsymbol{R}(\tau) + m\boldsymbol{v}, & (2.58)\\ \boldsymbol{U}'(\tau) &= R(\boldsymbol{\mu})\boldsymbol{U}(\tau), & (2.59)\\ \boldsymbol{W}'(\tau) &= R(\boldsymbol{\mu})\boldsymbol{W}(\tau). & (2.60)\end{aligned}$$

The Lagrangian depends on up to the second derivative of $\boldsymbol{r}$, and therefore there will be two conjugate momenta of these variables and another conjugate momentum associated to variables $\boldsymbol{\rho}$. The generalized coordinates will be $\boldsymbol{q}_1 \equiv \boldsymbol{r}$, $\boldsymbol{q}_2 \equiv \boldsymbol{u} = d\boldsymbol{r}/dt$, and $\boldsymbol{q}_3 \equiv \boldsymbol{\rho}$.

The canonical conjugate momenta of the independent degrees of freedom $r_i(t)$ and $\rho_i(t)$ are, according to (1.26):

$$\begin{aligned} \boldsymbol{p}_{r(1)} &= \boldsymbol{R} - d\boldsymbol{U}/dt, & (2.61)\\ \boldsymbol{p}_{r(2)} &= \boldsymbol{U}, & (2.62)\\ \boldsymbol{p}_{\rho} &= \boldsymbol{V}, & (2.63)\end{aligned}$$

where $\boldsymbol{p}_{r(1)}$ is the conjugate momentum of $\boldsymbol{q}_1$, $\boldsymbol{p}_{r(2)}$ is the conjugate momentum of $\boldsymbol{q}_2$, and $\boldsymbol{p}_\rho$ that of $\boldsymbol{q}_3$.

Noether's theorem defines the following constants of the motion:

a) Under time translation the action function is invariant and as usual we call the corresponding conserved quantity, the **total energy** of the system $H$. Since $\delta t = \delta b$ and $\delta q_i^{(s)} = 0$, $M = 1$ and $M_i^{(s)} = 0$, by applying (1.55) we have:

$$H = -(L - p_{(s)}^i q_i^{(s)})M = -(\hat{L}/\dot{t} - p_{(s)}^i q_i^{(s)}) = -T - \boldsymbol{R}\cdot\boldsymbol{u} - \boldsymbol{U}\cdot\dot{\boldsymbol{u}}/\dot{t} - \boldsymbol{W}\cdot\boldsymbol{\omega}/\dot{t}$$

$$+(\boldsymbol{R} - d\boldsymbol{U}/dt)\cdot\boldsymbol{u} + \boldsymbol{U}\cdot\dot{\boldsymbol{u}}/\dot{t} + \boldsymbol{V}\cdot\dot{\boldsymbol{\rho}}/\dot{t},$$

and since $\boldsymbol{W} \cdot \boldsymbol{\omega} = \boldsymbol{V} \cdot \dot{\boldsymbol{\rho}}$, it turns out that

$$H = -T - \frac{d\boldsymbol{U}}{dt} \cdot \boldsymbol{u}. \tag{2.64}$$

**b)** Under spatial translation, $A(x_1, x_2)$ is invariant and this defines the **total linear momentum** of the system. We have now:

$$\delta t = 0,\ M = 0,\ \delta r_i = \delta a_i,\ M_{ij}^{(0)} = \delta_{ij},\ \delta u_i = 0,\ M_{ij}^{(1)} = 0,$$

$$\delta \rho_i = 0,\ M_{ij}^{(\rho)} = 0,$$

and then

$$\boldsymbol{P} = \boldsymbol{R} - \frac{d\boldsymbol{U}}{dt}. \tag{2.65}$$

**c)** Under a pure Galilei transformation of velocity $\delta \boldsymbol{v}$, $A(x_1, x_2)$ is no longer invariant but taking into account (1.60) and the gauge function (2.55), it transforms as $\delta A = m\boldsymbol{r}_2 \cdot \delta \boldsymbol{v} - m\boldsymbol{r}_1 \cdot \delta \boldsymbol{v}$ and this defines the **total kinematical momentum** $\boldsymbol{K}$, in the following way:

$$\delta t = 0,\ M = 0,\ \delta r_i = \delta v_i t,\ M_{ij}^{(0)} = \delta_{ij} t,\ \delta u_i = \delta v_i,\ M_{ij}^{(1)} = \delta_{ij},$$

$$\delta \rho_i = 0,\ M_{ij}^{(\rho)} = 0,$$

and thus

$$\boldsymbol{K} = m\boldsymbol{r} - \boldsymbol{P}\, t - \boldsymbol{U}. \tag{2.66}$$

From $\dot{\boldsymbol{K}} = 0$, this leads to $\boldsymbol{P} = m\boldsymbol{u} - d\boldsymbol{U}/dt$, and thus by identification with (2.65), the function $\boldsymbol{R} = m\boldsymbol{u}$ irrespective of the particular Lagrangian. The total linear momentum does not lie along the velocity of point $\boldsymbol{r}$.

**d)** Finally, under rotations $A(x_1, x_2)$ remains invariant and the corresponding constant of the motion, the **total angular momentum** of the system, comes from the infinitesimal transformation of value $\delta \mu_i = \delta \alpha_i / 2$, i.e., half of the rotated infinitesimal angle, and then

$$\delta t = 0,\ M_i = 0, \delta r_i = \epsilon_{ijk} \delta \alpha_k r_j,\ M_{ij}^{(0)} = \epsilon_{ijk} r_k,$$

$$\delta u_i = \epsilon_{ijk} \delta \alpha_k u_j, \quad M_{ij}^{(1)} = \epsilon_{ijk} u_k,$$

$$\delta \rho_i = \delta \alpha^k (\delta_{ik} + \epsilon_{ijk} \rho^j + \rho_i \rho_k)/2, \quad M_{ij}^{(\rho)} = (\delta_{ij} + \epsilon_{ijk} \rho^k + \rho_i \rho_j)/2,$$

which leads to

$$V_i M_{ij}^{(\rho)} = \frac{\partial L}{\partial \omega^k} \frac{\partial \omega^k}{\partial \dot{\rho}_i} M_{ij}^{(\rho)} = W_j,$$

and therefore

$$\boldsymbol{J} = \boldsymbol{r} \times \boldsymbol{P} + \boldsymbol{u} \times \boldsymbol{U} + \boldsymbol{W} = \boldsymbol{r} \times \boldsymbol{P} + \boldsymbol{Z}. \tag{2.67}$$

We are tempted to consider $\boldsymbol{Z}$ as the spin of the system. Since $\dot{\boldsymbol{J}} = 0$, this function $\boldsymbol{Z}$ satisfies $d\boldsymbol{Z}/dt = \boldsymbol{P} \times \boldsymbol{u}$ and is not a constant of the motion for a free particle. We shall define a little later the spin as a constant angular momentum for a free particle, once we accurately identify the center of mass of the particle.

The center of mass observer is defined, as before, as that inertial observer for which $\boldsymbol{P} = 0$ and $\boldsymbol{K} = 0$. These six conditions do not define uniquely an inertial observer but rather a class of them up to a rotation and an arbitrary time translation. In fact, the condition $\boldsymbol{P} = 0$ establishes the class of observers for which the center of mass is at rest, and $\boldsymbol{K} = 0$ is the additional condition to locate it at the origin of coordinates, as we have seen for the point particle case. This comes from the analysis of (2.66), where $\boldsymbol{k} = \boldsymbol{U}/m$ is an observable with dimensions of length, and taking the derivative with respect to $\tau$ of both sides, taking into account that $\dot{\boldsymbol{P}} = 0$, we have:

$$\dot{\boldsymbol{K}} = 0 = m\dot{\boldsymbol{r}} - \boldsymbol{P}\dot{t} - m\dot{\boldsymbol{k}}, \qquad \textit{i.e.,} \qquad \boldsymbol{P} = m\frac{d(\boldsymbol{r} - \boldsymbol{k})}{dt}. \tag{2.68}$$

Then the point $\boldsymbol{q} = \boldsymbol{r} - \boldsymbol{k}$ is moving at constant speed and we say that it represents the position of the center of mass of the system. Thus, the observable $\boldsymbol{k} = \boldsymbol{r} - \boldsymbol{q}$ is just the relative position of point $\boldsymbol{r}$ with respect to the center of mass. Therefore $\boldsymbol{P} = 0$ and $\boldsymbol{K} = 0$ give rise to $d\boldsymbol{q}/dt = 0$, and $\boldsymbol{r} = \boldsymbol{k}$, *i.e.*, $\boldsymbol{q} = 0$, as we pointed out. With this definition, the kinematical momentum can be written as $\boldsymbol{K} = m\boldsymbol{q} - \boldsymbol{P}t$, similarly as for the point particle case, in terms of the center of mass position $\boldsymbol{q}$ and the total linear momentum $\boldsymbol{P}$.

The spin of the system is defined as the difference between the total angular momentum $\boldsymbol{J}$ and the orbital angular momentum of the center of mass motion $\boldsymbol{q} \times \boldsymbol{P}$, and thus

$$\boldsymbol{S} = \boldsymbol{J} - \boldsymbol{q} \times \boldsymbol{P} = \boldsymbol{J} - \frac{1}{m}\boldsymbol{K} \times \boldsymbol{P} = \boldsymbol{Z} + \boldsymbol{k} \times \boldsymbol{P} = -m\boldsymbol{k} \times \frac{d\boldsymbol{k}}{dt} + \boldsymbol{W}. \tag{2.69}$$

The spin $\boldsymbol{S}$, expressed in terms of the constants of the motion $\boldsymbol{J}$, $\boldsymbol{K}$ and $\boldsymbol{P}$, is also a constant of the motion.

It is the sum of two terms, one $\boldsymbol{Z} = \boldsymbol{u} \times \boldsymbol{U} + \boldsymbol{W}$, coming from the new degrees of freedom and another $\boldsymbol{k} \times \boldsymbol{P}$, which is the angular momentum of the linear momentum located at point $\boldsymbol{r}$ with respect to the center of mass. Alternatively we can describe the spin according to the last expression in which the term $-\boldsymbol{k} \times m d\boldsymbol{k}/dt$ suggests a contribution of (anti)orbital type coming from the motion around the center of mass. It is related to the zitterbewegung or more precisely to the function $\boldsymbol{U} = m\boldsymbol{k}$ which reflects the dependence of the Lagrangian on the acceleration. The other term $\boldsymbol{W}$ comes from the dependence on the other three degrees of freedom $\rho_i$, and thus on the angular velocity. This zitterbewegung is the motion of the center of charge around the center of mass as we shall see in the next section. Point $\boldsymbol{r}$, as representing the position of the center of charge, has been also suggested in previous works for the relativistic electron. [5]

To obtain the dynamical equations from $L$ we have to add the three differential constraints $\boldsymbol{u} = \dot{\boldsymbol{r}}/\dot{t}$, so that dynamical equations are obtained from the Lagrangian function $L + \boldsymbol{\lambda}(\tau) \cdot (\boldsymbol{u}\dot{t} - \dot{\boldsymbol{r}})$, where the three unknown functions $\lambda_i(\tau)$ are the Lagrange multipliers. If we use those constraints and replace in $L$ all terms containing the $\boldsymbol{u}$ variable by $\dot{\boldsymbol{r}}/\dot{t}$, then the variational derivative with respect to kinematical variable $u_i$ leads to:

$$\frac{\partial L}{\partial u_i} + \lambda_i \dot{t} - \frac{d}{d\tau}\left(\frac{\partial L}{\partial \dot{u}_i}\right) = 0 = \lambda_i \dot{t} - \frac{d}{d\tau} U_i,$$

because $L$ is now explicitly $u_i$ independent, so that the Lagrange multipliers are $\boldsymbol{\lambda} = d\boldsymbol{U}/dt$.

Variational derivatives with respect to time and position give rise to the corresponding Euler-Lagrange equations:

$$\frac{\partial L}{\partial t} - \frac{d}{d\tau}\left(\frac{\partial L}{\partial \dot{t}} + \boldsymbol{\lambda} \cdot \boldsymbol{u}\right) = 0,$$

$$\frac{\partial L}{\partial r_i} - \frac{d}{d\tau}\left(\frac{\partial L}{\partial \dot{r}_i} - \lambda_i\right) = 0.$$

Since $L$ is invariant under translations, because in this case the gauge function (2.55) vanishes, $\partial L/\partial t = \partial L/\partial r_i = 0$. This implies the existence of four constants of the motion that are interpreted as the total energy and linear momentum respectively

$$H = -T - \boldsymbol{\lambda} \cdot \boldsymbol{u}, \quad \boldsymbol{P} = \boldsymbol{R} - \boldsymbol{\lambda},$$

and with the identification of the Lagrange multipliers and the function $\boldsymbol{R} = m\boldsymbol{u}$, we get again

$$H = -T - \boldsymbol{u} \cdot \frac{d\boldsymbol{U}}{dt}, \quad \boldsymbol{P} = m\boldsymbol{u} - \frac{d\boldsymbol{U}}{dt}. \tag{2.70}$$

Finally, variational derivatives with respect to the orientation variables lead to

$$\frac{\partial L}{\partial \rho_i} - \frac{d}{d\tau}\left(\frac{\partial L}{\partial \dot{\rho}_i}\right) = 0. \tag{2.71}$$

If the Lagrangian is a function of the orientation variables only through its dependence on the angular velocity $\boldsymbol{\omega}(\boldsymbol{\rho}, \dot{\boldsymbol{\rho}})$, this leads to

$$\frac{d\boldsymbol{W}}{d\tau} = \boldsymbol{\omega} \times \boldsymbol{W}. \tag{2.72}$$

Indeed, from (2.71) we get

$$\frac{\partial L}{\partial \omega_j}\frac{\partial \omega_j}{\partial \rho_i} - \frac{d}{d\tau}\left(\frac{\partial L}{\partial \omega_j}\frac{\partial \omega_j}{\partial \dot{\rho}_i}\right) = 0,$$

i.e.,

$$W_j\,\frac{\partial \omega_j}{\partial \rho_i} - \frac{dW_j}{d\tau}\,\frac{\partial \omega_j}{\partial \dot{\rho}_i} - W_j\,\frac{d}{d\tau}\left(\frac{\partial \omega_j}{\partial \dot{\rho}_i}\right) = 0, \quad i = 1, 2, 3. \tag{2.73}$$

If we multiply each term of (2.73) by $\dot{\rho}_i$ and add all together for $i = 1, 2, 3$, since $W_j = \partial L/\partial \omega_j$, and $\boldsymbol{\omega}$ is a linear function of $\dot{\boldsymbol{\rho}}$, and therefore $\omega_j = (\partial \omega_j/\partial \dot{\rho}_i)\dot{\rho}_i$ holds, we get

$$W_j\,\frac{\partial \omega_j}{\partial \rho_i}\dot{\rho}_i - \frac{dW_j}{d\tau}\left(\frac{\partial \omega_j}{\partial \dot{\rho}_i}\right)\dot{\rho}_i - W_j\,\frac{d}{d\tau}\left(\frac{\partial \omega_j}{\partial \dot{\rho}_i}\right)\dot{\rho}_i = 0,$$

$$W_j\left(\frac{d\omega_j}{d\tau} - \frac{\partial \omega_j}{\partial \dot{\rho}_i}\,\ddot{\rho}_i\right) - \frac{dW_j}{d\tau}\,\omega_j - W_j\left(\frac{d\omega_j}{d\tau} - \frac{\partial \omega_j}{\partial \dot{\rho}_i}\,\ddot{\rho}_i\right) = 0,$$

so that $\omega_j\, dW_j/d\tau = 0$, i.e., $d\boldsymbol{W}/d\tau$ is orthogonal to $\boldsymbol{\omega}$ and since the matrix $\partial \omega_j/\partial \dot{\rho}_i$ is invertible we can separate the $d\boldsymbol{W}/d\tau$ term in (2.73) in terms of $\boldsymbol{W}$ and get (2.72). The different matrix terms of (2.73) are explicitly given by

$$\frac{\partial \omega_j}{\partial \rho_i} \equiv \frac{2}{1+\rho^2}\begin{pmatrix} -2\rho_1\omega_1 & -2\rho_2\omega_1 + \dot{\rho}_3 & -2\rho_3\omega_1 - \dot{\rho}_2 \\ -2\rho_1\omega_2 - \dot{\rho}_3 & -2\rho_2\omega_2 & -2\rho_3\omega_2 + \dot{\rho}_1 \\ -2\rho_1\omega_3 + \dot{\rho}_2 & -2\rho_2\omega_3 - \dot{\rho}_1 & -2\rho_3\omega_3 \end{pmatrix},$$

$$\frac{\partial \omega_j}{\partial \dot{\rho}_i} \equiv V = \frac{2}{1+\rho^2}\begin{pmatrix} 1 & -\rho_3 & \rho_2 \\ \rho_3 & 1 & -\rho_1 \\ -\rho_2 & \rho_1 & 1 \end{pmatrix},$$

with $j$ and $i$ as a row and column index, respectively. Thus

$$V^{-1} = \frac{1}{2}\begin{pmatrix} 1+\rho_1^2 & \rho_1\rho_2 + \rho_3 & \rho_1\rho_3 - \rho_2 \\ \rho_1\rho_2 - \rho_3 & 1+\rho_2^2 & \rho_2\rho_3 + \rho_1 \\ \rho_1\rho_3 + \rho_2 & \rho_2\rho_3 - \rho_1 & 1+\rho_3^2 \end{pmatrix},$$

$$\frac{dV}{d\tau} = -\frac{2(\boldsymbol{\rho}\cdot\dot{\boldsymbol{\rho}})}{1+\rho^2}\,V + \frac{2}{1+\rho^2}\begin{pmatrix} 0 & -\dot{\rho}_3 & \dot{\rho}_2 \\ \dot{\rho}_3 & 0 & -\dot{\rho}_1 \\ -\dot{\rho}_2 & \dot{\rho}_1 & 0 \end{pmatrix}.$$

Because $\dot{\boldsymbol{J}} = 0$, using (2.72) and the expression of $\boldsymbol{P}$, (2.65), this implies the general relation for a free particle

$$\dot{\boldsymbol{r}} \times \boldsymbol{R} + \dot{\boldsymbol{u}} \times \boldsymbol{U} + \boldsymbol{\omega} \times \boldsymbol{W} = 0, \tag{2.74}$$

which reflects the fact that velocity, acceleration and angular velocity are not independent magnitudes, and taking into account that $\boldsymbol{R}$ and $\dot{\boldsymbol{r}}$ have the same direction, it reduces to

$$\dot{\boldsymbol{u}} \times \boldsymbol{U} + \boldsymbol{\omega} \times \boldsymbol{W} = 0. \tag{2.75}$$

This relation shows that for a free particle, if the function $\boldsymbol{W}$ is in the direction of the angular velocity, $\boldsymbol{W} \sim \boldsymbol{\omega}$, then, necessarily $\boldsymbol{U}$ is lying along the direction of the acceleration $\boldsymbol{U} \sim \dot{\boldsymbol{u}}$. Conversely, if $\boldsymbol{W} = a\dot{\boldsymbol{u}}$, then $\boldsymbol{U} = a\boldsymbol{\omega}$, with the same proportionality coefficient $a$. The internal motion of the center of charge $\boldsymbol{r}$ and the rotation of the particle are not completely independent motions. We see that the analysis of (2.75) shows the possibility of at least two different kinds of particles: One in which $\boldsymbol{W} \sim \boldsymbol{\omega}$ and $\boldsymbol{U} \sim \dot{\boldsymbol{u}}$ or another in which conversely $\boldsymbol{U} = a\boldsymbol{\omega}$ and $\boldsymbol{W} = a\dot{\boldsymbol{u}}$. More general systems can be obtained by assuming a more complex structure of both $\boldsymbol{U}$ and $\boldsymbol{W}$ compatible with (2.75).

One important conclusion is that for spinning models in which the Lagrangian $L$ does not depend on $\boldsymbol{\omega}$, then necessarily $\boldsymbol{U}$ lies along $\dot{\boldsymbol{u}}$. On the other hand, if $L$ is independent of $\dot{\boldsymbol{u}}$, then $\boldsymbol{W}$ and $\boldsymbol{\omega}$ must be collinear.

It is the presence of functions $\boldsymbol{U}$ and $\boldsymbol{W}$ in (2.54) that distinguishes the structure of this particle when compared with the point particle (2.23) and consequently the spin structure is directly related to these functions.

## 4. GALILEI FREE PARTICLE WITH (ANTI)ORBITAL SPIN

To analyze the spin structure of the particle, and therefore the different contributions to spin coming from these functions $\boldsymbol{U}$ and $\boldsymbol{W}$, let us consider the following simpler example.

Consider a Galilei particle whose kinematical space is $X = \mathcal{G}/SO(3)$, so that any point $x \in X$ can be characterized by the seven variables $x \equiv (t, \boldsymbol{r}, \boldsymbol{u})$, $\boldsymbol{u} = d\boldsymbol{r}/dt$, which are interpreted as time, position and velocity of the particle respectively. In this example we have no orientation variables. The Lagrangian will also depend on the next order derivatives, *i.e.*, on the velocity which is already considered as a kinematical variable and on the acceleration of the particle. Rotation and translation invariance implies that $L$ will be a function of only $\boldsymbol{u}^2$, $(d\boldsymbol{u}/dt)^2$ and

$\boldsymbol{u}\cdot d\boldsymbol{u}/dt = d(u^2/2)/dt$, but this last term is a total time derivative and it will not be considered here.

Since from condition (2.75) $\boldsymbol{U} \sim \dot{\boldsymbol{u}}$, let us assume that our elementary system is represented by the following Lagrangian, which when written in terms of the three degrees of freedom and their derivatives is expressed as

$$L = \frac{m}{2}\left(\frac{d\boldsymbol{r}}{dt}\right)^2 - \frac{m}{2\omega^2}\left(\frac{d^2\boldsymbol{r}}{dt^2}\right)^2. \tag{2.76}$$

Parameter $m$ is the mass of the particle because the first term is gauge variant in terms of the gauge function (2.55) defined by this constant $m$, while parameter $\omega$ of dimensions of time$^{-1}$ represents an internal frequency. It is the frequency of the internal zitterbewegung. In terms of the kinematical variables and their derivatives, and in terms of some group invariant evolution parameter $\tau$, the Lagrangian can also be written as

$$L = \frac{m}{2}\frac{\dot{\boldsymbol{r}}^2}{\dot{t}} - \frac{m}{2\omega^2}\frac{\dot{\boldsymbol{u}}^2}{\dot{t}}, \tag{2.77}$$

where the dot means $\tau$-derivative. If we consider that the evolution parameter is dimensionless, all terms in the Lagrangian have dimensions of action. Because the Lagrangian is a homogeneous function of first degree in terms of the derivatives of the kinematical variables, $L$ can also be written as

$$L = T\dot{t} + \boldsymbol{R}\cdot\dot{\boldsymbol{r}} + \boldsymbol{U}\cdot\dot{\boldsymbol{u}}, \tag{2.78}$$

where the functions accompanying the derivatives of the kinematical variables are defined and explicitly given by

$$T = \frac{\partial L}{\partial \dot{t}} = -\frac{m}{2}\left(\frac{d\boldsymbol{r}}{dt}\right)^2 + \frac{m}{2\omega^2}\left(\frac{d^2\boldsymbol{r}}{dt^2}\right)^2,$$

$$\boldsymbol{R} = \frac{\partial L}{\partial \dot{\boldsymbol{r}}} = m\frac{d\boldsymbol{r}}{dt}, \tag{2.79}$$

$$\boldsymbol{U} = \frac{\partial L}{\partial \dot{\boldsymbol{u}}} = -\frac{m}{\omega^2}\frac{d^2\boldsymbol{r}}{dt^2}. \tag{2.80}$$

Dynamical equations obtained from Lagrangian (2.76) are:

$$\frac{1}{\omega^2}\frac{d^4\boldsymbol{r}}{dt^4} + \frac{d^2\boldsymbol{r}}{dt^2} = 0, \tag{2.81}$$

whose general solution is:

$$\boldsymbol{r}(t) = \boldsymbol{A} + \boldsymbol{B}t + \boldsymbol{C}\cos\omega t + \boldsymbol{D}\sin\omega t, \tag{2.82}$$

in terms of the 12 integration constants $\boldsymbol{A}$, $\boldsymbol{B}$, $\boldsymbol{C}$ and $\boldsymbol{D}$.

When applying Noether's theorem to the invariance of dynamical equations under the Galilei group, the corresponding constants of the motion can be written in terms of the above functions in the form:

$$\text{Energy} \quad H = -T - \boldsymbol{u} \cdot \frac{d\boldsymbol{U}}{dt}, \tag{2.83}$$

$$\text{linear momentum} \quad \boldsymbol{P} = \boldsymbol{R} - \frac{d\boldsymbol{U}}{dt} = m\boldsymbol{u} - \frac{d\boldsymbol{U}}{dt}, \tag{2.84}$$

$$\text{kinematical momentum} \quad \boldsymbol{K} = m\boldsymbol{r} - \boldsymbol{P}t - \boldsymbol{U}, \tag{2.85}$$

$$\text{angular momentum} \quad \boldsymbol{J} = \boldsymbol{r} \times \boldsymbol{P} + \boldsymbol{u} \times \boldsymbol{U}. \tag{2.86}$$

It is the presence of the $\boldsymbol{U}$ function that distinguishes the features of this system with respect to the point particle case. We find that the total linear momentum is not lying along the direction of the velocity $\boldsymbol{u}$, and the spin structure is directly related to the dependence of the Lagrangian on the acceleration.

If we substitute the general solution (2.82) in (2.83-2.86) we see in fact that the integration constants are related to the above conserved quantities

$$H = \frac{m}{2}\boldsymbol{B}^2 - \frac{m\omega^2}{2}(\boldsymbol{C}^2 + \boldsymbol{D}^2), \tag{2.87}$$

$$\boldsymbol{P} = m\boldsymbol{B}, \tag{2.88}$$

$$\boldsymbol{K} = m\boldsymbol{A}, \tag{2.89}$$

$$\boldsymbol{J} = \boldsymbol{A} \times m\boldsymbol{B} - m\omega\boldsymbol{C} \times \boldsymbol{D}. \tag{2.90}$$

We see that the kinematical momentum $\boldsymbol{K}$ in (2.85) differs from the point particle case (2.31) in the term $-\boldsymbol{U}$, such that if we define the vector $\boldsymbol{k} = \boldsymbol{U}/m$, with dimensions of length, then $\boldsymbol{K} = 0$ leads from (2.85) to the equation:

$$\boldsymbol{P} = m\frac{d(\boldsymbol{r} - \boldsymbol{k})}{dt},$$

and $\boldsymbol{q} = \boldsymbol{r} - \boldsymbol{k}$, defines the position of the center of mass of the particle that is a different point than $\boldsymbol{r}$ and using (2.80) is given by

$$\boldsymbol{q} = \boldsymbol{r} - \frac{1}{m}\boldsymbol{U} = \boldsymbol{r} + \frac{1}{\omega^2}\frac{d^2\boldsymbol{r}}{dt^2}. \tag{2.91}$$

In terms of it, dynamical equations (2.81) can be separated into the form:

$$\frac{d^2\boldsymbol{q}}{dt^2} = 0, \tag{2.92}$$

$$\frac{d^2\boldsymbol{r}}{dt^2} + \omega^2(\boldsymbol{r} - \boldsymbol{q}) = 0, \tag{2.93}$$

where (2.92) is just eq. (2.81) after twice differentiating (2.91), and Equation (2.93) is (2.91) after collecting all terms on the left hand side.

From (2.92) we see that point $\boldsymbol{q}$ moves in a straight trajectory at constant velocity while the motion of point $\boldsymbol{r}$, given in (2.93), is an isotropic harmonic motion of angular frequency $\omega$ around point $\boldsymbol{q}$.

The spin of the system $\boldsymbol{S}$ is defined as

$$\boldsymbol{S} = \boldsymbol{J} - \boldsymbol{q} \times \boldsymbol{P} = \boldsymbol{J} - \frac{1}{m}\boldsymbol{K} \times \boldsymbol{P}, \tag{2.94}$$

and since it is written in terms of constants of the motion it is clearly a constant of the motion, and its magnitude $S^2$ is also a Galilei invariant quantity that characterizes the system. In terms of the integration constants it is expressed as

$$\boldsymbol{S} = -m\omega\,\boldsymbol{C} \times \boldsymbol{D}. \tag{2.95}$$

From its definition we get

$$\boldsymbol{S} = \boldsymbol{u} \times \boldsymbol{U} + \boldsymbol{k} \times \boldsymbol{P} = -m(\boldsymbol{r} - \boldsymbol{q}) \times \frac{d}{dt}(\boldsymbol{r} - \boldsymbol{q}) = -\boldsymbol{k} \times m\frac{d\boldsymbol{k}}{dt}, \tag{2.96}$$

which appears as the (anti)orbital angular momentum of the relative motion of point $\boldsymbol{r}$ around the center of mass position $\boldsymbol{q}$ at rest, so that the total angular momentum can be written as

$$\boldsymbol{J} = \boldsymbol{q} \times \boldsymbol{P} + \boldsymbol{S} = \boldsymbol{L} + \boldsymbol{S}. \tag{2.97}$$

It is the sum of the orbital angular momentum $\boldsymbol{L}$ associated to the motion of the center of mass and the spin part $\boldsymbol{S}$. For a free particle both $\boldsymbol{L}$ and $\boldsymbol{S}$ are separately constants of the motion. We use the term (anti)orbital to suggest that if vector $\boldsymbol{k}$ represents the position of a point mass $m$, the angular momentum of this motion is in the opposite direction as the obtained spin observable. But as we shall see in a moment, vector $\boldsymbol{k}$ does not represent the position of the mass $m$ but rather the position of the charge $e$ of the particle.

By using the dynamical equations (2.92-2.93), the total energy can be cast into the form:

$$H = \frac{\boldsymbol{P}^2}{2m} + H_0, \tag{2.98}$$

where the internal energy

$$H_0 = -\frac{m\omega^2}{2}(\boldsymbol{r} - \boldsymbol{q})^2 - \frac{m}{2}\left(\frac{d}{dt}(\boldsymbol{r} - \boldsymbol{q})\right)^2 = -\frac{m\omega^2}{2}\boldsymbol{k}^2 - \frac{m}{2}\left(\frac{d\boldsymbol{k}}{dt}\right)^2, \tag{2.99}$$

is also expressed in terms of the relative vector $\boldsymbol{k}$, and is also a constant of the motion. It is negative definite as it corresponds to the bounded motion of point $\boldsymbol{r}$ around the center of mass $\boldsymbol{q}$. It is in fact the total mechanical energy of the harmonic oscillator that represents the internal zitterbewegung.

In general the zitterbewegung is a plane and closed elliptic orbit and we can also obtain circular trajectories. In the relativistic case we shall obtain more complex trajectories that can be interpreted as a kind of precessing ellipses in the trajectory plane.

Once we introduce the classical free path into the free Lagrangian and integrate from the initial to the final point, we obtain the action function for the free system along this path that can be written in terms of the boundary end point kinematical variables in the form

$$\begin{aligned} A(t_1, \boldsymbol{r}_1, \boldsymbol{u}_1; t_2, \boldsymbol{r}_2, \boldsymbol{u}_2) &= \frac{m}{2\Delta}\left[\frac{1}{\omega}(\boldsymbol{u}_2 - \boldsymbol{u}_1)^2 \sin\omega(t_2 - t_1)\right. \\ &+ \omega(\boldsymbol{r}_2 - \boldsymbol{r}_1)^2 \sin\omega(t_2 - t_1) \\ &- (t_2 - t_1)(\boldsymbol{u}_2 - \boldsymbol{u}_1)^2 \cos\omega(t_2 - t_1) \\ &+ 2(t_2 - t_1)(\boldsymbol{u}_2 \cdot \boldsymbol{u}_1)(1 - \cos\omega(t_2 - t_1)) \\ &- \left. 2(\boldsymbol{r}_2 - \boldsymbol{r}_1)\cdot(\boldsymbol{u}_1 + \boldsymbol{u}_2)(1 - \cos\omega(t_2 - t_1))\right], \end{aligned}$$

where $\Delta = 2(1 - \cos\omega(t_2 - t_1)) - \omega(t_2 - t_1)\sin\omega(t_2 - t_1)$. This is the action function that must be considered when analyzing Feynman's kernel in the path integral approach of this system. We observe in this case that the additional kinematical variables $\boldsymbol{u}$, with respect to the point particle case, are non-compact variables. The variational boundary value problem and the Newtonian boundary values at initial time $t_1$ can be expressed in terms of each other.

## 4.1 INTERACTING WITH AN EXTERNAL ELECTROMAGNETIC FIELD

But if $\boldsymbol{q}$ represents the center of mass position, then what position does point $\boldsymbol{r}$ represent? Point $\boldsymbol{r}$ represents the position of the charge of the particle. This can be seen by considering some interaction with an external field. The homogeneity condition of the Lagrangian in terms of the derivatives of the kinematical variables leads us to consider an interaction term of the form

$$L_I = -e\phi(t, \boldsymbol{r})\dot{t} + e\boldsymbol{A}(t, \boldsymbol{r}) \cdot \dot{\boldsymbol{r}}, \qquad (2.100)$$

which is linear in the derivatives of the kinematical variables $t$ and $\boldsymbol{r}$ and where the external potentials are only functions of $t$ and $\boldsymbol{r}$. We can also consider more general interaction terms of the form $\boldsymbol{N}(t, \boldsymbol{r}, \boldsymbol{u}) \cdot \dot{\boldsymbol{u}}$, and also more general terms in which functions $\phi$ and $\boldsymbol{A}$ also depend on $\boldsymbol{u}$ and $\dot{\boldsymbol{u}}$. But this will be something different than an interaction with an external electromagnetic field. At this stage of the formalism our challenge is to describe at least the electromagnetic properties of matter related to the spin structure so that those more general interaction terms will not be considered here.

Dynamical equations obtained from $L + L_I$ are

$$\frac{1}{\omega^2}\frac{d^4\boldsymbol{r}}{dt^4} + \frac{d^2\boldsymbol{r}}{dt^2} = \frac{e}{m}\left(\boldsymbol{E}(t,\boldsymbol{r}) + \boldsymbol{u} \times \boldsymbol{B}(t,\boldsymbol{r})\right), \qquad (2.101)$$

where the electric field $\boldsymbol{E}$ and magnetic field $\boldsymbol{B}$ are expressed in terms of the potentials in the usual form, $\boldsymbol{E} = -\nabla\phi - \partial\boldsymbol{A}/\partial t$, $\boldsymbol{B} = \nabla \times \boldsymbol{A}$. Because the interaction term does not modify the dependence of the Lagrangian on $\dot{\boldsymbol{u}}$, the function $\boldsymbol{U} = m\boldsymbol{k}$ has the same expression as in the free particle case. Therefore the spin and the center of mass definitions, (2.96) and (2.91) respectively, remain the same as in the previous case. Dynamical equations (2.101) can again be separated into the form

$$\frac{d^2\boldsymbol{q}}{dt^2} = \frac{e}{m}\left(\boldsymbol{E}(t,\boldsymbol{r}) + \boldsymbol{u} \times \boldsymbol{B}(t,\boldsymbol{r})\right), \qquad (2.102)$$

$$\frac{d^2\boldsymbol{r}}{dt^2} + \omega^2(\boldsymbol{r} - \boldsymbol{q}) = 0, \qquad (2.103)$$

where the center of mass $\boldsymbol{q}$ satisfies Newton's equations under the action of the total external Lorentz force, while point $\boldsymbol{r}$ still satisfies the isotropic harmonic motion of angular frequency $\omega$ around point $\boldsymbol{q}$. But the external force and the fields are defined at point $\boldsymbol{r}$ and not at point $\boldsymbol{q}$. It is the velocity $\boldsymbol{u}$ of point $\boldsymbol{r}$ that appears in the magnetic term of the Lorentz force. Point $\boldsymbol{r}$ clearly represents the position of the charge. In fact, this minimal coupling we have considered is the coupling of the electromagnetic potentials with the particle current, that in the relativistic case can be written as $j_\mu A^\mu$, but the current $j_\mu$ is associated to the motion of a charge $e$ at point $\boldsymbol{r}$.

This charge has an oscillatory motion of very high frequency $\omega$ that, in the case of the relativistic electron (see Sec.4.2 of chapter 3), is $\omega = 2mc^2/\hbar \simeq 1.55 \times 10^{21}\mathrm{s}^{-1}$. The average position of the charge is the center of mass, but it is this internal orbital motion, usually known as the zitterbewegung, that gives rise to the spin structure and also to the magnetic properties of the particle, as we shall see later.

When analyzed in the center of mass frame (see Fig. 2.1), $\boldsymbol{q} = 0$, $\boldsymbol{r} = \boldsymbol{k}$, the system reduces to a point charge whose motion is in general an ellipse, but if we choose $C = D$, and $\boldsymbol{C} \cdot \boldsymbol{D} = 0$, it reduces to a circle

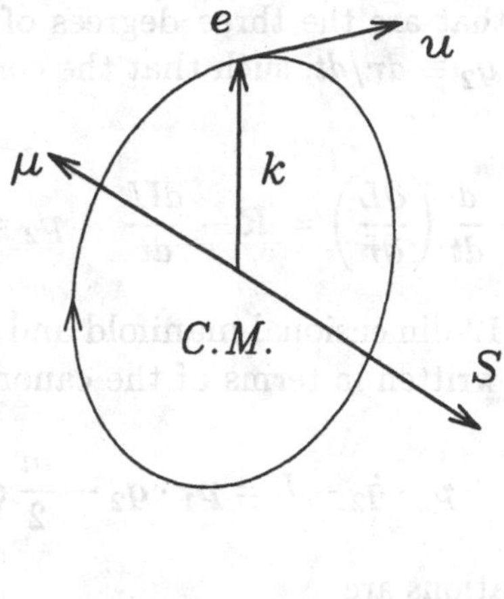


*Figure 2.1.* Charge motion in the C.M. frame.

of radius $a = C = D$, orthogonal to the spin. Then if the particle has charge $e$, it has a magnetic moment that according to the usual classical definition is: [6]

$$\boldsymbol{\mu} = \frac{1}{2}\int \boldsymbol{r} \times \boldsymbol{j}\, d^3\boldsymbol{r} = \frac{e}{2}\boldsymbol{k} \times \frac{d\boldsymbol{k}}{dt} = -\frac{e}{2m}\boldsymbol{S}, \tag{2.104}$$

where $\boldsymbol{j} = e\delta^3(\boldsymbol{r} - \boldsymbol{k})d\boldsymbol{k}/dt$ is the current associated to the motion of a charge $e$ located at point $\boldsymbol{k}$. The magnetic moment is orthogonal to the zitterbewegung plane and opposite to the spin if $e > 0$. It also has a non-vanishing oscillating electric dipole $\boldsymbol{d} = e\boldsymbol{k}$, orthogonal to $\boldsymbol{\mu}$ and therefore to $\boldsymbol{S}$ in the center of mass frame, such that its time average value vanishes for times larger than the natural period of this internal motion. Although this is a nonrelativistic example it is interesting to point out and compare with Dirac's relativistic analysis of the electron, [7] in which both momenta $\boldsymbol{\mu}$ and $\boldsymbol{d}$ appear, giving rise to two possible interacting terms in Dirac's Hamiltonian. We shall come back to this analysis later when we study the elementary relativistic particles.

## 4.2 CANONICAL ANALYSIS OF THE SYSTEM

Although the Lagrangian (2.76) depends on second order derivatives we can develop the corresponding canonical formalism. Starting from the Lagrangian (2.76)

$$L = \frac{m}{2}\,\dot{\boldsymbol{r}}^2 - \frac{m}{2\omega^2}\,\ddot{\boldsymbol{r}}^2,$$

where now the dot means time derivative, we have in this case six generalized coordinates that are the three degrees of freedom $\boldsymbol{q}_1 \equiv \boldsymbol{r}$ and their first derivatives $\boldsymbol{q}_2 \equiv d\boldsymbol{r}/dt$, such that the conjugate momenta are, according to (1.26),

$$\boldsymbol{p}_1 = \frac{\partial L}{\partial \dot{\boldsymbol{r}}} - \frac{d}{dt}\left(\frac{\partial L}{\partial \ddot{\boldsymbol{r}}}\right) = \boldsymbol{R} - \frac{d\boldsymbol{U}}{dt}, \quad \boldsymbol{p}_2 = \frac{\partial L}{\partial \ddot{\boldsymbol{r}}} = \boldsymbol{U}.$$

The phase space is a 12-dimensional manifold and the Hamiltonian is in fact the total energy written in terms of the canonical variables

$$H = \boldsymbol{p}_1 \cdot \dot{\boldsymbol{q}}_1 + \boldsymbol{p}_2 \cdot \dot{\boldsymbol{q}}_2 - L = \boldsymbol{p}_1 \cdot \boldsymbol{q}_2 - \frac{m}{2}\boldsymbol{q}_2^2 - \frac{\omega^2}{2m}\boldsymbol{p}_2^2.$$

Hamilton-Jacobi equations are

$$\dot{\boldsymbol{q}}_1 = \frac{\partial H}{\partial \boldsymbol{p}_1} = \boldsymbol{q}_2, \quad \dot{\boldsymbol{p}}_1 = -\frac{\partial H}{\partial \boldsymbol{q}_1} = 0,$$

$$\dot{\boldsymbol{q}}_2 = \frac{\partial H}{\partial \boldsymbol{p}_2} = -\frac{\omega^2}{m}\boldsymbol{p}_2, \quad \dot{\boldsymbol{p}}_2 = -\frac{\partial H}{\partial \boldsymbol{q}_2} = -\boldsymbol{p}_1 + m\boldsymbol{q}_2.$$

The ten Noether constants of the motion become in this formalism the generating functions of the corresponding canonical transformations of time and space translations, pure Galilei transformations and rotations. They are explicitly given by

$$\begin{aligned} H &= \boldsymbol{p}_1 \cdot \boldsymbol{q}_2 - \frac{m}{2}\boldsymbol{q}_2^2 - \frac{\omega^2}{2m}\boldsymbol{p}_2^2, \\ \boldsymbol{P} &= \boldsymbol{p}_1, \\ \boldsymbol{K} &= m\boldsymbol{q}_1 - \boldsymbol{p}_1 t - \boldsymbol{p}_2, \\ \boldsymbol{J} &= \boldsymbol{q}_1 \times \boldsymbol{p}_1 + \boldsymbol{q}_2 \times \boldsymbol{p}_2. \end{aligned}$$

Since the Poisson bracket of two constants of the motion is again a constant of the motion, we obtain that the above constants of motion satisfy the following commutation relations

$$\{J_i, J_k\} = \epsilon_{ikl}J_l, \quad \{J_i, P_k\} = \epsilon_{ikl}P_l, \quad \{J_i, K_k\} = \epsilon_{ikl}K_l, \quad \{J_i, H\} = 0,$$

$$\{P_i, P_k\} = 0, \quad \{P_i, H\} = 0, \quad \{K_i, K_k\} = 0, \quad \{K_i, H\} = P_i,$$
$$\{K_i, P_j\} = m\delta_{ij},$$

where $\{.,.\}$ is the corresponding Poisson bracket. Because $\{K_i, P_j\} \neq 0$ they are not the commutations relations of the Galilei group but rather those of the extended Galilei group. [8] It is interesting to compare them with the commutation relations (2.12), (2.13) and (2.16). Although $K_i$ satisfies $\{K_i, H\} = P_i$, it is a constant of the motion because, as we see in (2.85), $\boldsymbol{K}$ is time dependent and its total time derivative in the canonical approach is

$$\frac{d\boldsymbol{K}}{dt} = \{\boldsymbol{K}, H\} + \frac{\partial \boldsymbol{K}}{\partial t} = \boldsymbol{p}_1 - \boldsymbol{p}_1 = 0.$$

The center of mass position defined by $\boldsymbol{q} \equiv (\boldsymbol{K} + \boldsymbol{P}t)/m$ satisfies

$$\{q_i, q_j\} = 0, \quad \{q_i, P_j\} = \delta_{ij}, \quad \text{but} \quad \{q_i, q_{2j}\} = \frac{1}{m}\delta_{ij} \neq 0.$$

Therefore, the $\boldsymbol{q}$ are not canonical variables. It is not possible to find in general a canonical transformation that changes the center of charge position $\boldsymbol{q}_1$ for the center of mass $\boldsymbol{q}$ as new canonical variables. To do so we shall need also to replace the $\boldsymbol{q}_2$ variables. This feature will be analyzed later in Chapter 6 when considering the description of the position operator in quantum mechanics.

The spin observable $\boldsymbol{S} = \boldsymbol{J} - \boldsymbol{K} \times \boldsymbol{P}/m = \boldsymbol{q}_2 \times \boldsymbol{p}_2 + \boldsymbol{p}_2 \times \boldsymbol{p}_1/m$ satisfies the Poisson bracket commutation relations

$$\{S_i, S_k\} = \epsilon_{ikl} S_l, \quad \{J_i, S_k\} = \epsilon_{ikl} S_l,$$

$$\{S_i, P_k\} = 0, \quad \{S_i, K_j\} = 0, \quad \{S_i, H\} = 0,$$

showing respectively that it is an angular momentum, that transforms like a vector under rotations, and it is invariant under space translations and pure Galilei transformations and is a constant of the motion. If we had taken in expression (2.86) the angular momentum $\boldsymbol{J} = \boldsymbol{r} \times \boldsymbol{P} + \boldsymbol{Z}$, with $\boldsymbol{Z} = \boldsymbol{u} \times \boldsymbol{U} = \boldsymbol{q}_2 \times \boldsymbol{p}_2$, then this observable satisfies

$$\{Z_i, Z_k\} = \epsilon_{ikl} Z_l, \quad \{J_i, Z_k\} = \epsilon_{ikl} Z_l, \quad \{Z_i, P_k\} = 0.$$

But

$$\{Z_i, K_j\} = -\epsilon_{ijl} p_{2l}, \quad \{Z_i, H\} = (\boldsymbol{p}_1 \times \boldsymbol{q}_2)_i.$$

It satisfies the Poisson brackets of an angular momentum, transforms like a vector under rotations, and is invariant under space translations but not under pure Galilei transformations. It is not a constant of the motion. It satisfies the dynamical equation $d\boldsymbol{Z}/dt = \boldsymbol{P} \times \boldsymbol{u}$. That is why this observable cannot be considered as the spin of the system. We mention this here because in the quantum case this is precisely the spin observable equivalent to Dirac's spin operator of the electron. Dirac's spin observable satisfies the dynamical equation $d\boldsymbol{S}/dt = \boldsymbol{P} \times c\boldsymbol{\alpha}$, where $c\boldsymbol{\alpha}$ matrices play the role of the velocity operator $\boldsymbol{u}$ in Dirac's theory.

We thus see again that the two constants of the motion

$$H_0 = H - \frac{\boldsymbol{P}^2}{2m}, \quad \text{and} \quad S^2 = \left(\boldsymbol{J} - \frac{1}{m}\boldsymbol{K}\times\boldsymbol{P}\right)^2, \tag{2.105}$$

commute with the above ten generators and they are Galilei invariant properties of the particle, that together with the mass $m$ completely characterize the structure of this particle. They are the three functionally independent Casimir invariants of the extended Galilei group. In fact $H_0$ is the Galilei internal energy of the particle and $S^2$ the square of the spin.

## 4.3 SPINNING PARTICLE IN A UNIFORM MAGNETIC FIELD

Let us consider in detail the interaction of this model of particle with spin of orbital nature in an external uniform magnetic field $\boldsymbol{B}$. It is an exercise that can be solved explicitly. The advantage of a model defined in terms of a Lagrangian function is that we do not need to state any dynamical equation for spin, because the spin is a function of the independent degrees of freedom and therefore its dynamics can be obtained from them. The result is that we shall obtain as a first order approximation a torque equation of the usual form $d\boldsymbol{S}/dt = \boldsymbol{\mu}\times\boldsymbol{B}$, when the magnetic moment $\boldsymbol{\mu}$ is properly interpreted in terms of the charge motion.

In this case, the system of equations (2.102-2.103) reduces to

$$\frac{d^2\boldsymbol{q}}{dt^2} = \frac{e}{m}\boldsymbol{u}\times\boldsymbol{B}, \quad \frac{d^2\boldsymbol{r}}{dt^2} + \omega^2(\boldsymbol{r}-\boldsymbol{q}) = 0.$$

With the definition of the variables $\boldsymbol{v} = d\boldsymbol{q}/dt$, it is equivalent to a linear system of twelve differential equations of first order for the components of $\boldsymbol{r}$, $\boldsymbol{u}$, $\boldsymbol{q}$ and $\boldsymbol{v}$. If we define a new dimensionless time variable $\tau = \omega t$, then the above system depends only on the dimensionless parameter $a = eB/m\omega$ which is the quotient between the cyclotron frequency $|\omega_c| = eB/m$ and $\omega$, the natural frequency of the internal motion.

By taking the direction of the uniform magnetic field along the $OZ$ axis, the external force is orthogonal to it. Then if we call $q_3$ and $r_3$ the corresponding coordinates along that axis of the centre of mass and center of charge, they satisfy

$$\frac{d^2q_3}{dt^2} = 0, \quad \frac{d^2q_3}{dt^2} + \omega^2(r_3 - q_3) = 0 \tag{2.106}$$

whose general solution in terms of the initial data $q_3(0)$, $r_3(0)$, $v_3(0)$ and $u_3(0)$ is

$$q_3(t) = q_3(0) + v_3(0)t, \tag{2.107}$$

$$r_3(t) = (r_3(0) - q_3(0))\cos\omega t + \frac{1}{\omega}(u_3(0) - v_3(0))\sin\omega t + q_3(0) + v_3(0)t. \tag{2.108}$$

Similarly, the other components of the center of mass in terms of the new time variable are

$$\frac{d^2q_1}{d\tau^2} = a\frac{dr_2}{d\tau}, \quad \frac{d^2q_2}{d\tau^2} = -a\frac{dr_1}{d\tau},$$

and once integrated we get

$$\frac{dq_1}{d\tau} = ar_2 + b_1, \quad \frac{dq_2}{d\tau} = -ar_1 + b_2, \tag{2.109}$$

where $b_1$ and $b_2$ are two integration constants with dimensions of length. Thus we are left with the integration of a first order system formed by these two last equations (2.109) and the equations for the other two components of the center of charge that can be written as

$$\frac{dr_1}{d\tau} = u_1, \quad \frac{dr_2}{d\tau} = u_2, \tag{2.110}$$

$$\frac{du_1}{d\tau} = q_1 - r_1, \quad \frac{du_2}{d\tau} = q_2 - r_2. \tag{2.111}$$

The matrix of this linear system in terms of the variables $q_1$, $q_2$, $r_1$, $r_2$, $u_1$ and $u_2$, taken in this order, is just

$$M = \begin{pmatrix} 0 & 0 & 0 & a & 0 & 0 \\ 0 & 0 & -a & 0 & 0 & 0 \\ 0 & 0 & 0 & 0 & 1 & 0 \\ 0 & 0 & 0 & 0 & 0 & 1 \\ 1 & 0 & -1 & 0 & 0 & 0 \\ 0 & 1 & 0 & -1 & 0 & 0 \end{pmatrix},$$

whose characteristic equation is $\lambda^6 + 2\lambda^4 + \lambda^2 + a^2 = 0$. It is shown that it has six different roots, corresponding to the normal modes of the system. If we call $\lambda = iz$, these new variables verify $z^2(1-z^2)^2 = a^2$, and thus by solving the cubic equation $z(1-z^2) = a$, the six solutions of the form $\pm iz$ will be the six eigenvalues of the system. If we define

$$k = \frac{1}{3}\arcsin\left(\frac{3\sqrt{3}a}{2}\right), \tag{2.112}$$

then the six eigevalues are $\pm i\omega_j$, $j = 1, 2, 3$, where:

$$\omega_1 = \frac{2}{\sqrt{3}}\sin k, \quad \omega_2 = -\cos k - \frac{1}{\sqrt{3}}\sin k, \quad \omega_3 = \cos k - \frac{1}{\sqrt{3}}\sin k. \tag{2.113}$$

If $3\sqrt{3}|a|/2 \leq 1$ then the six roots are purely imaginary and the motion is three-periodic with these three frequencies. Otherwise, if there exist real roots, the corresponding solution will be exponential. In general, for the electron, as we shall see in the next chapter, the zitterbewegung frequency is $\omega = 2mc^2/\hbar$, and thus

$$a/B = e/m\omega = e\hbar/2m^2c^2 = 1.13 \times 10^{-10}\text{Tesla}^{-1},$$

so that even with very strong magnetic fields the parameter $a$ is very small and the usual solution will be oscillatory.

The general solution of the complete system will be a linear combination of these three oscillations and it will depend on twelve integration constants that will be expressed in terms of the initial position and velocity of the center of mass and center of charge. The general form for the evolution of the center of charge is:

$$\begin{aligned}
r_1(\tau) &= A\cos\omega_1\tau + B\sin\omega_1\tau + C\cos\omega_2\tau + D\sin\omega_2\tau + E\cos\omega_3\tau \\
&+ F\sin\omega_3\tau + b_2/a, \\
r_2(\tau) &= B\cos\omega_1\tau - A\sin\omega_1\tau + D\cos\omega_2\tau - C\sin\omega_2\tau \\
&+ F\cos\omega_3\tau - E\sin\omega_3\tau - b_1/a, \\
r_3(t) &= (r_3(0) - q_3(0))\cos\omega t \\
&+ \frac{1}{\omega}(u_3(0) - v_3(0))\sin\omega t + q_3(0) + v_3(0)t,
\end{aligned}$$

where

$$b_1/a = v_1(0)/a\omega - r_2(0), \quad b_2/a = v_2(0)/a\omega + r_1(0).$$

For the center of mass coordinates we get

$$\begin{aligned}
q_1(\tau) &= (1-\omega_1^2)(A\cos\omega_1\tau + B\sin\omega_1\tau) \\
&+ (1-\omega_2^2)(C\cos\omega_2\tau + D\sin\omega_2\tau) \\
&+ (1-\omega_3^2)(E\cos\omega_3\tau + F\sin\omega_3\tau) + b_2/a, \\
q_2(\tau) &= (1-\omega_1^2)(B\cos\omega_1\tau - A\sin\omega_1\tau) \\
&+ (1-\omega_2^2)(D\cos\omega_2\tau - C\sin\omega_2\tau) \\
&+ (1-\omega_3^2)(F\cos\omega_3\tau - E\sin\omega_3\tau) - b_1/a, \\
q_3(t) &= q_3(0) + v_3(0)t.
\end{aligned}$$

The six unknown constants $A, B, C, D, E$, and $F$ are of dimensions of length and satisfy the linear system

$$\begin{pmatrix} 1 & 1 & 1 \\ \omega_1 & \omega_2 & \omega_3 \\ \omega_1^2 & \omega_2^2 & \omega_3^2 \end{pmatrix} \begin{pmatrix} A \\ C \\ E \end{pmatrix} = \begin{pmatrix} -v_2(0)/a\omega \\ -u_2(0)/\omega \\ r_1(0) - q_1(0) \end{pmatrix},$$

and

$$\begin{pmatrix} 1 & 1 & 1 \\ \omega_1 & \omega_2 & \omega_3 \\ \omega_1^2 & \omega_2^2 & \omega_3^2 \end{pmatrix} \begin{pmatrix} B \\ D \\ F \end{pmatrix} = \begin{pmatrix} v_1(0)/a\omega \\ u_1(0)/\omega \\ r_2(0) - q_2(0) \end{pmatrix},$$

where $\boldsymbol{q}(0)$, $\boldsymbol{v}(0)$ and $\boldsymbol{r}(0)$, $\boldsymbol{u}(0)$, are respectively the position and velocity of the center of mass and center of charge at time $t = 0$.

If we call $N$ the inverse of the matrix containing the frequencies of the above equations, it is:

$$N = \frac{1}{\Delta} \begin{pmatrix} \omega_2\omega_3(\omega_3 - \omega_2) & \omega_2^2 - \omega_3^2 & \omega_3 - \omega_2 \\ \omega_1\omega_3(\omega_1 - \omega_3) & \omega_3^2 - \omega_1^2 & \omega_1 - \omega_3 \\ \omega_1\omega_2(\omega_2 - \omega_1) & \omega_1^2 - \omega_2^2 & \omega_2 - \omega_1 \end{pmatrix},$$

where $\Delta = (\omega_1 - \omega_2)(\omega_2 - \omega_3)(\omega_3 - \omega_1)$, in such a way that we can obtain the final expression of the integration constants in terms of the initial conditions.

To lowest order in $a$, since $k \approx \sqrt{3}a/2$, the normal modes are:

$$\omega_1 = a + O(a^3), \quad \omega_2 = -1 - \frac{a}{2} + \frac{3a^2}{8} + O(a^3), \quad \omega_3 = 1 - \frac{a}{2} - \frac{3a^2}{8} + O(a^3). \tag{2.114}$$

These frequencies will be compared in Section 2.5 of Chapter 5 with those of Barut-Zanghi relativistic model. In terms of the physical parameters and in the time evolution description, these normal frequencies are to lowest order:

$$\omega_1 = \omega_c, \quad \omega_2 = \omega - \frac{\omega_c}{2} - \frac{3\omega_c^2}{8\omega}, \quad \omega_2 = \omega + \frac{\omega_c}{2} - \frac{3\omega_c^2}{8\omega}, \tag{2.115}$$

where $\omega_c = eB/m$ and $\omega$ are the cyclotron and zitterbewegung frequency, respectively.

To properly characterize these initial values in terms of physical parameters, like the radius of the internal motion $R_0$, the cyclotron radius $R_c$, the center of mass velocity $v$ and the zitterbewegung frequency $\omega$, let us consider an electron that is sent with a velocity $\boldsymbol{v}$ orthogonal to the external uniform magnetic field $\boldsymbol{B}$. We take the $XOY$ plane such that the initial position of the center of mass is on the $OX$ axis at the coordinate $R_c = -vm/eB$, and the initial velocity $v$ along the positive direction of the $OY$ axis. With this convention, the center of mass will have a precession around the $OZ$ axis with cyclotron angular velocity $|\omega_c|$ in the positive direction while for a positive charged particle the initial position will be chosen as $-|R_c|$ on the $OX$ axis and the angular velocity will point in the negative $OZ$ axis.

The initial position of the center of charge is characterized by the three parameters $\phi$, $\theta$ and $\psi$, where $\theta$ and $\phi$ represent the initial orientation of the internal angular velocity $\omega$, and parameter $\psi$ is the initial phase position of the center of charge as shown in Figure 2.2. If all these three parameters are zero, $\boldsymbol{\omega}$ is pointing along $OZ$ and the initial position of the charge is at point $R_c + R_0$ on the $OX$ axis.

We thus have as initial conditions for our system, written in column matrix form:

$$\boldsymbol{q}(0) = \begin{pmatrix} R_c \\ 0 \\ 0 \end{pmatrix}, \quad \boldsymbol{r}(0) = \begin{pmatrix} R_c \\ 0 \\ 0 \end{pmatrix} + \mathcal{R}_{oz}(\phi)\mathcal{R}_{oy}(\theta)\mathcal{R}_{oz}(\psi) \begin{pmatrix} R_0 \\ 0 \\ 0 \end{pmatrix},$$

$$\boldsymbol{v}(0) = \begin{pmatrix} 0 \\ v \\ 0 \end{pmatrix}, \quad \boldsymbol{u}(0) = \begin{pmatrix} 0 \\ v \\ 0 \end{pmatrix} + \mathcal{R}_{oz}(\phi)\mathcal{R}_{oy}(\theta)\mathcal{R}_{oz}(\psi) \begin{pmatrix} 0 \\ \omega R_0 \\ 0 \end{pmatrix},$$

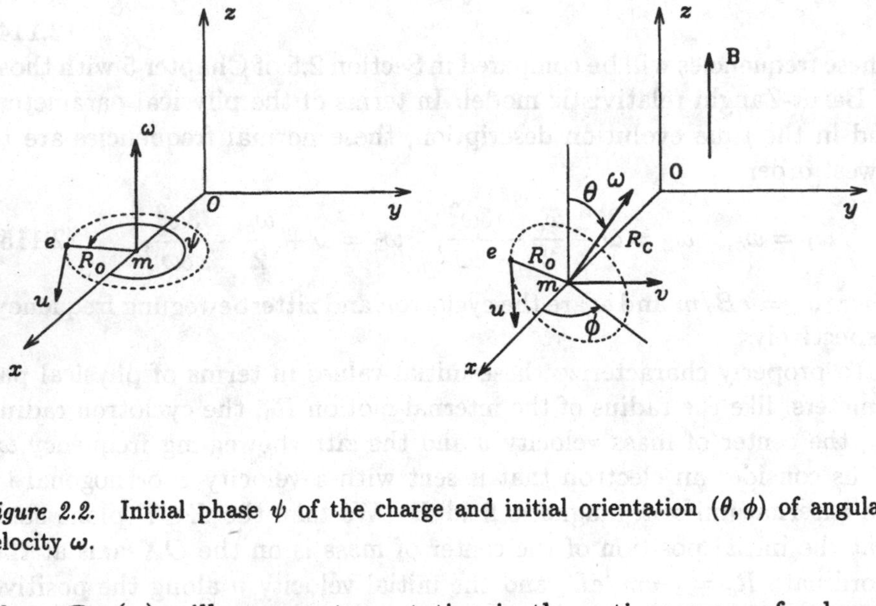


*Figure 2.2.* **Initial phase $\psi$ of the charge and initial orientation $(\theta, \phi)$ of angular velocity $\omega$.**

where $\mathcal{R}_{oz}(\alpha)$ will represent a rotation in the active sense, of value $\alpha$ around the $OZ$ axis. Since the spin is opposite to the internal angular velocity, its initial value is

$$\boldsymbol{S}(0) = \mathcal{R}_{oz}(\phi)\mathcal{R}_{oy}(\theta) \begin{pmatrix} 0 \\ 0 \\ -S \end{pmatrix}, \tag{2.116}$$

where $S = m\omega R_0^2$. Thus the initial conditions to determine the coefficients of the general solution are:

$$\begin{pmatrix} -v_2(0)/a\omega \\ -u_2(0)/\omega \\ r_1(0)-q_1(0) \end{pmatrix} = \begin{pmatrix} R_c \\ aR_c-\alpha R_0 \\ \beta R_0 \end{pmatrix}, \quad \begin{pmatrix} v_1(0)/a\omega \\ u_1(0)/\omega \\ r_2(0)-q_2(0) \end{pmatrix} = \begin{pmatrix} 0 \\ \gamma R_0 \\ \delta R_0 \end{pmatrix},$$

where $R_c = -vm/eB$, $\omega_c = -eB/m = -a\omega$, as before and the constant parameters:

$$\begin{aligned}
\alpha &= -\sin\phi\cos\theta\sin\psi + \cos\phi\cos\psi, \\
\beta &= \cos\phi\cos\theta\cos\psi - \sin\phi\sin\psi, \\
\gamma &= -\cos\phi\cos\theta\sin\psi - \sin\phi\cos\psi, \\
\delta &= \sin\phi\cos\theta\cos\psi + \cos\phi\sin\psi.
\end{aligned}$$

To lowest order in $a$, the frequencies become:

$$\omega_1-\omega_2 = 1+\frac{3}{2}a, \quad \omega_2-\omega_3=-2, \quad \omega_3-\omega_1 = 1-\frac{3}{2}a,$$

$$\omega_1+\omega_2 = -1+\frac{a}{2}, \quad \omega_2+\omega_3=-a, \quad \omega_3+\omega_1 = 1+\frac{a}{2},$$

$$\omega_1\omega_2 = -a\left(1+\frac{a}{2}\right), \quad \omega_2\omega_3 = -\left(1-\frac{a^2}{4}\right), \quad \omega_3\omega_1 = a\left(1-\frac{a}{2}\right),$$

and thus the inverse matrix $N$ to order $O(a^2)$ is

$$N = \begin{pmatrix} 1+2a^2 & -a & -1-9a^2/4 \\ a/2-a^2 & -1/2+a/2-3a^2/4 & 1/2-3a/4+9a^2/8 \\ -a/2-a^2 & 1/2+a/2+3a^2/4 & 1/2+3a/4+9a^2/8 \end{pmatrix}.$$

In this way the coefficients of the general solution, to first order in $a$, are:

$$\begin{aligned}
A &= R_c - \beta R_0 + aR_0\alpha, \\
B &= -R_0(a\gamma+\delta), \\
C &= \frac{R_0}{2}(\alpha+\beta) - \frac{aR_0}{4}(2\alpha+3\beta), \\
D &= \frac{R_0}{2}(\delta-\gamma) + \frac{aR_0}{4}(2\gamma-3\delta), \\
E &= \frac{R_0}{2}(\beta-\alpha) + \frac{aR_0}{4}(3\beta-2\alpha), \\
F &= \frac{R_0}{2}(\delta+\gamma) + \frac{aR_0}{4}(2\gamma+3\delta),
\end{aligned}$$

and the coefficients

$$b_1/a = -\delta R_0, \quad b_2/a = \beta R_0.$$

This motion depends on the cyclotron radius $R_c$, only through the parameter $A$, and the remaining terms depend on the internal radius $R_0$.

The general solution, neglecting terms of the order $aR_0$, can be written in a vector form as:

$$\begin{aligned} \boldsymbol{r}(t) &= \mathcal{R}_{\mathrm{oz}}(\omega_c t)\begin{pmatrix} R_c \\ 0 \\ 0 \end{pmatrix} + (\mathbb{I} - \mathcal{R}_{\mathrm{oz}}(\omega_c t))\,\mathcal{R}(\phi,\theta,\psi)\begin{pmatrix} R_0 \\ 0 \\ 0 \end{pmatrix} \\ &+ \mathcal{R}_{\mathrm{oz}}\left(-\frac{\omega_c t}{2}\right)\mathcal{R}(\phi,\theta,\psi+\omega t)\begin{pmatrix} R_0 \\ 0 \\ 0 \end{pmatrix} + O(aR_0), \end{aligned}$$

where $\mathbb{I}$ is the $3\times 3$ unit matrix and $\mathcal{R}(\phi,\theta,\psi) \equiv \mathcal{R}_{\mathrm{oz}}(\phi)\mathcal{R}_{\mathrm{oy}}(\theta)\mathcal{R}_{\mathrm{oz}}(\psi)$. The first two terms represent the center of mass motion to this order of approximation, while the third is precisely the relative motion of the center of charge around the center of mass. The neglected contribution of order $aR_0$ can be written as

$$\begin{aligned} &O(aR_0) \\ &= -J_z\left[\mathcal{R}_{\mathrm{oz}}(\omega_c t)\mathcal{R}(\phi,\theta,\psi) - \mathcal{R}_{\mathrm{oz}}\left(-\frac{\omega_c t}{2}\right)\mathcal{R}(\phi,\theta,\psi+\omega t)\right]\begin{pmatrix} 0 \\ aR_0 \\ 0 \end{pmatrix} \\ &\quad -J_z\left[\frac{\sin(\omega t)}{2}\mathcal{R}_{\mathrm{oz}}\left(-\frac{\omega_c t}{2}\right)\mathcal{R}(\phi,\theta,\psi)\begin{pmatrix} aR_0 \\ 0 \\ 0 \end{pmatrix}\right], \end{aligned}$$

where

$$J_z = \begin{pmatrix} 0 & -1 & 0 \\ 1 & 0 & 0 \\ 0 & 0 & 0 \end{pmatrix},$$

is the $3\times 3$ generator of rotations around the $OZ$ axis. The first two terms represent the correction to this order of the center of mass motion and the third is the correction of the internal relative motion. The presence of the generator $J_z$ in this term means that this correction does not make any contribution to the motion along the $OZ$ axis. The solution along $OZ$ is exactly:

$$q_3(t) = 0, \quad r_3(t) = -R_0 \sin\theta\,\cos(\omega t + \psi), \tag{2.117}$$

*i.e.*, a harmonic motion of amplitude $R_0 \sin\theta$, and frequency $\omega$.

The relative position of the center of charge with respect to the center of mass verifies:

$$\boldsymbol{k}(t) = \mathcal{R}_{oz}\left(-\frac{\omega_c t}{2}\right) \mathcal{R}(\phi, \theta, \psi + \omega t) \begin{pmatrix} R_0 \\ 0 \\ 0 \end{pmatrix}$$

$$-J_z \left[\frac{\sin(\omega t)}{2} \mathcal{R}_{oz}\left(-\frac{\omega_c t}{2}\right) \mathcal{R}(\phi, \theta, \psi) \begin{pmatrix} aR_0 \\ 0 \\ 0 \end{pmatrix}\right], \tag{2.118}$$

and if we neglect contributions to order $aR_0$, it just reduces to the first term

$$\boldsymbol{k}(t) \approx \mathcal{R}_{oz}\left(-\frac{\omega_c t}{2}\right) \mathcal{R}(\phi, \theta, \psi + \omega t) \begin{pmatrix} R_0 \\ 0 \\ 0 \end{pmatrix}, \tag{2.119}$$

that represents an oscillation with the natural frequency $\omega$ of the zitterbewegung around the initial spin axis, with a backwards precession with an angular velocity $\omega_c/2$.

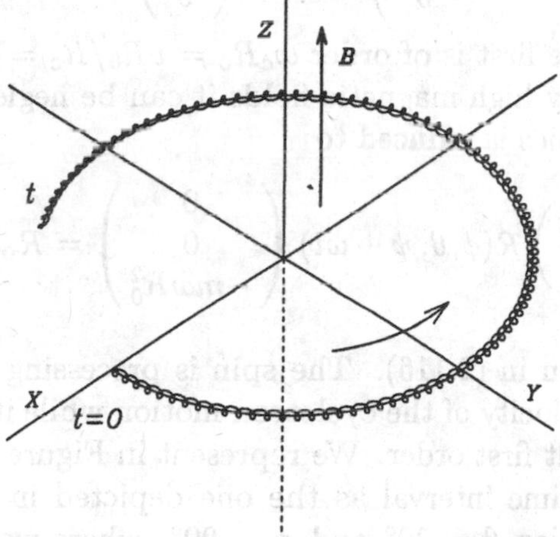


*Figure 2.3.* Motion of the center of charge and center of mass of a negative charged particle in a uniform magnetic field. The velocity of the center of mass is orthogonal to the field.

The center of charge and center of mass trajectory is depicted in Figure 2.3, where the curly trajectory is the motion of the charge.

To study the spin dynamics, we just substitute the general solution in its analytical definition

$$\boldsymbol{S}(t) = -m\boldsymbol{k}(t) \times \frac{d\boldsymbol{k}(t)}{dt}, \tag{2.120}$$

where we need to calculate the derivative of (2.119). To calculate this derivative, we have to take into account that

$$\mathcal{R}_{oz}(\omega t) = \exp(J_z \omega t),$$

and therefore

$$\dot{\mathcal{R}}_{oz}(\omega t) = \exp(J_z\omega t)J_z\omega = \mathcal{R}_{oz}(\omega t)J_z\omega = J_z\omega\,\mathcal{R}_{oz}(\omega t).$$

By taking the derivative of (2.119) we get the following terms:

$$\begin{aligned}\frac{d\boldsymbol{k}}{dt} &= \mathcal{R}_{oz}\left(-\frac{\omega_c t}{2}\right) J_z\,\mathcal{R}(\phi,\theta,\psi+\omega t)\begin{pmatrix}-\omega_c R_0/2\\ 0\\ 0\end{pmatrix}\\ &+ \mathcal{R}_{oz}\left(-\frac{\omega_c t}{2}\right)\mathcal{R}(\phi,\theta,\psi+\omega t)\begin{pmatrix}0\\ \omega R_0\\ 0\end{pmatrix},\end{aligned} \tag{2.121}$$

where

$$\begin{pmatrix}0\\ \omega R_0\\ 0\end{pmatrix} = \omega J_z\begin{pmatrix}R_0\\ 0\\ 0\end{pmatrix}. \tag{2.122}$$

Of these terms, the first is of order $\omega_c R_0 = vR_0/R_c = a\omega R_0 = ac$, and thus even with very high magnetic fields it can be neglected.

The spin dynamics is reduced to

$$\boldsymbol{S}(t) = \mathcal{R}_{oz}\left(-\frac{\omega_c t}{2}\right)\mathcal{R}(\phi,\theta,\psi+\omega t)\begin{pmatrix}0\\ 0\\ -m\omega R_0^2\end{pmatrix} = \mathcal{R}_{oz}\left(-\frac{\omega_c t}{2}\right)\boldsymbol{S}(0), \tag{2.123}$$

where $\boldsymbol{S}(0)$ is given in (2.116). The spin is precessing backwards with half the angular velocity of the cyclotron motion while its absolute value remains constant at first order. We represent in Figure 2.4 its evolution during the same time interval as the one depicted in Figure 2.3 with the initial orientation $\theta = 30°$ and $\phi = 90°$, where we can observe, in addition to the precession of constant absolute value, a tiny oscillation of the next order contribution.

The energy of the system is

$$H = -T - \boldsymbol{u}\cdot\frac{dU}{dt}, \tag{2.124}$$

that can be expressed as:

$$H = \frac{m}{2}\left(\frac{d\boldsymbol{r}}{dt}\right)^2 - \frac{m}{2\omega^2}\left(\frac{d^2\boldsymbol{r}}{dt^2}\right)^2 + \frac{m}{\omega^2}\frac{d\boldsymbol{r}}{dt}\cdot\frac{d^3\boldsymbol{r}}{dt^3} + eV(\boldsymbol{r},t),$$

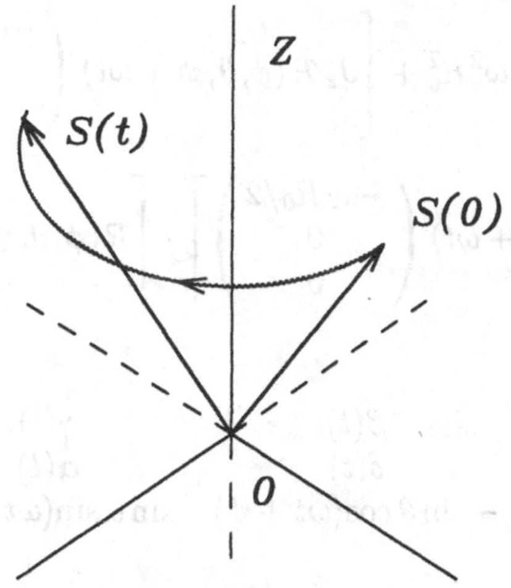


*Figure 2.4.* Precession of spin around the $OZ$ axis.

and, since function $V(\boldsymbol{r},t)=0$ in the presence of a constant magnetic field, it becomes:

$$H=\frac{m}{2}\left(\frac{d\boldsymbol{q}}{dt}\right)^2-\frac{m}{2}\left(\frac{d\boldsymbol{k}}{dt}\right)^2-\frac{m\omega^2}{2}\boldsymbol{k}^2=\frac{(\boldsymbol{P}-e\boldsymbol{A})^2}{2m}+H_0. \quad (2.125)$$

To lowest order the contribution comes from

$$\boldsymbol{q}(t)=\mathcal{R}_{oz}(\omega_c t)\begin{pmatrix}R_c\\0\\0\end{pmatrix}+(\mathbb{I}-\mathcal{R}_{oz}(\omega_c t))\,\mathcal{R}(\phi,\theta,\psi)\begin{pmatrix}R_0\\0\\0\end{pmatrix}.$$

Thus

$$\frac{d\boldsymbol{q}}{dt}=\mathcal{R}_{oz}(\omega_c t)\begin{pmatrix}0\\v\\0\end{pmatrix}-\mathcal{R}_{oz}(\omega_c t)\,J_z\,\mathcal{R}(\phi,\theta,\psi)\begin{pmatrix}\omega_c R_0\\0\\0\end{pmatrix},$$

in such a way that taking into account (2.118) and (2.121)

$$\begin{aligned}\left(\frac{d\boldsymbol{q}}{dt}\right)^2 &= v^2+\left[J_z\,\mathcal{R}(\phi,\theta,\psi)\begin{pmatrix}\omega_c R_0\\0\\0\end{pmatrix}\right]^2\\ &- 2\begin{pmatrix}0\\v\\0\end{pmatrix}\cdot\left[J_z\,\mathcal{R}(\phi,\theta,\psi)\begin{pmatrix}\omega_c R_0\\0\\0\end{pmatrix}\right],\end{aligned}$$

$$\left(\frac{d\boldsymbol{k}}{dt}\right)^2 = \omega^2 R_0^2 + \left[ J_z\,\mathcal{R}(\phi,\theta,\psi+\omega t) \begin{pmatrix} -\omega_c R_0/2 \\ 0 \\ 0 \end{pmatrix} \right]^2$$

$$+2\left[ J_z\,\mathcal{R}(\phi,\theta,\psi+\omega t) \begin{pmatrix} -\omega_c R_0/2 \\ 0 \\ 0 \end{pmatrix} \right] \cdot \left[ \mathcal{R}(\phi,\theta,\psi+\omega t) \begin{pmatrix} 0 \\ \omega R_0 \\ 0 \end{pmatrix} \right].$$

Since

$$\mathcal{R}(\phi,\theta,\psi{+}\omega t) = \begin{pmatrix} \beta(t) & \gamma(t) & \cos\phi\sin\theta \\ \delta(t) & \alpha(t) & \sin\phi\sin\theta \\ -\sin\theta\cos(\omega t+\psi) & \sin\theta\sin(\omega t+\psi) & \cos\theta \end{pmatrix},$$

$$J_z\,\mathcal{R}(\phi,\theta,\psi+\omega t) = \begin{pmatrix} -\delta(t) & -\alpha(t) & -\sin\phi\sin\theta \\ \beta(t) & \gamma(t) & \cos\phi\sin\theta \\ 0 & 0 & 0 \end{pmatrix},$$

where

$$\begin{aligned} \alpha(t) &= -\sin\phi\cos\theta\sin(\psi+\omega t) + \cos\phi\cos(\psi+\omega t), \\ \beta(t) &= \cos\phi\cos\theta\cos(\psi+\omega t) - \sin\phi\sin(\psi+\omega t), \\ \gamma(t) &= -\cos\phi\cos\theta\sin(\psi+\omega t) - \sin\phi\cos(\psi+\omega t), \\ \delta(t) &= \sin\phi\cos\theta\cos(\psi+\omega t) + \cos\phi\sin(\psi+\omega t) \end{aligned}$$

then

$$J_z\,\mathcal{R}(\phi,\theta,\psi+\omega t) \begin{pmatrix} \omega_c R_0 \\ 0 \\ 0 \end{pmatrix} = \omega_c R_0 \begin{pmatrix} -\delta(t) \\ \beta(t) \\ 0 \end{pmatrix}.$$

Consequently

$$\left(\frac{d\boldsymbol{q}}{dt}\right)^2 = v^2 + \omega_c^2 R_0^2(\delta(0)^2 + \beta(0)^2) - 2v\omega_c R_0 \beta(0),$$

$$\left(\frac{d\boldsymbol{k}}{dt}\right)^2 = \omega^2 R_0^2 + \frac{\omega_c^2 R_0^2}{4}(\delta(t)^2 + \beta(t)^2) + \omega\omega_c R_0^2(\delta(t)\gamma(t) - \beta(t)\alpha(t)).$$

Because

$$\delta(t)\gamma(t) - \beta(t)\alpha(t) = -\cos\theta,$$

$$\delta(0)^2 + \beta(0)^2 = 1 - \sin^2\theta\cos^2\psi,$$

$$\delta(t)^2 + \beta(t)^2 = 1 - \sin^2\theta\cos^2(\psi+\omega t),$$

if we write $\omega_c$ in terms of the parameter $a$, $\omega_c = -a\omega$, in the case of the electron $\omega R_0 = c$, the energy of this system to lower order of approximation in $a$ is:

$$H = H_0 - a\left(\frac{mc^2\cos\theta}{2} - mvc\beta(0)\right)$$

$$+a^2\frac{mc^2}{2}\left(\delta(0)^2 + \beta(0)^2 - \frac{1}{4}(\delta(t)^2 + \beta(t)^2)\right).$$

The lowest order of the interaction energy can be expressed as:

$$H_I = -\frac{1}{2}amc^2\cos\theta = -\frac{eB}{2m}\frac{mc^2}{\omega}\cos\theta = -\boldsymbol{\mu}\cdot\boldsymbol{B}, \tag{2.126}$$

and since $S = m\omega R_0^2 = mc^2/\omega$, $S_z = -S\cos\theta$, it implies

$$\mu_z = \frac{eS\cos\theta}{2m} = -\frac{eS_z}{2m}, \tag{2.127}$$

or

$$\boldsymbol{\mu} = -\frac{e}{2m}\boldsymbol{S}. \tag{2.128}$$

The interaction energy can also be written as

$$H_I = -\frac{eB}{2m}S\cos\theta = \frac{e}{2m}\boldsymbol{B}\cdot\boldsymbol{S} = \frac{-\boldsymbol{\omega}_c}{2}\cdot\boldsymbol{S}, \tag{2.129}$$

*i.e.*, as the scalar product of the spin and the angular velocity of precession of the spin.

From a simpler method, if assumed that the relation between the spin and magnetic moment is given by (2.128), and that the variation of the intrinsic angular momentum is governed by the torque equation

$$\frac{d\boldsymbol{S}}{dt} = \boldsymbol{\mu}\times\boldsymbol{B} = -\frac{e}{2m}\boldsymbol{S}\times\boldsymbol{B} = \boldsymbol{\Omega}\times\boldsymbol{S}.$$

The constant angular velocity of precession of spin is Larmor's angular frequency

$$\boldsymbol{\Omega} = \frac{e\boldsymbol{B}}{2m} = -\frac{\boldsymbol{\omega}_c}{2},$$

because $\boldsymbol{\omega}_c = -e\boldsymbol{B}/m$, *i.e.*, half and opposite to the cyclotron angular velocity. This produces the first order contribution because the spin conserves its absolute value. However, this simpler assumption does not contain the additional terms or corrections to the normal modes $\omega_i$, which can be relevant in high energy processes, and can be obtained from the general solution.

## 4.4 SPINNING PARTICLE IN A UNIFORM ELECTRIC FIELD

If the external field is a uniform electric field $\boldsymbol{E}$, pointing for instance along the $OZ$ axis, the dynamical equations are:

$$\frac{d^2\boldsymbol{q}}{dt^2} = \frac{e}{m}\boldsymbol{E}, \tag{2.130}$$

$$\frac{d^2\boldsymbol{r}}{dt^2} + \omega^2(\boldsymbol{r} - \boldsymbol{q}) = 0, \tag{2.131}$$

whose solution in terms of the initial boundary conditions for the position and velocity of both center of mass and center of charge is for the center of mass variables

$$\boldsymbol{q}(t) = \boldsymbol{q}(0) + \boldsymbol{v}(0)t + \frac{e\boldsymbol{E}}{2m}t^2.$$

Here, as before, we write $\boldsymbol{v} = d\boldsymbol{q}/dt$ and $\boldsymbol{u} = d\boldsymbol{r}/dt$, and thus

$$\begin{aligned} r_1(t) &= (r_1(0) - q_1(0))\cos\omega t + \frac{u_1(0) - v_1(0)}{\omega}\sin\omega t + q_1(0) + v_1(0)t, \\ r_2(t) &= (r_2(0) - q_2(0))\cos\omega t + \frac{u_2(0) - v_2(0)}{\omega}\sin\omega t + q_2(0) + v_2(0)t, \\ r_3(t) &= (r_3(0) - q_3(0) + \frac{eE}{m\omega^2})\cos\omega t + \frac{u_3(0) - v_3(0)}{\omega}\sin\omega t + q_3(0) \\ &- \frac{eE}{m\omega^2} + v_3(0)t + \frac{eE}{2m}t^2. \end{aligned}$$

The center of mass has a motion with uniform acceleration along the field direction while the center of charge experiences a harmonic motion of frequency $\omega$ around the center of mass.

The zitterbewegung is governed by the dynamical equation

$$\frac{d^2\boldsymbol{k}}{dt^2} + \omega^2\boldsymbol{k} = -\frac{e}{m}\boldsymbol{E},$$

obtained by substracting (2.130) from (2.131). It has the general solution

$$\boldsymbol{k}(t) = \boldsymbol{k}(0)\cos\omega t + \frac{\dot{\boldsymbol{k}}(0)}{\omega}\sin\omega t - \frac{e\boldsymbol{E}}{m\omega^2}(1 - \cos\omega t),$$

and therefore is a motion not contained in the initial plane spanned by the vectors $\boldsymbol{k}(0)$ and $\dot{\boldsymbol{k}}(0)$, but also includes a contribution in the direction of the external field.

The spin dynamical equation is

$$\frac{d\boldsymbol{S}}{dt} = -m\boldsymbol{k} \times \frac{d^2\boldsymbol{k}}{dt^2} = \boldsymbol{d} \times \boldsymbol{E}.$$

It is the torque of the electric dipole $\boldsymbol{d} = e\boldsymbol{k}$ with the electric field. It reduces to

$$\frac{d\boldsymbol{S}}{dt} = e\boldsymbol{k}(0) \times \boldsymbol{E} \cos\omega t + \frac{\dot{\boldsymbol{k}}(0)}{\omega} \times \boldsymbol{E} \sin\omega t.$$

The end point of the spin vector describes in general an ellipse contained in a plane orthogonal to the external field $\boldsymbol{E}$, and generated by the two fixed directions $\boldsymbol{k}(0) \times \boldsymbol{E}$ and $\dot{\boldsymbol{k}}(0) \times \boldsymbol{E}$. Therefore the absolute value of spin is no longer a constant of the motion but it has a bounded average value.

## 4.5 CIRCULAR ZITTERBEWEGUNG

Another example of particles with the same kinematical space $X = \mathcal{G}/SO(3)$ and therefore kinematical variables $t$, $\boldsymbol{r}$ and $\boldsymbol{u}$ is given by the following Lagrangian

$$L = \frac{1}{2}m\,\frac{\dot{\boldsymbol{r}}^2}{\dot{t}} - mR\,|\dot{\boldsymbol{u}}|, \tag{2.132}$$

which satisfies the homogeneity condition of first order in the derivatives of the kinematical variables and that $\boldsymbol{U} \sim \dot{\boldsymbol{u}}$. The constant parameter $R$ is going to be the constant radius of the circular zitterbewegung. The first term in the Lagrangian (2.132) is a gauge variant term with the same gauge function as in the previous Galilei examples, while the second term is Galilei invariant.

In this case, the center of mass position is defined, as in (2.91), by

$$\boldsymbol{q} = \boldsymbol{r} - \frac{1}{m}\boldsymbol{U} = \boldsymbol{r} + R\,\frac{\dot{\boldsymbol{u}}}{|\dot{\boldsymbol{u}}|},$$

that satisfies $d^2\boldsymbol{q}/dt^2 = 0$, while for the relative position $\boldsymbol{k} = \boldsymbol{r} - \boldsymbol{q}$ we get the dynamical equation

$$\boldsymbol{k} = -R\frac{\dot{\boldsymbol{u}}}{|\dot{\boldsymbol{u}}|}, \tag{2.133}$$

opposite to the acceleration and of absolute value $|\boldsymbol{k}| = R$. The internal motion is of constant radius. Therefore the internal motion for the center of mass observer is a circle of radius $R$ at a constant speed. The spin is also orthogonal to this plane with the same relative orientation as an (anti)-orbital angular momentum $\boldsymbol{S} = \boldsymbol{u} \times \boldsymbol{U} = -mR\boldsymbol{u} \times \dot{\boldsymbol{u}}/|\dot{\boldsymbol{u}}|$, and where $\dot{\boldsymbol{u}}/|\dot{\boldsymbol{u}}|$ is a Galilei invariant unit vector in the direction of the acceleration. Once the spin of the system is fixed, the internal velocity is just $|d\boldsymbol{k}/dt| = S/mR$.

The total energy and linear momentum are still given by (2.83) and (2.84) respectively, so that the internal energy of this system reduces to

$$H_0 = H - \frac{\boldsymbol{P}^2}{2m} = -\frac{1}{2m}\left(\frac{d\boldsymbol{U}}{dt}\right)^2 = -\frac{m}{2}\left(\frac{d\boldsymbol{k}}{dt}\right)^2 = -\frac{S^2}{2mR^2},$$

in terms of the internal velocity and a difference with the other example (2.99) is the absence of the 'elastic term' proportional to $\boldsymbol{k}^2$ that in this case is a true constant. The internal energy of this Galilei system does not supply any additional intrinsic property to the basic ones of spin $S$, internal radius $R$ and mass $m$, and the zitterbewegung frequency is given by $|d\boldsymbol{k}/dt|/R = S/mR^2$.

When comparing this system with the other one (2.76) it looks simpler because it is not necessary to fix the lengths of the two axes of the internal elliptic motion as in the mentioned example.

## 5. SPINNING GALILEI PARTICLE WITH ORIENTATION

Another simple example of spinning particles is the one in which the spin is related only to the angular variables that describe orientation.

Let us assume now a dynamical system whose kinematical space is $X = \mathcal{G}/\mathbb{R}^3_v$, where $\mathbb{R}^3_v \equiv \{\mathbb{R}^3, +\}$ is the 3-parameter Abelian subgroup of pure Galilei transformations. Then, the kinematical variables are $x \equiv (t, \boldsymbol{r}, \boldsymbol{\rho})$, which are interpreted as the time, position and orientation respectively.

The Lagrangian for this model takes the general form

$$L = T\dot{t} + \boldsymbol{R}\cdot\dot{\boldsymbol{r}} + \boldsymbol{W}\cdot\boldsymbol{\omega}.$$

Because of the structure of the exponent (2.14), the gauge function for this system can be taken the same as before. The general relationship (2.75) leads to $\boldsymbol{W}\times\boldsymbol{\omega} = 0$, because the Lagrangian is independent of $\dot{\boldsymbol{u}}$, and therefore $\boldsymbol{W}$ and $\boldsymbol{\omega}$ must be collinear. According to the transformation properties of the Lagrangian, the third term $\boldsymbol{W}\cdot\boldsymbol{\omega}$ is Galilei invariant and since $\boldsymbol{W}$ and $\boldsymbol{\omega}$ are collinear, we can take $\boldsymbol{W}\sim\boldsymbol{\omega}$ and one possible Lagrangian that describes this model is of the form:

$$L = \frac{m}{2}\frac{\dot{\boldsymbol{r}}^2}{\dot{t}} + \frac{I}{2}\frac{\boldsymbol{\omega}^2}{\dot{t}}. \tag{2.134}$$

The different Noether's constants are

$$H = \frac{m}{2}\left(\frac{d\boldsymbol{r}}{dt}\right)^2 + \frac{I}{2}\boldsymbol{\Omega}^2, \quad \boldsymbol{P} = m\boldsymbol{u},$$

$$K = mr - Pt, \quad J = r \times P + W,$$

where $u = dr/dt$ is the velocity of point $r$, and $\Omega = \omega/\dot{t}$ is the time evolution angular velocity. Point $r$ is moving at a constant speed and it also represents the position of the center of mass. The spin is just the observable $S \equiv W$ that satisfies the dynamical equation $dS/dt = \omega \times S = 0$, and thus the frame linked to the body rotates with a constant angular velocity $\Omega$.

The spin takes the constant value $S = I\Omega$, whose absolute value is independent of the inertial observer and also the angular velocity $\Omega = \omega/\dot{t}$ is constant. The parameter $I$ plays the role of a principal moment of inertia, suggesting a linear relationship between the spin and the angular velocity, which corresponds to a particle with spherical symmetry. The particle can also be considered as an extended object of gyration radius $R_0$, related to the other particle parameters by $I = mR_0^2$.

This system corresponds classically to a rigid body with spherical symmetry where the orientation variables $\rho$ can describe for instance, the orientation of its principal axes of inertia in a suitable parametrization of the rotation group. This is a system of six degrees of freedom. Three represent the position of the center of charge $r$ and the other three $\rho$, represent the orientation of a Cartesian frame linked to that point $r$. Since for this system there is no dependence on the acceleration, the centers of mass and charge will be represented by the same point. Spherically symmetric rigid bodies are particular cases of Galilei elementary particles, but as we shall see in next chapter they cannot be considered as elementary particles in the relativistic approach because this seven-dimensional kinematical space is no longer a homogeneous space of the Poincaré group.

In the center of mass frame there is no current associated to this particle and therefore it has neither magnetic nor electric dipole structure. As seen in previous examples, all magnetic properties seem therefore to be related to the zitterbewegung part of spin and are absent in this rigid body-like model.

## 6. GENERAL NONRELATIVISTIC SPINNING PARTICLE

We have seen in previous examples the partial structure of spin according to the dependence of the Lagrangian either on the acceleration or on the angular velocity. Coming back to the most general case, if the Lagrangian depends on both magnitudes, then one possibility is

$$L = \frac{m}{2}\frac{\dot{r}^2}{\dot{t}} + \frac{b}{2}\frac{1}{\dot{t}}\left(\omega^2 - \frac{1}{\beta^2}\dot{u}^2\right), \tag{2.135}$$

in terms of three parameters $m$, $b$ and $\beta$. Parameter $m$ represents the mass of the system because the first term is gauge variant in terms of the gauge function (2.55), which depends on this parameter while the other two terms are Galilei invariant. One expects, by comparing with the previous examples, that $b \sim I$ will be related to a moment of inertia and $b/\beta^2 \sim m/w^2$ with the zitterbewegung frequency $w$. In general these three parameters will be finally expressed in terms of the three invariants of the extended Galilei group mass $m$, spin $S$ and internal energy $H_0$.

Observables $\boldsymbol{U}$ and $\boldsymbol{W}$ will be

$$\boldsymbol{U} = \frac{b}{\beta^2}\frac{d\boldsymbol{u}}{dt}, \quad \boldsymbol{W} = b\boldsymbol{\Omega},$$

where $\boldsymbol{\Omega} = \boldsymbol{\omega}/\dot{t}$ is the angular velocity in a time evolution, while $\boldsymbol{\omega}$ is the angular velocity in the arbitrary $\tau$ evolution. The dynamics of $\boldsymbol{W}$ reduces to $d\boldsymbol{W}/d\tau = \boldsymbol{\omega} \times \boldsymbol{W}$, and thus $\boldsymbol{\Omega}$ is a constant vector. In the center of mass frame $\boldsymbol{U} = m\boldsymbol{k}$, so that the zitterbewegung equation in the center of mass frame is

$$\frac{d^2\boldsymbol{k}}{dt^2} + \frac{m\beta^2}{b}\boldsymbol{k} = 0. \qquad (2.136)$$

The spin of this system, according to (2.69) takes the form

$$\boldsymbol{S} = -m\boldsymbol{k} \times \frac{d\boldsymbol{k}}{dt} + \boldsymbol{W},$$

related to the zitterbewegung and to the angular velocity of the particle which is a constant arbitrary vector. In the center of mass frame the spin is a constant of motion and thus the above zitterbewegung trajectory (2.136) is a plane motion of angular frequency $(m\beta^2/b)^{1/2}$ orthogonal to the conserved vector $\boldsymbol{u} \times \boldsymbol{U}$. The zitterbewegung motion is independent of the rotation of the body frame. It turns out that, in addition to some internal elliptic trajectory, the body frame linked to the point $\boldsymbol{r}$ is rotating with a constant angular velocity. In a particular circular trajectory the different observables will have the relative orientation depicted in Figure 2.5, where vector $\boldsymbol{\omega}$ and thus $\boldsymbol{W}$ may have any orientation.

But we can also have Lagrangians of the form

$$L = \frac{m}{2}\frac{\dot{\boldsymbol{r}}^2}{\dot{t}} + b\frac{\boldsymbol{\omega}\cdot\dot{\boldsymbol{u}}}{\dot{t}}, \qquad (2.137)$$

where the parameter $b$ will be expressed in terms of the internal radius of the zitterbewegung $R_0$ and the internal velocity $\beta$ of this motion or in

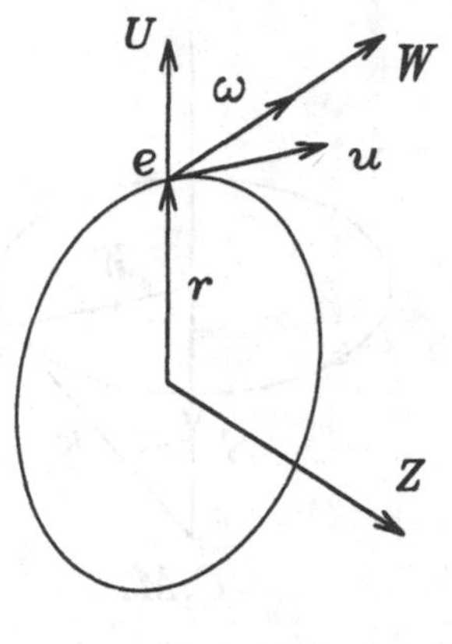


*Figure 2.5.* **Representation of observables $U$, $W$ and $Z$ in the center of mass frame.**

**terms of the spin and internal energy, thus suggesting a kind of system for which there is a fixed internal energy-spin ratio.**

**For this system**

$$U = a\frac{\omega}{t} = a\Omega, \quad W = a\frac{\dot{u}}{t} = a\frac{d^2 r}{dt^2},$$

**and $Z \equiv u \times U + W = a\,(u \times \Omega + du/dt)$.**

**Total linear and kinematical momentum are constant and take the values**

$$P = mu - a\frac{d\Omega}{dt}, \quad K = mr - Pt - a\Omega,$$

**in such a way that in the center of mass frame $P = K = 0$, the position of the charge with respect to the center of mass is $k = (a/m)\Omega$. But $dk/dt = (a/m)d\Omega/dt$, and $d^2k/dt^2 = (a/m)d^2\Omega/dt^2$, so that the constant spin vector (2.69) is**

$$S = \frac{a^2}{m}\left(\frac{d\Omega}{dt} \times \Omega + \frac{d^2\Omega}{dt^2}\right) = -mk \times \frac{dk}{dt} + \frac{a^2}{m}\frac{d^2\Omega}{dt^2} = Y + W. \quad (2.138)$$

**It also has two parts, one of (anti)orbital nature related to the zitterbewegung $Y$, which is not orthogonal to the trajectory, and another $W$ related to the variation of the angular velocity which is not a constant of the motion. The angular velocity of the body is directed along the relative position vector $k$ between the centers of mass and charge.**

**If we take the constant spin in the center of mass frame along the negative direction of the $OZ$ axis, solutions of equations (2.138) can be**

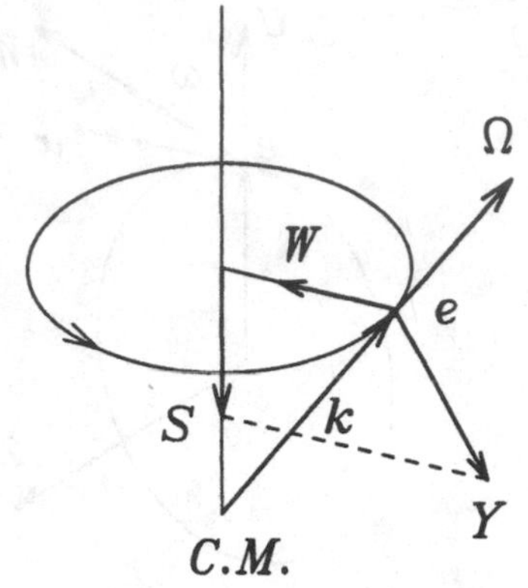


*Figure 2.6.* Zitterbewegung of system described by Lagrangian (2.137), showing the relation between the spin observables $\boldsymbol{S}$, $\boldsymbol{Y}$ and $\boldsymbol{W}$.

found such that $\boldsymbol{\Omega}$ is of constant modulus, with a constant component $\Omega_z$ which is precisely the angular velocity of precession. This motion is the same as the motion of a free rigid body with a symmetry axis.

The structure of the above Lagrangians is suggested by the corresponding relativistic version to be explored in the next chapter, where possible Poincaré invariant terms are of the form $\omega^2 - \alpha^2$ and $\omega \cdot \alpha$, and $\alpha$ is a vector function of the acceleration $\dot{\boldsymbol{u}}$.

## 6.1 CIRCULAR ZITTERBEWEGUNG

As we have seen we can also form first order Galilei invariant terms of the form $\sqrt{\omega^2 - \dot{\boldsymbol{u}}^2/\beta^2}$ and $\sqrt{\omega \cdot \dot{\boldsymbol{u}}}$, where in the first case parameter $\beta$ plays the role of the constant velocity of internal zitterbewegung. We can thus consider a Lagrangian of the form

$$L = \frac{m}{2}\frac{\dot{\boldsymbol{r}}^2}{\dot{t}} + b\sqrt{\omega^2 - \dot{\boldsymbol{u}}^2/\beta^2}, \qquad (2.139)$$

and because $\omega$ is linear in $\dot{\boldsymbol{\rho}}$, it is also homogeneous of first order in terms of the derivatives. It depends on two arbitrary parameters $m$ and $b$ and a third one $\beta$ that will be identified later with the velocity of the charge in the center of mass frame.

Observables $\boldsymbol{U}$ and $\boldsymbol{W}$ are given by

$$\boldsymbol{U} = -\frac{b}{\beta^2}\frac{\dot{\boldsymbol{u}}}{\sqrt{\omega^2 - \dot{\boldsymbol{u}}^2/\beta^2}}, \quad \boldsymbol{W} = b\frac{\omega}{\sqrt{\omega^2 - \dot{\boldsymbol{u}}^2/\beta^2}}.$$

Because dynamical equations for the orientation variables reduce to $d\boldsymbol{W}/dt = \boldsymbol{\omega} \times \boldsymbol{W} = 0$, $\boldsymbol{W}$ is a vector which is constant in time. Therefore, by squaring this vector we get that $\dot{\boldsymbol{u}}^2/\omega^2$ is also a constant of the motion. Dynamical equations for the position in the center of mass frame are $m\boldsymbol{r} - \boldsymbol{U} = 0$, and thus

$$m\mathbf{r} = -\frac{b}{\beta^2}\frac{\dot{\boldsymbol{u}}}{\sqrt{\omega^2 - \dot{\boldsymbol{u}}^2/\beta^2}},$$

and by squaring anew this variable we also reach the conclusion that $r^2$ is also a constant of the motion in the center of mass frame and thus it represents a circular zitterbewegung of radius $R_0$. Therefore if we choose $|\boldsymbol{u}| = \beta$, $|d\boldsymbol{u}/dt| = \beta^2/R_0$ are constant magnitudes and $\boldsymbol{\omega}$ is also a constant vector. Then we also get a circular zitterbewegung, even with a spin contribution coming from the rotation of the body frame, that in the limit when $\boldsymbol{\omega} \to 0$ gives rise to the model explored in Section 4.5.

Since in this frame the spin is also a constant of the motion it is expressed as the sum of two conserved vectors, one orthogonal to the zitterbewegung plane $\boldsymbol{Y} = \boldsymbol{u} \times \boldsymbol{U}$ and another $\boldsymbol{W}$ that, being a constant vector, can have any arbitrary orientation as the one depicted in Figure 2.5.

Another example with internal circular zitterbewegung which does not reduce to the previous system analyzed in Section 4.5 in the limit $\boldsymbol{\omega} \to 0$, is given by the Lagrangian

$$L = \frac{m}{2}\frac{\dot{\boldsymbol{r}}^2}{\dot{t}} + b\sqrt{\boldsymbol{\omega} \cdot \dot{\boldsymbol{u}}}.$$

Observables $\boldsymbol{U}$ and $\boldsymbol{W}$ are given by

$$\boldsymbol{U} = \frac{b\boldsymbol{\omega}}{2\sqrt{\boldsymbol{\omega} \cdot \dot{\boldsymbol{u}}}}, \quad \boldsymbol{W} = \frac{b\dot{\boldsymbol{u}}}{2\sqrt{\boldsymbol{\omega} \cdot \dot{\boldsymbol{u}}}}.$$

In the center of mass frame the position of the charge is

$$\boldsymbol{k} = \frac{1}{m}\boldsymbol{U} = \frac{b\boldsymbol{\omega}}{2m\sqrt{\boldsymbol{\omega} \cdot \dot{\boldsymbol{u}}}}, \tag{2.140}$$

and since the dependence on the orientation variables is only through the angular velocity we have $d\boldsymbol{W}/d\tau = \boldsymbol{\omega} \times \boldsymbol{W}$, which gives rise to

$$\frac{d}{d\tau}\left(\frac{\dot{\boldsymbol{u}}}{\sqrt{\boldsymbol{\omega} \cdot \dot{\boldsymbol{u}}}}\right) = \frac{\boldsymbol{\omega} \times \dot{\boldsymbol{u}}}{\sqrt{\boldsymbol{\omega} \cdot \dot{\boldsymbol{u}}}} = \frac{2m}{b}\boldsymbol{k} \times \dot{\boldsymbol{u}}, \tag{2.141}$$

after replacing $\boldsymbol{\omega}$ with its expression in terms of $\boldsymbol{k}$ from (2.140). Also from (2.140), by taking the scalar product with $\dot{\boldsymbol{u}}$ we can express

$$\sqrt{\boldsymbol{\omega} \cdot \dot{\boldsymbol{u}}} = \frac{2m}{b}\boldsymbol{k} \cdot \dot{\boldsymbol{u}},$$

and if we replace this on the left-hand side of (2.141) and differentiate with respect to $\tau$ we get

$$\frac{\ddot{\boldsymbol{u}}}{\boldsymbol{k}\cdot\dot{\boldsymbol{u}}} - \dot{\boldsymbol{u}}\,\frac{\boldsymbol{u}\cdot\dot{\boldsymbol{u}} + \boldsymbol{k}\cdot\ddot{\boldsymbol{u}}}{(\boldsymbol{k}\cdot\dot{\boldsymbol{u}})^2} = \frac{4m^2}{b^2}\boldsymbol{k}\times\dot{\boldsymbol{u}},$$

and making the scalar product of both sides with $\boldsymbol{k}$ we reach $\boldsymbol{u}\cdot\dot{\boldsymbol{u}} = 0$. Therefore the internal motion in the center of mass frame is a circle at a constant velocity $u$. In this frame the spin is also constant and reduces to the function $\boldsymbol{S} = \boldsymbol{u}\times\boldsymbol{U} + \boldsymbol{W}$ that takes the expression

$$\boldsymbol{S} = \frac{b}{2m\sqrt{\boldsymbol{\omega}\cdot\dot{\boldsymbol{u}}}}\left(\dot{\boldsymbol{u}} + \boldsymbol{u}\times\boldsymbol{\omega}\right),$$

which is orthogonal to the zitterbewegung plane, and in this case the center of mass is not exactly the center of the circle, as can be seen in the figure.

In this example we also have that the angular velocity is of constant modulus and precess around the spin direction with constant angular velocity, as in the previous example.

## 6.2 CLASSICAL NON-RELATIVISTIC GYROMAGNETIC RATIO

Particular but interesting examples are those in which total spin and magnetic moment have the same direction. This amounts to the existence of a linear relation between the zitterbewegung part $\boldsymbol{Y}$ and the total spin $\boldsymbol{S}$, so that when analyzed in the center of mass frame $\boldsymbol{Y}$ and $\boldsymbol{W}$ are parallel or antiparallel vectors. From the Lagrangian

$$L_1 = \frac{m}{2}\frac{\dot{r}^2}{\dot{t}} + b_1\sqrt{\omega^2 - \dot{u}^2/\beta^2},$$

a particular solution is the one depicted in Figure 2.7($a$), where $W > Y$, and thus total spin is orthogonal to the zitterbewegung plane but opposite to the zitterbewegung spin $\boldsymbol{Y}$. The Lagrangian

$$L_2 = \frac{m}{2}\frac{\dot{r}^2}{\dot{t}} - b_2\sqrt{\dot{u}^2/\beta^2 - \omega^2},$$

which is defined whenever $Y > W$, gives rise to a spin directed along $\boldsymbol{Y}$ as in Figure 2.7($b$).

If, as we have seen in previous examples, the magnetic moment is produced by the particle current, then it is related to the zitterbewegung part of the spin by

$$\boldsymbol{\mu} = -\frac{e}{2m}\boldsymbol{Y}.$$

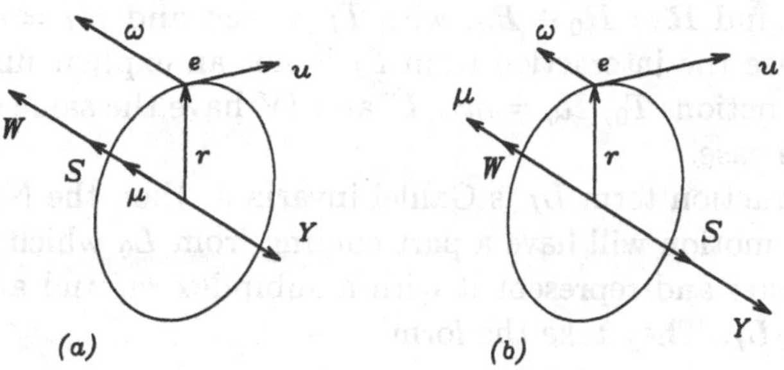


*Figure 2.7.* **Positive charged particles with parallel (*a*) and antiparallel (*b*), spin and magnetic moment.**

But when measuring the angular momentum of the particle it is not possible to separate the two spin parts, so that what is measured is the total spin part $\boldsymbol{S}$. If we call $g = Y/S$ the ratio between the absolute value of these two spin observables, then for Lagrangian $L_1$ and for a positive charged particle $e > 0$, the magnetic moment is written as

$$\boldsymbol{\mu}_1 = g\frac{e}{2m}\boldsymbol{S},$$

while for Lagrangian $L_2$

$$\boldsymbol{\mu}_2 = -g\frac{e}{2m}\boldsymbol{S}.$$

It is easy to see that the invariant parameters $b_i$ and $\beta$ for these examples reduce to

$$b_1 = S\sqrt{g^2+1}, \quad b_2 = S\sqrt{g^2-1}, \quad H_0 = -\frac{m}{2}\beta^2.$$

## 7. INTERACTION WITH AN EXTERNAL FIELD

For a general Lagrangian system with some external interaction, we shall assume that the total Lagrangian is of the form $L = L_0 + L_I$, where $L_0(\dot{t}, \dot{\boldsymbol{r}}, \dot{\boldsymbol{u}}, \boldsymbol{\omega})$ is a general Lagrangian for a free system, and we shall consider as a particular case the one analyzed in the example (2.77). It does not depend explicitly on $t$ and $\boldsymbol{r}$; we shall assume that the dependence on orientation variables is only through its dependence on angular velocity $\boldsymbol{\omega}$ and that whenever the variable $\boldsymbol{u}$ appears it is understood to be replaced by $\dot{\boldsymbol{r}}/\dot{t}$. For the interaction term we shall consider the same minimal coupling $L_I(t, \boldsymbol{r}, \dot{t}, \dot{\boldsymbol{r}})$ as in (2.100).

The total Lagrangian can be written as

$$L = T\dot{t} + \boldsymbol{R}\cdot\dot{\boldsymbol{r}} + \boldsymbol{U}\cdot\dot{\boldsymbol{u}} + \boldsymbol{W}\cdot\boldsymbol{\omega}.$$

The only difference with the free particle case is contained in the terms $T = T_0 + T_I$ and $\boldsymbol{R} = \boldsymbol{R}_0 + \boldsymbol{R}_I$, with $T_I = -e\phi$ and $\boldsymbol{R}_I = e\boldsymbol{A}$, respectively. Because the interaction term $L_I$ is not an explicit function of $\dot{\boldsymbol{u}}$ and $\boldsymbol{\omega}$, the functions $T_0$, $\boldsymbol{R}_0 = m\boldsymbol{u}$, $\boldsymbol{U}$ and $\boldsymbol{W}$ have the same expressions as in the free case.

If the interaction term $L_I$ is Galilei invariant, then the Noether constants of the motion will have a part coming from $L_0$ which we call the mechanical part and represent it with a subindex $m$, and another part coming from $L_I$. They take the form

$$\begin{aligned} H &= H_m + e\phi, \\ \boldsymbol{P} &= \boldsymbol{P}_m + e\boldsymbol{A}, \\ \boldsymbol{K} &= m\boldsymbol{r} - \boldsymbol{P}t - \boldsymbol{U} = \boldsymbol{K}_m - e\boldsymbol{A}t, \\ \boldsymbol{J} &= \boldsymbol{r} \times \boldsymbol{P} + \boldsymbol{u} \times \boldsymbol{U} + \boldsymbol{W} = \boldsymbol{J}_m + \boldsymbol{r} \times e\boldsymbol{A}. \end{aligned}$$

The corresponding 'mechanical' observables have the same analytical expressions in terms of the kinematical variables and their time derivatives as in the free particle case.

In general, the interaction term $L_I$ will not be translation invariant and therefore the above observables are no longer constants of the motion. Their time variation is obtained by the corresponding Euler-Lagrange equation, such that for the energy it gives

$$\frac{dH}{dt} = \frac{dH_m}{dt} + \frac{d(e\phi(t,\boldsymbol{r}))}{dt} = -\frac{1}{\dot{t}}\frac{\partial L_I}{\partial t}.$$

Thus we get

$$\frac{dH_m}{dt} = e\boldsymbol{u} \cdot \left( -\nabla\phi(t,\boldsymbol{r}) - \frac{\partial \boldsymbol{A}(t,\boldsymbol{r})}{\partial t} \right) = e\boldsymbol{u} \cdot \boldsymbol{E}, \qquad (2.142)$$

and therefore the variation of the mechanical energy is the work of the external electric field along the charge trajectory.

For the linear momentum we get

$$\frac{d\boldsymbol{P}}{dt} = \frac{d\boldsymbol{P}_m}{dt} + e\frac{d\boldsymbol{A}}{dt} = \frac{1}{\dot{t}}\nabla L_I,$$

so that

$$\frac{d\boldsymbol{P}_m}{dt} = e(\boldsymbol{E} + \boldsymbol{u} \times \boldsymbol{B}) = m\frac{d^2\boldsymbol{q}}{dt^2}, \qquad (2.143)$$

because the definition of the center of mass position (2.91) is unchanged. We get again Newton's dynamical equations for the center of mass position $\boldsymbol{q}$ but with the fields defined at the charge position $\boldsymbol{r}$.

Dynamical equations for the orientation variables are the same as in the free case (2.72) and have the form

$$\frac{d\boldsymbol{W}}{d\tau} = \boldsymbol{\omega} \times \boldsymbol{W}. \tag{2.144}$$

As seen in (1.17) all these dynamical equations are not independent. They are related by:

$$\frac{dH_m}{dt} - e\boldsymbol{u}\cdot\boldsymbol{E} = \boldsymbol{u}\cdot\left(\frac{d\boldsymbol{P}_m}{dt} - e(\boldsymbol{E}+\boldsymbol{u}\times\boldsymbol{B})\right) + \frac{\boldsymbol{\omega}}{t}\cdot\left(\frac{d\boldsymbol{W}}{dt} - \frac{\boldsymbol{\omega}}{t}\times\boldsymbol{W}\right)$$

suggesting that

$$\frac{dH_m}{dt} = \boldsymbol{u}\cdot\frac{d\boldsymbol{P}_m}{dt} + \boldsymbol{\Omega}\cdot\frac{d\boldsymbol{W}}{dt},$$

and where $\boldsymbol{\Omega} = \boldsymbol{\omega}/t$ is the angular velocity in a time evolution description instead of the evolution in terms of parameter $\tau$. For those systems for which $\boldsymbol{W}$ satisfies (2.144) we have that $\boldsymbol{\omega}\cdot d\boldsymbol{W}/dt = 0$, and thus

$$\frac{dH_m}{dt} = \boldsymbol{u}\cdot\frac{d\boldsymbol{P}_m}{dt}.$$

For the relative position vector $\boldsymbol{k} = \boldsymbol{r} - \boldsymbol{q}$, we get

$$\frac{d^2\boldsymbol{k}}{dt^2} + \omega^2\boldsymbol{k} = -\frac{1}{m}\boldsymbol{F}(t,\boldsymbol{r}), \tag{2.145}$$

in terms of the external Lorentz force. The spin is expressed as in (2.69) by

$$\boldsymbol{S} = -m\boldsymbol{k}\times\frac{d\boldsymbol{k}}{dt} + \boldsymbol{W},$$

and taking into account (2.145) its dynamics is

$$\frac{d\boldsymbol{S}}{dt} = \boldsymbol{k}\times\boldsymbol{F} + \boldsymbol{\Omega}\times\boldsymbol{W}.$$

In the particular case where $\boldsymbol{W} \sim \boldsymbol{\omega}$ it reduces to the torque of the external force applied at point $\boldsymbol{r}$, with respect to the center of mass, and therefore, both electric and magnetic fields produce a change in the spin of the particle.

If the electric field is conservative, then the change of the mechanical energy (2.142) between two arbitrary points of the charge trajectory is

$$H_m(t_2) - H_m(t_1) = e\phi(\boldsymbol{r}_1) - e\phi(\boldsymbol{r}_2)$$

$$\simeq (e\phi(\boldsymbol{q}_1) - e\boldsymbol{k}_1\cdot\boldsymbol{E}(\boldsymbol{q}_1)) - (e\phi(\boldsymbol{q}_2) - e\boldsymbol{k}_2\cdot\boldsymbol{E}(\boldsymbol{q}_2)),$$

after expanding the potentials around the center of mass position, and considering that $|\boldsymbol{k}|$ is very small. By rearranging terms we see that the sum of the mechanical and potential energy is conserved, as it corresponds to a time-independent interaction. The particle, in the presence of an electrostatic field, has a potential energy of value $e\phi(\boldsymbol{q}) - \boldsymbol{d}\cdot\boldsymbol{E}(\boldsymbol{q})$ and therefore it behaves like a point particle of charge $e$ and an intrinsic electric dipole moment $\boldsymbol{d} = e\boldsymbol{k}$ at the center of mass $\boldsymbol{q}$, although this dipole is oscillating with very high frequency and its time-average value vanishes.

The time variation of the total angular momentum is

$$\frac{d\boldsymbol{J}}{dt} = \boldsymbol{q}\times\boldsymbol{F} + \boldsymbol{k}\times\boldsymbol{F} + \frac{d\boldsymbol{W}}{dt} + \frac{d}{dt}(\boldsymbol{r}\times e\boldsymbol{A}).$$

Let us define the center of mass observer by $\boldsymbol{q} = 0$ and $d\boldsymbol{q}/dt = 0$. Consider the simpler case where $d\boldsymbol{W}/dt = 0$ or the analyzed example with (anti)orbital spin where $\boldsymbol{W} = 0$. If the external fields are smooth enough, at least in a region larger than the radius of the zitterbewegung, we take for the vector potential the expression $\boldsymbol{A} = \boldsymbol{B}\times\boldsymbol{r}/2$ in terms of a constant vector $\boldsymbol{B}$, and the above equation in the center of mass frame is written as

$$\frac{d\boldsymbol{J}}{dt} = e\boldsymbol{k}\times\left(\boldsymbol{E} + \frac{d\boldsymbol{k}}{dt}\times\boldsymbol{B}\right) + \frac{e}{2}\frac{d\boldsymbol{k}}{dt}\times(\boldsymbol{B}\times\boldsymbol{k}) + \frac{e}{2}\boldsymbol{k}\times\left(\boldsymbol{B}\times\frac{d\boldsymbol{k}}{dt}\right).$$

By arranging terms we arrive at the torque equation

$$\frac{d\boldsymbol{J}}{dt} = \boldsymbol{d}\times\boldsymbol{E} + \boldsymbol{\mu}\times\boldsymbol{B}.$$

The particle, in addition to the electric dipole $\boldsymbol{d} = e\boldsymbol{k}$ will therefore behave as though it has also associated a magnetic moment $\boldsymbol{\mu}$ which is expressed in terms of the internal zitterbewegung as in (2.104) by

$$\boldsymbol{\mu} = \frac{e}{2}\boldsymbol{k}\times\frac{d\boldsymbol{k}}{dt}.$$

We thus see that under smooth external fields the spinning particle behaves like a point charge $e$ placed at the center of mass $\boldsymbol{q}$, with the addition of a magnetic moment $\boldsymbol{\mu}$ related to the zitterbewegung part of the spin, and also an electric dipole $\boldsymbol{d}$ lying on the zitterbewegung plane, which is oscillating with the zitterbewegung frequency $\omega$.

We can make an alternative analysis of the structure of this particle. If we replace variable $\boldsymbol{r}$ by $\boldsymbol{q}+\boldsymbol{k}$ and make a Taylor expansion of the potentials around point $\boldsymbol{q}$, provided $|\boldsymbol{k}|$ is considered small, then,

$$\phi(t,\boldsymbol{r}) \simeq \phi(t,\boldsymbol{q}) + \boldsymbol{k}\cdot\nabla\phi(t,\boldsymbol{q}), \quad \boldsymbol{A}(t,\boldsymbol{r}) \simeq \boldsymbol{A}(t,\boldsymbol{q}) + (\boldsymbol{k}\cdot\nabla)\boldsymbol{A}(t,\boldsymbol{q}).$$

The interaction Lagrangian (2.100) can be written as

$$\begin{aligned} L_I &= -e\phi(t,q)\dot{t} + e\dot{q}\cdot A(t,q) \\ &- ek\cdot\nabla\phi(t,q)\dot{t} \quad (2.146) \\ &+ e\dot{k}\cdot A(t,q) \quad (2.147) \\ &+ e\dot{q}\cdot(k\cdot\nabla)A(t,q) \\ &+ ek\cdot(k\cdot\nabla)A(t,q). \end{aligned}$$

We then have the expression

$$e\frac{d}{d\tau}(k\cdot A(t,q)) = e\dot{k}\cdot A(t,q) + ek\cdot\left(\frac{\partial A(t,q)}{\partial t}\dot{t} + \frac{\partial A(t,q)}{\partial q_i}\dot{q}_i\right).$$

The first term on the right-hand side is (2.147) and therefore can be expressed as a function of this total $\tau$-derivative, which can be withdrawn from the Lagrangian, and the remaining two additional terms. The first one with term (2.146) gives rise to $(d\cdot E)\dot{t}$. Then

$$\begin{aligned} L_I &= -e\phi(t,q)\dot{t} + e\dot{q}\cdot A(t,q) + d\cdot E(t,q)\dot{t} \\ &+ e\dot{q}\cdot(k\cdot\nabla)A(t,q) - ek\cdot\left(\frac{\partial A(t,q)}{\partial q_i}\dot{q}_i\right) \quad (2.148) \\ &+ e\dot{k}\cdot(k\cdot\nabla)A(t,q). \quad (2.149) \end{aligned}$$

The two terms in expression (2.148) are written as $e\,(k\times\dot{q})\cdot B$, where $B = \nabla\times A$. In the case of smooth fields, if we consider for the vector potential its expression in terms of a uniform magnetic field $A = B\times r/2$, the last term (2.149) is

$$\frac{e}{2}\left(k\times\dot{k}\right)\cdot B(t,q) = (\mu\cdot B)\dot{t}.$$

Finally we get for the interaction Lagrangian the expression

$$L_I = -e\phi(t,q)\dot{t} + e\dot{q}\cdot A(t,q) + d\cdot E(t,q)\dot{t} + \left(\mu + d\times\frac{dq}{dt}\right)\cdot B(t,q)\dot{t}.$$

It clearly can be interpreted in this approximation as a nonrelativistic model of a point particle with charge $e$ at point $q$, moving with velocity $\dot{q}$ and an electric and magnetic dipole $d$ and $\mu$ respectively, located at the center of mass like some 'intrinsic' properties. Because the motion of an electric dipole with velocity $\dot{q}$ produces a magnetic dipole $d\times\dot{q}$, it supplies an additional term in the Lagrangian. The motion of a magnetic dipole produces an electric dipole $\dot{q}\times\mu/c^2$ which is negligible in this nonrelativistic approach.

## 8. TWO-PARTICLE SYSTEMS

Let us consider a nonrelativistic system such that its kinematical space $X$ can be written as the Cartesian product $X = X_1 \times X_2$ in terms of two manifolds $X_a$, $a = 1,2$, such that both are homogeneous spaces of the Galilei group $\mathcal{G}$ and therefore can be taken as the kinematical spaces of two elementary particles. Let $x_a \in X_a$ be the corresponding kinematical variables for each elementary particle and $x \equiv (x_1, x_2) \in X$. Can we say that $X$ represents the kinematical space of a two-particle system? In general an elementary particle is not only characterized by its kinematical space $X$ but also by the corresponding gauge function $\alpha(g; x)$ defined on $G \times X$.

Let us go further and assume that the gauge function for our system $\alpha(g; x_1, x_2)$ can be written in terms of gauge functions for elementary particles in the form

$$\alpha(g; x) \equiv \alpha(g; x_1, x_2) = \alpha_1(g; x_1) + \alpha_2(g; x_2), \tag{2.150}$$

in terms of the single particle gauge functions $\alpha_a$, $a = 1,2$, each one defined on the corresponding $G \times X_a$ manifold. In this case we shall say that $X$ represents the kinematical space of a decomposable two-particle system, with a gauge function defined by (2.150).

For instance let us consider two Galilei point particles of masses $m_a$ with kinematical variables $t_a$ and $\boldsymbol{r}_a$, and gauge functions

$$\alpha_a(g; x_a) = m_a \left(\boldsymbol{v}^2 t_a/2 + \boldsymbol{v} \cdot R(\boldsymbol{\mu})\boldsymbol{r}_a\right), \quad \forall g \in \mathcal{G}, \quad \forall x_a \in X_a, a = 1,2.$$

From (1.62) we see that our total gauge function satisfies

$$\alpha(g'; gx) + \alpha(g; x) - \alpha(g'g; x) = (m_1 + m_2)\left(\boldsymbol{v}'^2 b/2 + \boldsymbol{v}' \cdot R(\boldsymbol{\mu}')\boldsymbol{a}\right)$$

$$= \xi_m(g', g), \quad \forall g', g \in \mathcal{G},$$

in terms of an exponent $\xi_m$ of the Galilei group that depends on the parameter $m = m_1 + m_2$, interpreted as the total mass of the compound system.

The homogeneity condition on the derivatives of the kinematical variables leads us to write the general Lagrangian in the form

$$L = T_1 \dot{t}_1 + \boldsymbol{R}_1 \cdot \dot{\boldsymbol{r}}_1 + T_2 \dot{t}_2 + \boldsymbol{R}_2 \cdot \dot{\boldsymbol{r}}_2, \tag{2.151}$$

where $T_a = \partial L/\partial \dot{t}_a$ and $\boldsymbol{R}_a = \partial L/\partial \dot{\boldsymbol{r}}_a$, $a = 1,2$, respectively. All these functions depend on the eight kinematical variables and their first order derivatives with respect to some dimensionless evolution parameter $\tau$, being homogeneous functions of zero degree of the derivatives. We have two time observables $t_1$ and $t_2$, one for each particle. In a synchronous description, as we shall see later, we shall establish the constraint $t_1 = t_2 = t$ to obtain a synchronous single time evolution of the system.

According to the transformation equations of the kinematical variables and their derivatives under the Galilei group $\mathcal{G}$ we get for the different functions that appear in (2.151) the transformation equations:

$$T'_a = T_a - \boldsymbol{v}\cdot R(\boldsymbol{\mu})\boldsymbol{R}_a - \frac{m_a}{2}\boldsymbol{v}^2, \quad (2.152)$$

$$\boldsymbol{R}'_a = R(\boldsymbol{\mu})\boldsymbol{R}_a + m_a\boldsymbol{v}, \quad a = 1,2. \quad (2.153)$$

Invariance of dynamical equations under the Galilei group lead to the following constants of the motion, by means of Noether's theorem, named as usual, total energy, total linear momentum, total kinematical momentum and total angular momentum, respectively:

$$H = -T_1 - T_2, \quad (2.154)$$

$$\boldsymbol{P} = \boldsymbol{R}_1 + \boldsymbol{R}_2, \quad (2.155)$$

$$\boldsymbol{K} = m_1\boldsymbol{r}_1 + m_2\boldsymbol{r}_2 - \boldsymbol{R}_1 t_1 - \boldsymbol{R}_2 t_2, \quad (2.156)$$

$$\boldsymbol{J} = \boldsymbol{r}_1\times\boldsymbol{R}_1 + \boldsymbol{r}_2\times\boldsymbol{R}_2. \quad (2.157)$$

Total energy and linear momentum transform as:

$$H' = H + \boldsymbol{v}\cdot R(\boldsymbol{\mu})\boldsymbol{P} + \frac{m}{2}\boldsymbol{v}^2, \quad (2.158)$$

$$\boldsymbol{P}' = R(\boldsymbol{\mu})\boldsymbol{P} + m\boldsymbol{v}, \quad (2.159)$$

in terms of the total mass $m$ of the system and therefore the magnitude $H - \boldsymbol{P}^2/2m = H_0$ or internal energy of the system, is an invariant and a constant of the motion. The different observables are invariant under translations. Therefore they depend on the variables $\theta = t_2 - t_1$ and $\boldsymbol{r} = \boldsymbol{r}_2 - \boldsymbol{r}_1$, are homogeneous functions of zero degree in terms of the derivatives and thus are functions of $\lambda = \dot{t}_2/\dot{t}_1$ and $\boldsymbol{u}_a = \dot{\boldsymbol{r}}_a/\dot{t}_a$, $a = 1,2$. These variables transform under $\mathcal{G}$ in the form

$$\theta' = \theta, \quad \boldsymbol{r}' = R(\boldsymbol{\mu})\boldsymbol{r} + \boldsymbol{v}\theta, \quad \lambda' = \lambda, \quad \boldsymbol{u}'_a = R(\boldsymbol{\mu})\boldsymbol{u}_a + \boldsymbol{v}. \quad (2.160)$$

## 8.1 SYNCHRONOUS DESCRIPTION

If there exists an observer who is able to produce a synchronous description, *i.e.*, for all $\tau$ the time variables $t_1(\tau) = t_2(\tau) = t(\tau)$, then all other observables also produce a synchronous description. In this case

$\theta = 0$ and $\lambda = 1$. If we take the time derivative of the kinematical momentum (2.156) we get for the total linear momentum

$$\boldsymbol{P} = \boldsymbol{R}_1 + \boldsymbol{R}_2 = m_1\boldsymbol{u}_1 + m_2\boldsymbol{u}_2. \tag{2.161}$$

Under an infinitesimal pure Galilei transformation we get, for the functions $\boldsymbol{R}_a$, the differential conditions

$$\frac{\partial R_{ai}}{\partial u_{1j}} + \frac{\partial R_{ai}}{\partial u_{2j}} = m_a\delta_{ij},$$

whose general solution is of the form

$$\boldsymbol{R}_a = m_a\boldsymbol{u}_a + \boldsymbol{A}_a(\boldsymbol{r}, \boldsymbol{u}), \tag{2.162}$$

in which no addition on repeated index $a$ is performed, and where the $\boldsymbol{A}_a$ are arbitrary functions of $\boldsymbol{r}$ and $\boldsymbol{u} = \boldsymbol{u}_2 - \boldsymbol{u}_1$. Taking into account (2.161) we get

$$\boldsymbol{A}_1(\boldsymbol{r}, \boldsymbol{u}) + \boldsymbol{A}_2(\boldsymbol{r}, \boldsymbol{u}) = 0. \tag{2.163}$$

Since in this spinless system we can interpret each function $m_a\boldsymbol{u}_a$ as the mechanical linear momentum of particle $a$, then the time derivative (with opposite sign) of the function $\boldsymbol{A}_a$ is the force acting on particle $a$. From (2.163) we see that these systems in a synchronous description satisfy Newton's third law (action-reaction principle). In fact, the two indexes 1 and 2 being arbitrary, the functions $\boldsymbol{A}_a$ satisfy the symmetry properties

$$\boldsymbol{A}_2(\boldsymbol{r}, \boldsymbol{u}) = \boldsymbol{A}_1(-\boldsymbol{r}, -\boldsymbol{u}) = -\boldsymbol{A}_1(\boldsymbol{r}, \boldsymbol{u}) = -\boldsymbol{A}_2(-\boldsymbol{r}, -\boldsymbol{u}). \tag{2.164}$$

Due to the vector character of $\boldsymbol{A}_a(\boldsymbol{r}, \boldsymbol{u})$ under rotations it takes the general form

$$\boldsymbol{A}(\boldsymbol{r}, \boldsymbol{u}) = f(\boldsymbol{r}^2, \boldsymbol{u}^2, \boldsymbol{r}\cdot\boldsymbol{u})\boldsymbol{r} + g(\boldsymbol{r}^2, \boldsymbol{u}^2, \boldsymbol{r}\cdot\boldsymbol{u})\boldsymbol{u}, \tag{2.165}$$

where $f$ and $g$ are two arbitrary functions of its arguments and the symmetry conditions (2.164) eliminate a possible $\boldsymbol{r}\times\boldsymbol{u}$ term in (2.165).

In this synchronous description the center of mass position of the system can be defined, at any instant $\tau$, as

$$m\boldsymbol{q}(\tau) = m_1\boldsymbol{r}_1(\tau) + m_2\boldsymbol{r}_2(\tau),$$

and this leads for the total linear momentum to $\boldsymbol{P} = m d\boldsymbol{q}/dt$.

From now on we can consider this system in a synchronous description as a system of six degrees of freedom, with kinematical variables $t$, $\boldsymbol{r}$ and

$\boldsymbol{q}$, that transform under $\mathcal{G}$ as:

$$\begin{aligned} t'(\tau) &= t(\tau), \\ \boldsymbol{r}'(\tau) &= R(\boldsymbol{\mu})\boldsymbol{r}(\tau), \\ \boldsymbol{q}'(\tau) &= R(\boldsymbol{\mu})\boldsymbol{q}(\tau) + \boldsymbol{v}t(\tau) + \boldsymbol{a}, \end{aligned}$$

with a gauge function

$$\alpha(g; x) = m\left(\boldsymbol{v}^2 t(\tau)/2 + \boldsymbol{v} \cdot R(\boldsymbol{\mu})\boldsymbol{q}(\tau)\right).$$

It must be remarked here that the seven kinematical variables do not belong to a homogeneous space of $\mathcal{G}$, because in general the relative separation at two different instants $|\boldsymbol{r}(\tau_1)| \neq |\boldsymbol{r}(\tau_2)|$ and the way $\boldsymbol{r}(\tau)$ transforms does not define a group element that maps one into the other. It therefore represents a non-elementary system of total mass $m$, where the relative position $\boldsymbol{r}$ describes the internal motion.

In terms of $\boldsymbol{r}$ and $\boldsymbol{q}$ we have $\boldsymbol{r}_1 = \boldsymbol{q} - (m_2/m)\boldsymbol{r}$ and $\boldsymbol{r}_2 = \boldsymbol{q} + (m_1/m)\boldsymbol{r}$, so that the general Lagrangian in terms of these kinematical variables takes the form

$$L = -H\dot{t} + \boldsymbol{P} \cdot \dot{\boldsymbol{q}} + \left(\frac{m_1 m_2}{m}\,\boldsymbol{u} + \boldsymbol{A}_1(\boldsymbol{r}, \boldsymbol{u})\right) \cdot \dot{\boldsymbol{r}}.$$

Here $H = -\partial L/\partial \dot{t}$, $\boldsymbol{P} = \partial L/\partial \dot{\boldsymbol{q}}$, and observable $\boldsymbol{R} = (m_1 m_2/m)\boldsymbol{u} + \boldsymbol{A}_1$. The energy $H$ is translation invariant, and it is only a function of $\boldsymbol{r}$, $\boldsymbol{u} = \dot{\boldsymbol{r}}/\dot{t}$ and $\boldsymbol{w} = \dot{\boldsymbol{q}}/\dot{t}$. $\boldsymbol{P}$ is $m d\boldsymbol{q}/dt$ and $\boldsymbol{A}_1$ is the arbitrary function obtained in (2.162) whose general form is given in (2.165). Since the energy transforms under $\mathcal{G}$ as in equation (2.158), then under an infinitesimal pure Galilei transformation it follows that

$$\frac{\partial H(\boldsymbol{r}, \boldsymbol{u}, \boldsymbol{w})}{\partial w_i} = P_i = m\,\frac{dq_i}{dt}.$$

The general solution of this equation is of the form

$$H = \frac{1}{2}m\left(\frac{d\boldsymbol{q}}{dt}\right)^2 + V(\boldsymbol{r}, \boldsymbol{u}),$$

with $V$, an arbitrary function of its arguments, interpreted as the internal energy of the system. The final expression of the Lagrangian is

$$L = \frac{1}{2}m\,\dot{\boldsymbol{q}}^2/\dot{t} + \mu\dot{\boldsymbol{r}}^2/\dot{t} - V(\boldsymbol{r}, \boldsymbol{u})\,\dot{t} + \boldsymbol{A}(\boldsymbol{r}, \boldsymbol{u}) \cdot \dot{\boldsymbol{r}} = L_0 + L_I,$$

where $\mu = m_1 m_2/m$ is the reduced mass of the system. The arbitrary functions $V$ and $\boldsymbol{A}$ transform under $\mathcal{G}$ as

$$\begin{aligned} V' &= V, \\ \boldsymbol{A}' &= R(\boldsymbol{\mu})\boldsymbol{A}. \end{aligned}$$

Nevertheless $V$ and $\boldsymbol{A}$ are not independent functions, since the observable $\boldsymbol{R} = \mu\boldsymbol{u} + \boldsymbol{A} = \partial L/\partial\dot{\boldsymbol{r}}$. When making this partial derivative of the Lagrangian we get

$$\boldsymbol{R} = \mu u_i + A_i = 2\mu u_i - \frac{\partial V}{\partial u_i} + A_i + \frac{\partial \boldsymbol{A}}{\partial u_i} \cdot \boldsymbol{u},$$

and in consequence

$$\mu u_i = \frac{\partial V}{\partial u_i} - \frac{\partial \boldsymbol{A}}{\partial u_i} \cdot \boldsymbol{u}.$$

If the vector potential $\boldsymbol{A}$ does not depend on $\boldsymbol{u}$, then $V$ reduces to the general form $V(\boldsymbol{r}, \boldsymbol{u}) = \mu\boldsymbol{u}^2/2 + \phi(\boldsymbol{r})$ so that the final expression of the Lagrangian $L_I$ becomes

$$L_I = \left[\frac{1}{2}\mu\,\boldsymbol{u}^2 - \phi(\boldsymbol{r}) + \boldsymbol{A}(\boldsymbol{r}) \cdot \boldsymbol{u}\right]\dot{t}. \tag{2.166}$$

In a time evolution description this corresponds to the interaction of a point particle of mass $\mu$, located at point $\boldsymbol{r}$, under the action of an external field with scalar and vector potentials $\phi(\boldsymbol{r})$ and $\boldsymbol{A}(\boldsymbol{r})$, respectively.

Dynamical equations obtained from (2.166), with $\boldsymbol{A}$ of the form $\boldsymbol{A} = f(\boldsymbol{r})\boldsymbol{r}$, as deduced from its general expression (2.165), are

$$\mu\frac{d^2\boldsymbol{r}}{dt^2} = -\nabla\phi(\boldsymbol{r}),$$

since curl $\boldsymbol{A} = 0$.

The function $\boldsymbol{A}(\boldsymbol{r})$ is interpreted as the linear momentum transfer between the two particles but it plays no role in the dynamical equations. In the general case, for arbitrary $V$ and $\boldsymbol{A}$, we obtain the dynamical equations

$$\mu\frac{d^2\boldsymbol{r}}{dt^2} + \frac{\partial\boldsymbol{A}}{\partial u_j}\frac{d^2 r_j}{dt^2} = -\nabla V(\boldsymbol{r}, \boldsymbol{u}) + \boldsymbol{u} \times \mathrm{curl}\,\boldsymbol{A}(\boldsymbol{r}, \boldsymbol{u}),$$

where on the right-hand side we still have a Lorentz-like force but on the lef-hand side we have a term that can be interpreted as a mass transfer between particles. For instance, if $\boldsymbol{A}(\boldsymbol{r}, \boldsymbol{u}) = f(r)\boldsymbol{r} + g(r)\boldsymbol{u}$ then the above dynamical equations become

$$(\mu + g(r))\frac{d^2\boldsymbol{r}}{dt^2} = -\nabla V(\boldsymbol{r}, \boldsymbol{u}) + \boldsymbol{u} \times (\nabla g(r) \times \boldsymbol{u}),$$

where the $g(r)$ part can be interpreted as the mass transfer during the linear momentum transfer.

As a remark, in this case the radial part $f(r)\boldsymbol{r}$ of $\boldsymbol{A}(\boldsymbol{r},\boldsymbol{u})$ does not play any role in the dynamics because it contributes to the total Lagrangian through a term of the form $f(r)\boldsymbol{r}\cdot\boldsymbol{u}$ such that if we write $f(r)\boldsymbol{r} = \nabla h(r)$, then $f(r)\boldsymbol{r}\cdot\boldsymbol{u} = dh(r)/dt$, and since it is a total time derivative it can be withdrawn.

For point particles there are no spin effects and if the interaction is of electromagnetic nature, $\boldsymbol{A}$ does not depend on $\boldsymbol{u}$ and the interaction is characterized only by the static potential energy $\phi(\boldsymbol{r})$.

## 9. TWO INTERACTING SPINNING PARTICLES

A possible two-particle spinning system with a gauge function of the form (2.150) is given by the following Lagrangian

$$L = L_1 + L_2 - \phi(r) = L_1 + L_2 - \frac{e_1 e_2}{|\boldsymbol{r}_1 - \boldsymbol{r}_2|}, \tag{2.167}$$

in a synchronous description and in terms of a static potential energy. For each $L_a$, $a = 1, 2$, we can take for instance the corresponding free Lagrangian for a free spinning particle of mass $m_a$ and zitterbewegung frequency $\omega_a$ as described in Sec. 4.. The interaction term depends only on the relative distance $|\boldsymbol{r}_1 - \boldsymbol{r}_2|$ between the charges $e_a$ of the particles and therefore it is a Galilei invariant term.

We must be cautious about the choice of a static Coulomb-like interaction between particles, because in the spinning particle model which we have described in Sec. 4., the spin is related to the zitterbewegung so that the charges whose motion produces the spin are neither static nor in uniform motion. Therefore the corresponding Maxwell field associated to each one is no longer static and even far from a Coulomb-like behaviour. It is clear at this point that a thorough analysis of the electromagnetic structure of a spinning particle must be carried out before producing any analysis of possible electromagnetic interactions between them. We shall come back to the electromagnetic structure of the electron in Chapter 6. There, some hints about the static and Coulomb-like behaviour of the average Maxwell field associated to a point charge in motion, which describes circles at the speed of light, will be given. If we take this suggestion for granted, the interaction potential $\phi(r)$ of this example will be interpreted as the interaction of each particle in the average electrostatic field of the other. In this case, we have to consider also the magnetic interaction of each charge with the magnetic field produced by the static magnetic moments associated to the other particle. Nevertheless we shall avoid this analysis for the moment.

Since the interaction does not depend on the accelerations, the definition of the center of mass observable for each particle remains the usual one and the dynamical equations we obtain for system (2.167), are:

$$m_1\frac{d^2\boldsymbol{q}_1}{dt^2} = e_1e_2\frac{\boldsymbol{r}_1-\boldsymbol{r}_2}{|\boldsymbol{r}_1-\boldsymbol{r}_2|^3},\quad \frac{d^2\boldsymbol{r}_1}{dt^2}+\omega_1^2(\boldsymbol{r}_1-\boldsymbol{q}_1)=0, \quad (2.168)$$

$$m_2\frac{d^2\boldsymbol{q}_2}{dt^2} = e_1e_2\frac{\boldsymbol{r}_2-\boldsymbol{r}_1}{|\boldsymbol{r}_1-\boldsymbol{r}_2|^3},\quad \frac{d^2\boldsymbol{r}_2}{dt^2}+\omega_2^2(\boldsymbol{r}_2-\boldsymbol{q}_2)=0. \quad (2.169)$$

To simplify the problem let us assume that the particles are identical and they are electrons. Then, by defining a dimensionless time $\theta=\omega t$, as in Sec. 4.3 and by using the zitterbewegung radius $R=\hbar/2m_ec$ as unit of length, the only parameter that controls the dynamical equations is

$$a=2e^2/c\hbar=2\alpha\simeq 2/137=0.01459.$$

Here $\alpha$ is the fine structure constant, and the dynamical equations (2.168-2.169) become

$$\frac{d^2\boldsymbol{q}_i}{d\theta^2}=(-1)^i\,a\,\frac{\boldsymbol{r}_1-\boldsymbol{r}_2}{|\boldsymbol{r}_1-\boldsymbol{r}_2|^3},\quad \frac{d^2\boldsymbol{r}_i}{d\theta^2}+\boldsymbol{r}_i-\boldsymbol{q}_i=0,\quad i=1,2.$$

If we define the variables: $\boldsymbol{q}=(\boldsymbol{q}_1+\boldsymbol{q}_2)/2$, and $\boldsymbol{r}=(\boldsymbol{r}_1+\boldsymbol{r}_2)/2$, as center of mass and center of charge position, respectively, and the relative separation between the centers of mass and charge of each particle $\boldsymbol{k}=\boldsymbol{q}_1-\boldsymbol{q}_2$, $\boldsymbol{l}=\boldsymbol{r}_1-\boldsymbol{r}_2$, these variables satisfy:

$$\frac{d^2\boldsymbol{q}}{d\alpha^2}=0,\quad \frac{d^2\boldsymbol{r}}{d\alpha^2}+\boldsymbol{r}-\boldsymbol{q}=0,$$

$$\frac{d^2\boldsymbol{k}}{d\alpha^2}=-\frac{2a\boldsymbol{l}}{l^3},\quad \frac{d^2\boldsymbol{l}}{d\alpha^2}+\boldsymbol{l}-\boldsymbol{k}=0.$$

The center of mass and center of charge of the system become uncoupled with the other relative variables, so that the center of mass has a straight free motion while the center of charge of the system makes an isotropic harmonic motion of unit frequency around it.

Now about the other variables $\boldsymbol{k}$ and $\boldsymbol{l}$, their motion is the same as that of a single particle with a center of mass at $\boldsymbol{k}$ under the action of a Coulomb-like central force from the origin of coordinates, which is located at point $\boldsymbol{l}$. When the problem is solved in the center of mass frame, $\boldsymbol{q}_1=-\boldsymbol{q}_2$, and thus $\boldsymbol{k}=2\boldsymbol{q}_1$, the evolution of particle 1 in this reference frame is equivalent, up to a global factor 2, to the evolution of variable $\boldsymbol{k}$.

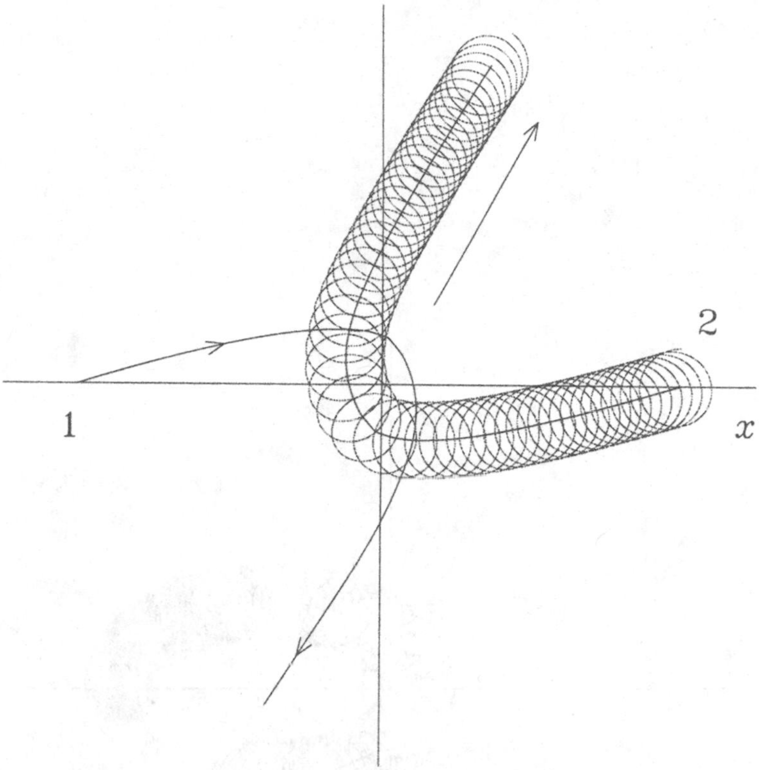


*Figure 2.8.* **Representation in the C.M. frame of the motion of the center of mass of two opposite charged particles with antiparallel spins. For particle 2, the motion of its center of charge is also represented. The interaction produces a chaotic scattering.**

In Figure 2.8, we represent the motion of an electron-positron system interacting according to the Lagrangian (2.167). Both particles are polarized orthogonal to the evolution plane in opposite directions and the motion is depicted in the center of mass frame. Initial positions for their center of masses are represented by points 1 and 2 on the $x$-axis, respectively, and they are sent into each other with a non-vanishing impact parameter. Particles approach each other and finally separate in a chaotic scattering process. The direction of the dispersion depends in a non-linear way on the relative initial phases of the internal motion.

For a different initial configuration we can obtain a bounded system, like the one depicted in Figure 2.9 in which we also represent the center of charge motion of particle 2.

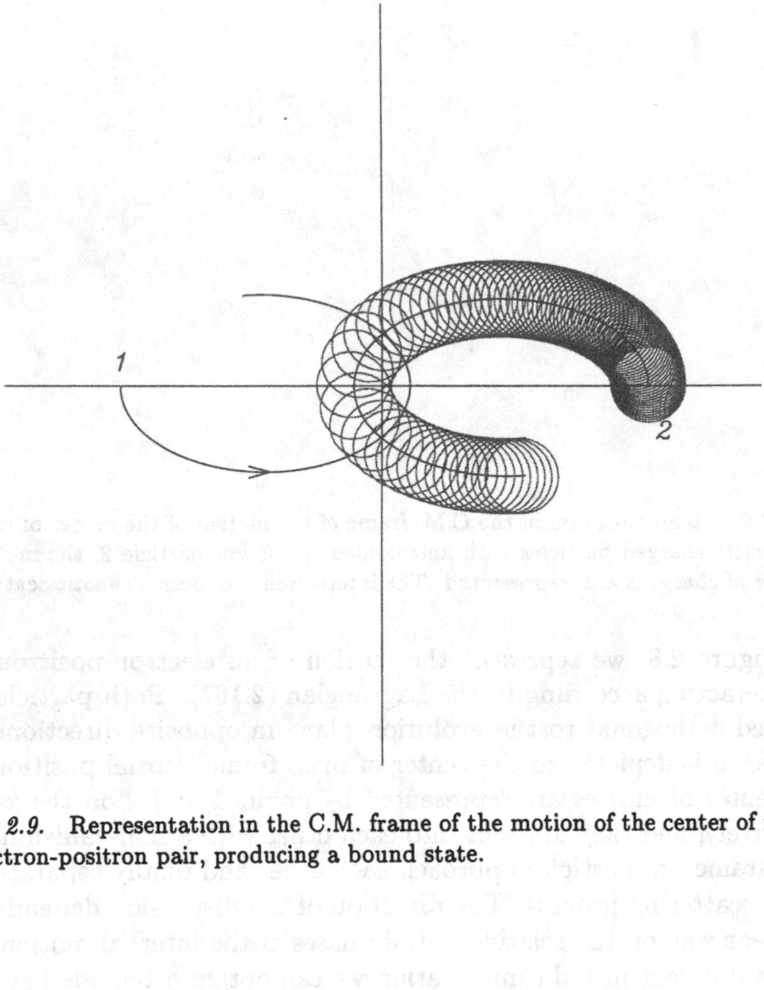


*Figure 2.9.* **Representation in the C.M. frame of the motion of the center of mass of an electron-positron pair, producing a bound state.**

## Notes


1 D. Hestenes, *Space-time algebra*, Gordon and Breach, NY (1966).

2 V. Bargmann, *Ann. Math.* **5**, 1 (1954).

3 J.M. Levy-Leblond, *Galilei Group and Galilean Invariance*, in E.M. Loebl, *Group Theory and its applications*, Acad. Press, NY (1971), vol. 2, p. 221.

4 M. Rivas, *J. Phys.* **A 18**, 1971 (1985).

5 A.O. Barut and A.J. Bracken, *Phys. Rev. D* **23**, 2454 (1981).

6 J.D. Jackson, *Classical Electrodynamics*, John Wiley & Sons, NY (1998), 3rd. ed. p.186.

7 P.A.M. Dirac, *The Principles of Quantum mechanics*, Oxford Univ. Press, 4th ed. (1967).

8 see ref. 2 and 3.

Chapter 3

# RELATIVISTIC ELEMENTARY PARTICLES

In this chapter, by analyzing different homogeneous spaces of the Poincaré group $\mathcal{P}$, we study some models of elementary relativistic particles that our formalism predicts. We shall consider first a particular Poincaré group parametrization that will be useful for the description of the different homogeneous spaces to define the particles. As a first example the relativistic point particle is analyzed. Later different kinds of particles arise and are clasified by the velocity of the charge either below, equal to or above the speed of light $c$. Except for the photon, the center of mass of the particles is moving with velocity below $c$, so that the possible tachyonic motion is related only to the motion of the point charge. The class of particles whose charge is moving at the speed of light is particularly important, giving rise to the Lagrangian description of the electron, although the electron structure will be unveiled after quantization of this model. These particles and tachyons have no non-relativistic analog so that the Poincaré group produces a larger catalogue of spinning objects.

The spin description we obtain here and the dipole structure of the elementary particles are equivalent to the nonrelativistic case, in terms of the zitterbewegung and rotation of the particle, although the explicit relativistic expressions in terms of kinematical variables are more complicated.

Instead of using a covariant notation for the different observables as tensor magnitudes on Minkowski's space-time, we prefer to use three-vector notation, as in the nonrelativistic case, to show the similarities between both approaches. Nevertheless, usually, at the end of each section, we produce the corresponding description in terms of covariant quantities.

# 1. POINCARÉ GROUP

The Poincaré group is the group of transformations of Minkowski's space-time that leave invariant the separation between any two close space-time events $ds^2 = \eta_{\mu\nu}dx^\mu dx^\nu$. We shall consider the contravariant components $x^\mu \equiv (ct, \boldsymbol{r})$, and $x' = gx$ is expressed as $x'^\mu = \Lambda^\mu{}_\nu x^\nu + a^\mu$, in terms of a constant matrix $\Lambda$ and a constant translation four-vector $a^\mu \equiv (cb, \boldsymbol{a})$. We take for the covariant components of Minkowski's metric tensor $\eta_{\mu\nu} \equiv \mathrm{diag}(1, -1, -1, -1)$. Then $dx'^\mu = \Lambda^\mu{}_\nu dx^\nu$ and $ds^2 = \eta_{\mu\nu}dx'^\mu dx'^\nu = \eta_{\sigma\rho}dx^\sigma dx^\rho$ implies for the matrix $\Lambda$

$$\eta_{\mu\nu}\Lambda^\mu{}_\sigma\Lambda^\nu{}_\rho = \eta_{\sigma\rho}. \tag{3.1}$$

Relations (3.1) represent ten conditions among the 16 components of the matrix $\Lambda$, so that each matrix depends on six essential parameters, which can be chosen in many ways. Throughout this book we shall take three of them as the components of the relative velocity $\boldsymbol{v}$ between inertial observers and the remaining three as the orientation $\boldsymbol{\alpha}$ of their Cartesian frames, expressed in a suitable parametrization of the rotation group.

Therefore, every element of the Poincaré group $\mathcal{P}$ will be represented, as in the previous case of the Galilei group, by the ten parameters $g \equiv (b, \boldsymbol{a}, \boldsymbol{v}, \boldsymbol{\alpha})$ and the group action on a space-time point $x \equiv (t, \boldsymbol{r})$ will be interpreted in the same way, *i.e.*, $x' = gx$:

$$x' = \exp(bH)\exp(\boldsymbol{a}\cdot\boldsymbol{P})\exp(\boldsymbol{\beta}\cdot\boldsymbol{K})\exp(\boldsymbol{\alpha}\cdot\boldsymbol{J})x, \tag{3.2}$$

as the action of a rotation followed by a boost or pure Lorentz transformation and finally a space and time translation. It is explicitly given on the space-time variables by

$$t' = \gamma t + \gamma(\boldsymbol{v}\cdot R(\boldsymbol{\mu})\boldsymbol{r})/c^2 + b, \tag{3.3}$$

$$\boldsymbol{r}' = R(\boldsymbol{\mu})\boldsymbol{r} + \gamma\boldsymbol{v}t + \gamma^2(\boldsymbol{v}\cdot R(\boldsymbol{\mu})\boldsymbol{r})\boldsymbol{v}/(1+\gamma)c^2 + \boldsymbol{a}. \tag{3.4}$$

Parameter $\boldsymbol{\beta}$ in (3.2) is the normal parameter for the pure Lorentz transformations, that in terms of the relative velocity among observers $\boldsymbol{v}$ is expressed as $\boldsymbol{\beta}/\beta \tanh\beta = \boldsymbol{v}/c$ as we shall see below. The dimensions and domains of the parameters $b$, $\boldsymbol{a}$ and $\boldsymbol{\mu}$ are the same as those of the Galilei group, and the parameter $\boldsymbol{v} \in \mathbb{R}^3$, with the upper bound $v < c$, has also dimensions of velocity. The physical meaning of these ten parameters, that relate any two inertial observers, is the same as in the Galilei case. The parameter $\boldsymbol{v}$ is the velocity of observer $O$, as measured by $O'$, and $R(\boldsymbol{\mu})$ represents the orientation of Cartesian frame $O$ relative to $O'$, once $O'$ is boosted with velocity $\boldsymbol{v}$. The factor $\gamma(v) = (1 - v^2/c^2)^{-1/2}$.

The composition law of the group is obtained from $x'' = \Lambda' x' + a' = \Lambda'(\Lambda x + a) + a'$ that by identification with $x'' = \Lambda'' x + a''$ reduces to $\Lambda'' = \Lambda'\Lambda$ and $a'' = \Lambda' a + a'$, *i.e.*, the composition law of the Lorentz transformations, that we will find in the next Section 1.1, and a Poincaré transformation $(\Lambda', a')$ of the four-vector $a^\mu$. In this parametrization $g'' = g'g$, is: [1]

$$b'' = \gamma' b + \gamma'(\boldsymbol{v}' \cdot R(\boldsymbol{\mu}')\boldsymbol{a})/c^2 + b', \qquad (3.5)$$

$$\boldsymbol{a}'' = R(\boldsymbol{\mu}')\boldsymbol{a} + \gamma'\boldsymbol{v}' b + \frac{\gamma'^2}{(1+\gamma')c^2}(\boldsymbol{v}' \cdot R(\boldsymbol{\mu}')\boldsymbol{a})\boldsymbol{v}' + \boldsymbol{a}', \qquad (3.6)$$

$$\boldsymbol{v}'' = \frac{R(\boldsymbol{\mu}')\boldsymbol{v} + \gamma'\boldsymbol{v}' + \frac{\gamma'^2}{(1+\gamma')c^2}(\boldsymbol{v}' \cdot R(\boldsymbol{\mu}')\boldsymbol{v})\boldsymbol{v}'}{\gamma'(1 + \boldsymbol{v}' \cdot R(\boldsymbol{\mu}')\,\boldsymbol{v}/c^2)}, \qquad (3.7)$$

$$\boldsymbol{\mu}'' = \frac{\boldsymbol{\mu}' + \boldsymbol{\mu} + \boldsymbol{\mu}' \times \boldsymbol{\mu} + \boldsymbol{F}(\boldsymbol{v}', \boldsymbol{\mu}', \boldsymbol{v}, \boldsymbol{\mu})}{1 - \boldsymbol{\mu}' \cdot \boldsymbol{\mu} + G(\boldsymbol{v}', \boldsymbol{\mu}', \boldsymbol{v}, \boldsymbol{\mu})}, \qquad (3.8)$$

where $\boldsymbol{F}(\boldsymbol{v}', \boldsymbol{\mu}', \boldsymbol{v}, \boldsymbol{\mu})$ and $G(\boldsymbol{v}', \boldsymbol{\mu}', \boldsymbol{v}, \boldsymbol{\mu})$ are the real analytic functions:

$$\begin{aligned} \boldsymbol{F}(\boldsymbol{v}', \boldsymbol{\mu}', \boldsymbol{v}, \boldsymbol{\mu}) &= \frac{\gamma\gamma'}{(1+\gamma)(1+\gamma')c^2}\left[\boldsymbol{v} \times \boldsymbol{v}' + \boldsymbol{v}(\boldsymbol{v}' \cdot \boldsymbol{\mu}') + \boldsymbol{v}'(\boldsymbol{v} \cdot \boldsymbol{\mu})\right. \\ &+ \boldsymbol{v} \times (\boldsymbol{v}' \times \boldsymbol{\mu}') + (\boldsymbol{v} \times \boldsymbol{\mu}) \times \boldsymbol{v}' + (\boldsymbol{v} \cdot \boldsymbol{\mu})(\boldsymbol{v}' \times \boldsymbol{\mu}') \\ &+ \left.(\boldsymbol{v} \times \boldsymbol{\mu})(\boldsymbol{v}' \cdot \boldsymbol{\mu}') + (\boldsymbol{v} \times \boldsymbol{\mu}) \times (\boldsymbol{v}' \times \boldsymbol{\mu}')\right], \end{aligned} \qquad (3.9)$$

$$\begin{aligned} G(\boldsymbol{v}', \boldsymbol{\mu}', \boldsymbol{v}, \boldsymbol{\mu}) &= \frac{\gamma\gamma'}{(1+\gamma)(1+\gamma')c^2}\left[\boldsymbol{v} \cdot \boldsymbol{v}' + \boldsymbol{v} \cdot (\boldsymbol{v}' \times \boldsymbol{\mu}') + \boldsymbol{v}' \cdot (\boldsymbol{v} \times \boldsymbol{\mu})\right. \\ &- \left.(\boldsymbol{v} \cdot \boldsymbol{\mu})(\boldsymbol{v}' \cdot \boldsymbol{\mu}') + (\boldsymbol{v} \times \boldsymbol{\mu}) \cdot (\boldsymbol{v}' \times \boldsymbol{\mu}')\right]. \end{aligned} \qquad (3.10)$$

The unit element of the group is $(0, \boldsymbol{0}, \boldsymbol{0}, \boldsymbol{0})$ and the inverse of any arbitrary element $(b, \boldsymbol{a}, \boldsymbol{v}, \boldsymbol{\mu})$ is

$$(-\gamma b + \gamma \boldsymbol{v} \cdot \boldsymbol{a}/c^2, -R(-\boldsymbol{\mu})(\boldsymbol{a} - \gamma \boldsymbol{v} b + \frac{\gamma^2}{(1+\gamma)c^2}(\boldsymbol{v} \cdot \boldsymbol{a})\boldsymbol{v}), -R(-\boldsymbol{\mu})\boldsymbol{v}, -\boldsymbol{\mu}).$$

The group generators in the realization (3.3, 3.4), and in terms of the normal parameters $(b, \boldsymbol{a}, \boldsymbol{\beta}, \boldsymbol{\alpha})$, are

$$H = \partial/\partial t,\ P_i = \partial/\partial x^i,\ K_i = ct\partial/\partial x^i + (x_i/c)\partial/\partial t,\ J_k = \varepsilon_{kl}{}^i x^l \partial/\partial x^i.$$

Thus, $\boldsymbol{K}$ and $\boldsymbol{J}$ are dimensionless and the commutation relations become

$$[\boldsymbol{J}, \boldsymbol{J}] = -\boldsymbol{J},\ [\boldsymbol{J}, \boldsymbol{P}] = -\boldsymbol{P},\ [\boldsymbol{J}, \boldsymbol{K}] = -\boldsymbol{K},\ [\boldsymbol{J}, H] = 0,\ [H, \boldsymbol{P}] = 0, \qquad (3.11)$$

$$[H, \boldsymbol{K}] = c\boldsymbol{P},\ [\boldsymbol{P}, \boldsymbol{P}] = 0,\ [\boldsymbol{K}, \boldsymbol{K}] = \boldsymbol{J},\ [\boldsymbol{K}, \boldsymbol{P}] = -H/c. \tag{3.12}$$

If, as usual, we call $x^0 = ct$, $P_0 = H/c$, and $K_i = J_{0i} = -J_{i0}$ and $J_k = \frac{1}{2}\epsilon_{klr}J_{lr}$, $x_\mu = \eta_{\mu\nu}x^\nu$, $\mu = 0,1,2,3$ and $\partial_\nu \equiv \partial/\partial x^\nu$, then,

$$P_\mu = \partial_\mu, \quad J_{\mu\nu} = -J_{\nu\mu} = x_\mu\partial_\nu - x_\nu\partial_\mu.$$

In covariant notation the commutation relations appear:

$$\begin{aligned} [P_\mu, P_\nu] &= 0, \\ [J_{\mu\nu}, P_\sigma] &= -\eta_{\mu\sigma}P_\nu + \eta_{\nu\sigma}P_\mu, \\ [J_{\mu\nu}, J_{\rho\sigma}] &= -\eta_{\mu\rho}J_{\nu\sigma} - \eta_{\nu\sigma}J_{\mu\rho} + \eta_{\nu\rho}J_{\mu\sigma} + \eta_{\mu\sigma}J_{\nu\rho}. \end{aligned}$$

The Poincaré group has two functionally independent Casimir invariants. One is interpreted as the squared mass of the system,

$$P^\mu P_\mu = (H/c)^2 - \boldsymbol{P}^2 = m^2c^2, \tag{3.13}$$

and the other is the square of the Pauli-Lubanski four-vector $W^\mu$. The Pauli-Lubanski four-vector is defined as

$$W^\mu = \frac{1}{2}\epsilon^{\mu\nu\sigma\lambda}P_\nu J_{\sigma\lambda} \equiv (\boldsymbol{P}\cdot\boldsymbol{J}, H\boldsymbol{J}/c - \boldsymbol{K}\times\boldsymbol{P}) \equiv (\boldsymbol{P}\cdot\boldsymbol{S}, H\boldsymbol{S}/c), \tag{3.14}$$

which is by construction orthogonal to $P_\mu$, i.e., $W^\mu P_\mu = 0$.

It is related to the spin of the system $\boldsymbol{S}$, defined through the relation

$$H\boldsymbol{S}/c = H\boldsymbol{J}/c - \boldsymbol{K}\times\boldsymbol{P}, \tag{3.15}$$

so that its time component $W^0 = \boldsymbol{P}\cdot\boldsymbol{S} = \boldsymbol{P}\cdot\boldsymbol{J}$ is the helicity of the particle, and the spatial part is the vector (3.15).

The other Casimir operator is thus

$$W^\mu W_\mu = (\boldsymbol{P}\cdot\boldsymbol{J})^2 - (H\boldsymbol{J}/c - \boldsymbol{K}\times\boldsymbol{P})^2 = -m^2c^2S^2, \tag{3.16}$$

where it depends on $S^2$, the absolute value squared of the spin. We see in the relativistic case that the two parameters $m$ and $S$ characterize the two Casimir invariants and therefore they are the intrinsic properties of the elementary particle the formalism provides. In the quantum case, since the representation must be irreducible $S^2 = s(s+1)\hbar^2$, for any $s = 0, 1/2, 1, \ldots$, depending on the value of the quantized spin of the particle, but in the classical case $S^2$ can take any continuous value.

These $W^\mu$ operators satisfy the commutation relations:

$$[W^\mu, W^\nu] = \epsilon^{\mu\nu\sigma\rho}W_\sigma P_\rho, \tag{3.17}$$

where we take $\epsilon^{0123} = +1$, and

$$[P^\mu, W^\nu] = 0, \qquad [M_{\mu\nu}, W_\sigma] = -\eta_{\mu\sigma}W_\nu + \eta_{\nu\sigma}W_\mu. \tag{3.18}$$

The Poincaré group has no non-trivial exponents, so that gauge functions when restricted to homogeneous spaces of $\mathcal{P}$ vanish.

## 1.1 LORENTZ GROUP

The Lorentz group $\mathcal{L}$ is the subgroup of transformations of the form $(0, 0, \boldsymbol{v}, \boldsymbol{\mu})$, and every Lorentz transformation $\Lambda(\boldsymbol{v}, \boldsymbol{\mu})$ will be interpreted as $\Lambda(\boldsymbol{v}, \boldsymbol{\mu}) = L(\boldsymbol{v})R(\boldsymbol{\mu})$, as mentioned before where $L(\boldsymbol{v})$ is a boost or pure Lorentz transformation and $R(\boldsymbol{\mu})$ a spatial rotation. Expressions (3.7, 3.8) come from $\Lambda(\boldsymbol{v}'', \boldsymbol{\mu}'') = \Lambda(\boldsymbol{v}', \boldsymbol{\mu}')\Lambda(\boldsymbol{v}, \boldsymbol{\mu})$. Expression (3.7) is the relativistic composition of velocities since

$$\begin{aligned} L(\boldsymbol{v}'')R(\boldsymbol{\mu}'') &= L(\boldsymbol{v}')R(\boldsymbol{\mu}')L(\boldsymbol{v})R(\boldsymbol{\mu}) \\ &= L(\boldsymbol{v}')R(\boldsymbol{\mu}')L(\boldsymbol{v})R(-\boldsymbol{\mu}')R(\boldsymbol{\mu}')R(\boldsymbol{\mu}), \end{aligned}$$

but the conjugate of the boost $R(\boldsymbol{\mu}')L(\boldsymbol{v})R(-\boldsymbol{\mu}') = L(R(\boldsymbol{\mu}')\boldsymbol{v})$ is another boost and thus

$$L(\boldsymbol{v}'')R(\boldsymbol{\mu}'') = L(\boldsymbol{v}')L(R(\boldsymbol{\mu}')\boldsymbol{v})R(\boldsymbol{\mu}')R(\boldsymbol{\mu}).$$

The product $L(\boldsymbol{v}')L(R(\boldsymbol{\mu}')\boldsymbol{v}) = L(\boldsymbol{v}'')R(\boldsymbol{w})$ where $\boldsymbol{v}''$ is the relativistic composition of the velocities $\boldsymbol{v}'$ and $R(\boldsymbol{\mu}')\boldsymbol{v}$, and $R(\boldsymbol{w})$ is the Thomas-Wigner rotation associated to the boosts $L(\boldsymbol{v}')$ and $L(R(\boldsymbol{\mu}')\boldsymbol{v})$.

Therefore, expression (3.7) is equivalent to

$$L(\boldsymbol{v}'') = L(\boldsymbol{v}')L(R(\boldsymbol{\mu}')\boldsymbol{v})R(-\boldsymbol{w}), \tag{3.19}$$

and (3.8) is

$$R(\boldsymbol{\mu}'') = R(\boldsymbol{w})R(\boldsymbol{\mu}')R(\boldsymbol{\mu}) \equiv R(\boldsymbol{\phi})R(\boldsymbol{\mu}). \tag{3.20}$$

The Thomas-Wigner rotation matrix $R(\boldsymbol{w})$ is:

$$R(\boldsymbol{w})_{ij} = \delta_{ij} + \frac{1}{1+\gamma''}\left(\frac{\gamma'^2}{c^2}\left(\frac{1-\gamma}{1+\gamma'}\right)v'_i v'_j + \frac{\gamma^2}{c^2}\left(\frac{1-\gamma'}{1+\gamma}\right)R'_{ik}v_k R'_{jl}v_l\right.$$

$$\left. + \frac{\gamma'\gamma}{c^2}(v'_i R'_{jk}v_k - v'_j R'_{ik}v_k) + \frac{2\gamma'^2\gamma^2(v'_k R'_{kl}v_l)}{(1+\gamma')(1+\gamma)c^2}v'_i R'_{jk}v_k\right),$$

and the factor

$$\gamma'' = \gamma'\gamma\left(1 + \frac{\boldsymbol{v}' \cdot R(\boldsymbol{\mu})\boldsymbol{v}}{c^2}\right).$$

Matrix $R(\boldsymbol{w})$ is written in terms of the vector parameter $\boldsymbol{w}$, which is a function of $\boldsymbol{v}'$, $\boldsymbol{\mu}'$ and $\boldsymbol{v}$, given by

$$\boldsymbol{w} = \frac{\boldsymbol{F}(\boldsymbol{v}', \boldsymbol{0}, R(\boldsymbol{\mu}')\boldsymbol{v}, \boldsymbol{0})}{1 + G(\boldsymbol{v}', \boldsymbol{0}, R(\boldsymbol{\mu}')\boldsymbol{v}, \boldsymbol{0})}, \tag{3.21}$$

and the parameter $\boldsymbol{\phi}$, such that $R(\boldsymbol{\phi}) = R(\boldsymbol{w})R(\boldsymbol{\mu}')$ is

$$\boldsymbol{\phi} = \frac{\boldsymbol{\mu}' + \boldsymbol{F}(\boldsymbol{v}', \boldsymbol{\mu}', \boldsymbol{v}, \boldsymbol{0})}{1 + G(\boldsymbol{v}', \boldsymbol{\mu}', \boldsymbol{v}, \boldsymbol{0})}. \tag{3.22}$$

If any one of the two velocities $\boldsymbol{v}$ or $\boldsymbol{v}'$ vanishes, $R(\boldsymbol{w})_{ij} = \delta_{ij}$.

The composition law is obtained by the homomorphism between the Lorentz group $\mathcal{L}$ and the group $SL(2,\mathbb{C})$ of $2 \times 2$ complex matrices of determinant +1. The Lie algebra of this group has as generators $\boldsymbol{J} = -i\boldsymbol{\sigma}/2$ and $\boldsymbol{K} = \boldsymbol{\sigma}/2$, where $\sigma_i$ are Pauli spin matrices. A rotation of angle $\alpha$ around a rotation axis given by the unit vector $\boldsymbol{n}$ is given by the $2 \times 2$ unitary matrix $\exp(\boldsymbol{\alpha} \cdot \boldsymbol{J})$, as in Sec.1. of Chapter 2,

$$R(\boldsymbol{\alpha}) = \cos(\alpha/2)\sigma_0 - i\boldsymbol{n} \cdot \boldsymbol{\sigma} \sin(\alpha/2). \tag{3.23}$$

In terms of the vector $\boldsymbol{\mu} = \tan(\alpha/2)\boldsymbol{n}$,

$$R(\boldsymbol{\mu}) = \frac{1}{\sqrt{1+\mu^2}}(\sigma_0 - i\boldsymbol{\mu} \cdot \boldsymbol{\sigma}), \tag{3.24}$$

where $\sigma_0$ is the $2 \times 2$ unit matrix. A pure Lorentz transformation of normal parameters $\beta_i$ is represented by the hermitian matrix $\exp(\boldsymbol{\beta} \cdot \boldsymbol{K})$. This matrix is:

$$L(\boldsymbol{\beta}) = \cosh(\beta/2)\sigma_0 + \frac{\boldsymbol{\sigma} \cdot \boldsymbol{\beta}}{\beta} \sinh(\beta/2). \tag{3.25}$$

In terms of the relative velocity parameters, taking into account the functions $\cosh\beta = \gamma(v)$, $\sinh\beta = \gamma v/c$ and the trigonometric relations $\cosh(\beta/2) = \sqrt{(\cosh\beta + 1)/2}$ and $\tanh(\beta/2) = \sinh\beta/(1 + \cosh\beta)$, the matrix can be written as

$$L(\boldsymbol{v}) = \sqrt{\frac{1+\gamma}{2}} \left( \sigma_0 + \frac{\gamma}{1+\gamma} \frac{\boldsymbol{\sigma} \cdot \boldsymbol{v}}{c} \right). \tag{3.26}$$

Then, every element of $SL(2,\mathbb{C})$ is parametrized by the six real numbers $(\boldsymbol{v}, \boldsymbol{\mu})$, and interpreted as

$$A(\boldsymbol{v}, \boldsymbol{\mu}) = L(\boldsymbol{v})R(\boldsymbol{\mu}). \tag{3.27}$$

We thus see that every $2 \times 2$ matrix $A \in SL(2,\mathbb{C})$ can be written in terms of a complex four-vector $a^\mu$ and the four Pauli matrices $\sigma_\mu$. As $A = a^\mu \sigma_\mu$, and $\det A = 1$ leads to $a^\mu a_\mu = 1$ or $(a^0)^2 - \boldsymbol{a}^2 = 1$. The general form of (3.27) is

$$A(\boldsymbol{v}, \boldsymbol{\mu}) = \sqrt{\frac{1+\gamma}{2(1+\mu^2)}} \left[ \sigma_0 \left( 1 - i\frac{\boldsymbol{\mu} \cdot \boldsymbol{u}}{1+\gamma} \right) + \boldsymbol{\sigma} \cdot \left( \frac{\boldsymbol{u} + \boldsymbol{u} \times \boldsymbol{\mu}}{1+\gamma} - i\boldsymbol{\mu} \right) \right], \tag{3.28}$$

here the dimensionless vector $\boldsymbol{u} = \gamma(v)\boldsymbol{v}/c$.

Conversely, since $\mathrm{Tr}\,(\sigma_\mu\sigma_\nu) = 2\delta_{\mu\nu}$, we obtain $a^\mu = (1/2)\mathrm{Tr}\,(A\sigma_\mu)$. If we express (3.28) in the form $A(\boldsymbol{v},\boldsymbol{\mu}) = a^\mu\sigma_\mu$ we can determine $\boldsymbol{\mu}$ and $\boldsymbol{v}$, and thus $\boldsymbol{u}$, from the components of the complex four-vector $a^\mu$ as:

$$\boldsymbol{\mu} = -\frac{\mathrm{Im}\,(\boldsymbol{a})}{\mathrm{Re}\,(a^0)}, \tag{3.29}$$

$$\boldsymbol{u} = 2\left[\mathrm{Re}\,(a^0)\mathrm{Re}\,(\boldsymbol{a}) + \mathrm{Im}\,(a^0)\mathrm{Im}\,(\boldsymbol{a}) + \mathrm{Re}\,(\boldsymbol{a}) \times \mathrm{Im}\,(\boldsymbol{a})\right], \tag{3.30}$$

where $\mathrm{Re}\,(a^\mu)$ and $\mathrm{Im}\,(a^\mu)$ are the real and imaginary parts of the corresponding components of the four-vector $a^\mu$. When $\mathrm{Re}\,(a^0) = 0$ expression (3.29) is defined and represents a rotation of value $\pi$ along the axis in the direction of vector $\mathrm{Im}\,(\boldsymbol{a})$.

If we represent every Lorentz transformation in terms of a rotation and a boost, *i.e.*, in the reverse order, $\Lambda(\boldsymbol{v},\boldsymbol{\mu}) = R(\boldsymbol{\mu})L(\boldsymbol{v})$, then the general expression of $A$ is the same as (3.28) with a change of sign in the cross product term $\boldsymbol{u} \times \boldsymbol{\mu}$. Therefore, the decomposition is also unique, the rotation $R(\boldsymbol{\mu})$ is the same as before but the Lorentz boost is given in terms of the variables $a^\mu$ by

$$\boldsymbol{u} = 2\left[\mathrm{Re}\,(a^0)\mathrm{Re}\,(\boldsymbol{a}) + \mathrm{Im}\,(a^0)\mathrm{Im}\,(\boldsymbol{a}) + \mathrm{Im}\,(\boldsymbol{a}) \times \mathrm{Re}\,(\boldsymbol{a})\right].$$

Note the difference in the third term which is reversed when compared with (3.30).

In the four-dimensional representation of the Lorentz group on Minkowski space-time, a boost is expressed as $L(\boldsymbol{\beta}) = \exp(\boldsymbol{\beta}\cdot\boldsymbol{K})$ in terms of the dimensionless normal parameters $\beta_i$ and the $4\times 4$ boost generators $K_i$ given by

$$K_1 = \begin{pmatrix} 0 & 1 & 0 & 0 \\ 1 & 0 & 0 & 0 \\ 0 & 0 & 0 & 0 \\ 0 & 0 & 0 & 0 \end{pmatrix}, \quad K_2 = \begin{pmatrix} 0 & 0 & 1 & 0 \\ 0 & 0 & 0 & 0 \\ 1 & 0 & 0 & 0 \\ 0 & 0 & 0 & 0 \end{pmatrix}, \quad K_3 = \begin{pmatrix} 0 & 0 & 0 & 1 \\ 0 & 0 & 0 & 0 \\ 0 & 0 & 0 & 0 \\ 1 & 0 & 0 & 0 \end{pmatrix}.$$

If we call $B = \boldsymbol{\beta}\cdot\boldsymbol{K} \equiv \sum_i \beta_i K_i$, we have

$$B = \begin{pmatrix} 0 & \beta_1 & \beta_2 & \beta_3 \\ \beta_1 & 0 & 0 & 0 \\ \beta_2 & 0 & 0 & 0 \\ \beta_3 & 0 & 0 & 0 \end{pmatrix}, \quad B^2 = \begin{pmatrix} \beta^2 & 0 & 0 & 0 \\ 0 & \beta_1\beta_1 & \beta_1\beta_2 & \beta_1\beta_3 \\ 0 & \beta_2\beta_1 & \beta_2\beta_2 & \beta_2\beta_3 \\ 0 & \beta_3\beta_1 & \beta_3\beta_2 & \beta_3\beta_3 \end{pmatrix},$$

with $\beta^2 = \beta_1^2 + \beta_2^2 + \beta_3^2$ and $B^3 = \beta^2 B$, and so on for the remaining powers of $B$, so that the final expression for $L(\boldsymbol{\beta}) = \exp(\boldsymbol{\beta} \cdot \boldsymbol{K})$ is

$$\begin{pmatrix} C & (\beta_1/\beta)S & (\beta_2/\beta)S & (\beta_3/\beta)S \\ (\beta_1/\beta)S & 1+\frac{\beta_1\beta_1}{\beta^2}(C-1) & \frac{\beta_1\beta_2}{\beta^2}(C-1) & \frac{\beta_1\beta_3}{\beta^2}(C-1) \\ (\beta_2/\beta)S & \frac{\beta_2\beta_1}{\beta^2}(C-1) & 1+\frac{\beta_2\beta_2}{\beta^2}(C-1) & \frac{\beta_2\beta_3}{\beta^2}(C-1) \\ (\beta_3/\beta)S & \frac{\beta_3\beta_1}{\beta^2}(C-1) & \frac{\beta_3\beta_2}{\beta^2}(C-1) & 1+\frac{\beta_3\beta_3}{\beta^2}(C-1) \end{pmatrix}$$

where $S = \sinh\beta$ and $C = \cosh\beta$. What is the physical interpretation of $\beta_i$? Let us assume that observers $O$ and $O'$ relate their space-time measurements $x$ and $x'$ by $x'^{\mu} = L(\boldsymbol{\beta})^{\mu}{}_{\nu}x^{\nu}$. Observer $O$ sends at time $t$ and at a later time $t + dt$ two light signals from a source placed at the origin of its Cartesian frame. These two signals when measured by $O'$ take place at points $\boldsymbol{r}'$ and $\boldsymbol{r}' + d\boldsymbol{r}'$ and at instants $t'$ and $t' + dt'$, respectively. Then they are related by

$$cdt' = L^0{}_0 cdt, \qquad dx'^i = L^i{}_0 cdt$$

because $dx^i = 0$. The quotient $dx'^i/dt'$ is just the velocity of the light source $v^i$, *i.e.*, of the origin of the $O$ frame as measured by observer $O'$, and then this velocity $v^i = cL^i{}_0/L^0{}_0 = c(\beta_i/\beta)S/C$, such that the relation between the normal parameters and the relative velocity between observers is

$$\frac{\boldsymbol{v}}{c} = \frac{\boldsymbol{\beta}}{\beta}\tanh\beta$$

and therefore $\tanh\beta = v/c$. Function $\cosh\beta \equiv \gamma(v) = (1 - v^2/c^2)^{-1/2}$ and when the transformation is expressed in terms of the relative velocity it takes the form of the symmetric matrix:

$$L(\boldsymbol{v}) = \begin{pmatrix} \gamma & \gamma v_x/c & \gamma v_y/c & \gamma v_z/c \\ \gamma v_x/c & 1+\frac{v_x^2}{c^2}\frac{\gamma^2}{\gamma+1} & \frac{v_x v_y}{c^2}\frac{\gamma^2}{\gamma+1} & \frac{v_x v_z}{c^2}\frac{\gamma^2}{\gamma+1} \\ \gamma v_y/c & \frac{v_y v_x}{c^2}\frac{\gamma^2}{\gamma+1} & 1+\frac{v_y^2}{c^2}\frac{\gamma^2}{\gamma+1} & \frac{v_y v_z}{c^2}\frac{\gamma^2}{\gamma+1} \\ \gamma v_z/c & \frac{v_z v_x}{c^2}\frac{\gamma^2}{\gamma+1} & \frac{v_z v_y}{c^2}\frac{\gamma^2}{\gamma+1} & 1+\frac{v_z^2}{c^2}\frac{\gamma^2}{\gamma+1} \end{pmatrix}. \tag{3.31}$$

The inverse transformation $L^{-1}(\boldsymbol{v}) = L(-\boldsymbol{v})$. The orthogonal $4 \times 4$ rotation matrix takes the block form

$$R(\boldsymbol{\mu}) = \begin{pmatrix} 1 & 0 \\ 0 & \widetilde{R}(\boldsymbol{\mu}) \end{pmatrix}, \tag{3.32}$$

where $\widetilde{R}(\boldsymbol{\mu})$ is the $3 \times 3$ orthogonal matrix (2.7). When a Lorentz transformation is expressed in the form $\Lambda(\boldsymbol{v}, \boldsymbol{\mu}) = L(\boldsymbol{v})R(\boldsymbol{\mu})$, then by construction the first column of $\Lambda(\boldsymbol{v}, \boldsymbol{\mu})$ is just the first column of (3.31) where the velocity parameters $\boldsymbol{v}$ are defined. Therefore, given the general Lorentz transformation $\Lambda(\boldsymbol{v}, \boldsymbol{\mu})$, from its first column we determine the parameters $\boldsymbol{v}$ and thus the complete $L(\boldsymbol{v})$ can be worked out. The rotation involved can be easily calculated as $L(-\boldsymbol{v})\Lambda(\boldsymbol{v}, \boldsymbol{\mu}) = R(\boldsymbol{\mu})$. If expressed in the reverse order $\Lambda(\boldsymbol{v}, \boldsymbol{\mu}) = R(\boldsymbol{\mu})L(\boldsymbol{v})$, then it is the first row of $\Lambda$ that coincides with the first row of (3.31). It turns out that, given any general Lorentz transformation $\Lambda(\boldsymbol{v}, \boldsymbol{\mu})$, then $\Lambda(\boldsymbol{v}, \boldsymbol{\mu}) = L(\boldsymbol{v})R(\boldsymbol{\mu}) = R(\boldsymbol{\mu})L(\boldsymbol{v}')$ with the same rotation in both sides as derived in (3.29) and $L(\boldsymbol{v}') = R(-\boldsymbol{\mu})L(\boldsymbol{v})R(\boldsymbol{\mu}) = L(R(-\boldsymbol{\mu})\boldsymbol{v})$, *i.e*, the velocity $\boldsymbol{v}' = R(-\boldsymbol{\mu})\boldsymbol{v}$. In any case, the decomposition of a general Lorentz transformation as a product of a rotation and a boost is a unique one, in terms of the same rotation $R(\boldsymbol{\mu})$ and a boost to be determined, depending on the order in which we take these two operations.

Matrix $\Lambda$ can be considered as a tetrad (*i.e.*, a set of four orthonormal four-vectors, one time-like and the other three space-like) attached by observer $O'$ to the origin of observer $O$. In fact, if the matrix is considered in the form $\Lambda(\boldsymbol{v}, \boldsymbol{\mu}) = L(\boldsymbol{v})R(\boldsymbol{\mu})$, then the first column of $\Lambda$ is the four-velocity of the origin of the $O$ Cartesian frame and the other three columns are just the three unit vectors of the $O$ reference frame, rotated with rotation $R(\boldsymbol{\mu})$ and afterwards boosted with $L(\boldsymbol{v})$. We shall consider this tetrad structure when analyzing a general relativistic spinning particle.

As an application of the composition law, let us consider the well-known Thomas effect [2] of the spin precession of an accelerated electron. Let $O_L$ be a laboratory inertial observer and $O$ an instantaneous inertial observer at rest with the electron at time $t$. Since the electron is moving in the laboratory with velocity $\boldsymbol{v}_L$, the relationship between these observers is $x_L = L(\boldsymbol{v}_L)x$, where the boost matrix $L(\boldsymbol{v}_L)$ is as given in (3.31). If the acceleration of the electron is $\boldsymbol{a}$ in the $O$ frame, then let $O'$ be another inertial observer at rest with the electron at time $t + dt$ which is moving with the velocity $\boldsymbol{a}dt$ with respect to $O$ and such that its Cartesian frame is just that of $O$, boosted with this velocity. Therefore their space-time measurements are related by $x = L(\boldsymbol{a}dt)x'$ and thus $x_L = L(\boldsymbol{v}_L)L(\boldsymbol{a}dt)x'$. The composition of these two boosts is just $L(\boldsymbol{v}_L)L(\boldsymbol{a}dt) = L(\boldsymbol{v}_L + \boldsymbol{a}_L dt_L)R(d\boldsymbol{\alpha}_L)$, where $\boldsymbol{v}_L + \boldsymbol{a}_L dt_L$ is the new velocity of the electron in the laboratory frame and $\boldsymbol{a}_L$ the acceleration also in this frame. $R(d\boldsymbol{\alpha}_L)$ is the infinitesimal Thomas-Wigner rotation associated to the composition of the boosts. By application of equation (3.8), since the rotation involved is infinitesimal, $\tan(d\alpha_L/2) \simeq d\alpha_L/2$, we get

$$d\boldsymbol{\mu}_L = \frac{d\boldsymbol{\alpha}_L}{2} = \frac{\boldsymbol{F}(\boldsymbol{v}_L, 0, \boldsymbol{a}dt, 0)}{1 + G(\boldsymbol{v}_L, 0, \boldsymbol{a}dt, 0)} = \frac{\gamma}{2(1+\gamma)c^2}\, \boldsymbol{a}dt \times \boldsymbol{v}_L,$$

where we have written $\gamma(\boldsymbol{v}_L)\equiv\gamma$, $\gamma(\boldsymbol{a}dt)\simeq 1$ and the infinitesimal function $G$ in the denominator is neglected as compared with 1. Now $dt_L=\gamma(\boldsymbol{v}_L)dt$ and the relation between the accelerations in the $O$ and $O_L$ frames, given later in (3.58), is

$$\boldsymbol{a}_L=\frac{1}{\gamma^2}\left(\boldsymbol{a}-\frac{\gamma}{1+\gamma}\frac{\boldsymbol{v}_L\cdot\boldsymbol{a}}{c^2}\boldsymbol{v}_L\right),\quad \text{and thus}\quad \boldsymbol{a}\times\boldsymbol{v}_L=\gamma^2\boldsymbol{a}_L\times\boldsymbol{v}_L.$$

We finally obtain that

$$\boldsymbol{\omega}_T=\frac{d\boldsymbol{\alpha}_L}{dt_L}=\frac{\gamma^2}{(1+\gamma)c^2}\boldsymbol{a}_L\times\boldsymbol{v}_L,$$

which is the angular velocity of rotation of the instantaneous frame located at rest with the particle, measured in the laboratory frame.

If the electron is under the action of external forces but no torques, $O$ and $O'$, at rest with the particle, agree that the spin remains constant in time in their frames and therefore for the laboratory observer the spin, in absence of torques, is precessing with Thomas angular velocity $\boldsymbol{\omega}_T$.

## 2. RELATIVISTIC POINT PARTICLE

As an example we shall analyze first the relativistic point particle, *i.e.*, that system for which the kinematical space is the quotient structure $X=\mathcal{P}/\mathcal{L}$, where $\mathcal{P}$ is the Poincaré group and the subgroup $\mathcal{L}$ is the Lorentz group. Then every point $x\in X$ is characterized by the variables $x\equiv(t(\tau),\boldsymbol{r}(\tau))$, with domains $t\in\mathbb{R}$, $\boldsymbol{r}\in\mathbb{R}^3$ as the corresponding group parameters, in such a way that under the action of a group element $g\equiv(b,\boldsymbol{a},\boldsymbol{v},\boldsymbol{\mu})$ of $\mathcal{P}$ they transform as:

$$t'(\tau) = \gamma t(\tau)+\gamma(\boldsymbol{v}\cdot R(\boldsymbol{\mu})\boldsymbol{r}(\tau))/c^2+b, \tag{3.33}$$

$$\boldsymbol{r}'(\tau) = R(\boldsymbol{\mu})\boldsymbol{r}(\tau)+\gamma\boldsymbol{v}t(\tau)+\frac{\gamma^2}{(1+\gamma)c^2}(\boldsymbol{v}\cdot R(\boldsymbol{\mu})\boldsymbol{r}(\tau))\boldsymbol{v}+\boldsymbol{a}, \tag{3.34}$$

and are interpreted as the time and position of the system. If, as usual, we assume that the evolution parameter $\tau$ is invariant under the group, taking the $\tau$-derivatives of (3.33) and (3.34) we get

$$\dot{t}'(\tau) = \gamma\dot{t}(\tau)+\gamma(\boldsymbol{v}\cdot R(\boldsymbol{\mu})\dot{\boldsymbol{r}}(\tau))/c^2, \tag{3.35}$$

$$\dot{\boldsymbol{r}}'(\tau) = R(\boldsymbol{\mu})\dot{\boldsymbol{r}}(\tau)+\gamma\boldsymbol{v}\dot{t}(\tau)+\frac{\gamma^2}{(1+\gamma)c^2}(\boldsymbol{v}\cdot R(\boldsymbol{\mu})\dot{\boldsymbol{r}}(\tau))\boldsymbol{v}. \tag{3.36}$$

The homogeneity condition of the Lagrangian, in terms of the derivatives of the kinematical variables, reduces to three the number of degrees of freedom of the system. This leads to the general expression

$$L=T\dot{t}+\boldsymbol{R}\cdot\dot{\boldsymbol{r}}, \tag{3.37}$$

where $T = \partial L/\partial\dot{t}$ and $R_i = \partial L/\partial\dot{r}_i$, will be functions of $t$ and $\boldsymbol{r}$ and homogeneous functions of zero degree of $\dot{t}(\tau)$ and $\dot{\boldsymbol{r}}(\tau)$. Because the Lagrangian is invariant under $\mathcal{P}$, the functions $T$ and $\boldsymbol{R}$ transform under the group $\mathcal{P}$ in the form:

$$T' = \gamma T - \gamma(\boldsymbol{v}\cdot R(\boldsymbol{\mu})\boldsymbol{R}), \tag{3.38}$$

$$\boldsymbol{R}' = R(\boldsymbol{\mu})\boldsymbol{R} - \gamma\boldsymbol{v}T/c^2 + \frac{\gamma^2}{1+\gamma}(\boldsymbol{v}\cdot R(\boldsymbol{\mu})\boldsymbol{R})\boldsymbol{v}/c^2. \tag{3.39}$$

We thus see that $T$ and $\boldsymbol{R}$ are invariant under translations and therefore they must be functions independent of $t$ and $\boldsymbol{r}$.

The conjugate momenta of the independent degrees of freedom $q_i = r_i$ are $p_i = \partial L/\partial\dot{r}_i$, and consequently Noether's theorem (1.55) leads to the following constants of the motion, that are calculated similarly as in the Galilei case except for the invariance under pure Lorentz transformations. We have now no gauge function and the variations are $\delta t = \boldsymbol{r}\cdot\delta\boldsymbol{v}/c^2$, $M_i = r_i/c^2$ and $\delta\boldsymbol{r} = t\delta\boldsymbol{v}$, $M_{ij} = t\delta_{ij}$ and thus we get:

$$\text{Energy} \quad H = -T, \tag{3.40}$$

$$\text{linear momentum} \quad \boldsymbol{P} = \boldsymbol{R} = \boldsymbol{p}, \tag{3.41}$$

$$\text{kinematical momentum} \quad \boldsymbol{K} = H\boldsymbol{r}/c^2 - \boldsymbol{P}t, \tag{3.42}$$

$$\text{angular momentum} \quad \boldsymbol{J} = \boldsymbol{r}\times\boldsymbol{P}. \tag{3.43}$$

The energy and the linear momentum transform as:

$$H'(\tau) = \gamma H(\tau) + \gamma(\boldsymbol{v}\cdot R(\boldsymbol{\mu})\boldsymbol{P}(\tau)), \tag{3.44}$$

$$\boldsymbol{P}'(\tau) = R(\boldsymbol{\mu})\boldsymbol{P}(\tau) + \frac{\gamma\boldsymbol{v}}{c^2}H(\tau) + \frac{\gamma^2}{(1+\gamma)c^2}(\boldsymbol{v}\cdot R(\boldsymbol{\mu})\boldsymbol{P}(\tau))\boldsymbol{v}. \tag{3.45}$$

They transform like the contravariant components of a four-vector $P^\mu \equiv (H/c, \boldsymbol{P})$. The observables $c\boldsymbol{K}$ and $\boldsymbol{J}$ are the essential components of the antisymmetric tensor $J^{\mu\nu} = -J^{\nu\mu} = x^\mu P^\nu - x^\nu P^\mu$, $cK_i = J^{i0}$ and $J_k = \epsilon_{kil}J^{il}/2$.

Taking the $\tau$ derivative of the kinematical momentum, $\dot{\boldsymbol{K}} = 0$, we get $\boldsymbol{P} = H\dot{\boldsymbol{r}}/c^2\dot{t} = (H/c^2)d\boldsymbol{r}/dt = H\boldsymbol{u}/c^2$, where $\boldsymbol{u}$ is the velocity of the particle and point $\boldsymbol{r}$ represents both the center of mass and center of charge position of the particle.

The six conditions $\boldsymbol{P} = 0$ and $\boldsymbol{K} = 0$, imply $\boldsymbol{u} = 0$ and $\boldsymbol{r} = 0$, so that the system is at rest and placed at the origin of the reference frame, similarly as in the nonrelativistic case. We again call this class of observers the center of mass observer.

From (3.44) and (3.45) we see that the magnitude $(H/c)^2 - \boldsymbol{P}^2 = m^2c^2$ is a Poincaré invariant and a constant of the motion that defines the mass of the particle. By using the expression of $\boldsymbol{P} = H\boldsymbol{u}/c^2$, we get

$$H = \pm mc^2(1 - u^2/c^2)^{-1/2},$$

and the energy can be either positive or negative and $u < c$. If $u > c$, then the invariant $(H/c)^2 - P^2 < 0$ and it is not possible to define the rest mass of the system. By substitution of the found expressions for $T$ and $\boldsymbol{R}$ in (3.37), the Lagrangian of the system is just

$$L = \mp mc\sqrt{c^2\dot{t}^2 - \dot{\boldsymbol{r}}^2}. \tag{3.46}$$

Expansion of this Lagrangian to lowest order in $u/c$, in the case of positive energy, we get

$$L = -mc^2\dot{t} + \frac{m}{2}\frac{\dot{\boldsymbol{r}}^2}{\dot{t}},$$

where the first term $-mc^2\dot{t}$ that can be withdrawn is just the equivalent to the Galilei internal energy term $-H_0\dot{t}$ of (2.38).

The spin of this system, defined similarly as in the nonrelativistic case,

$$\boldsymbol{S} \equiv \boldsymbol{J} - \boldsymbol{q} \times \boldsymbol{P} = \boldsymbol{J} - \frac{c^2}{H}\boldsymbol{K} \times \boldsymbol{P} = 0, \tag{3.47}$$

vanishes, so that the relativistic point particle is also a spinless system.

The ten constants of the motion (3.40-3.43) are the generating functions of the corresponding canonical transformations of the system, such that on phase space in terms of the canonical conjugate variables they take the form

$$H = \sqrt{m^2c^4 + p^2c^2}, \quad \boldsymbol{P} = \boldsymbol{p}, \quad \boldsymbol{K} = H\boldsymbol{q}/c^2 - \boldsymbol{p}t, \quad \boldsymbol{J} = \boldsymbol{q} \times \boldsymbol{p}, \tag{3.48}$$

and taking the Poisson bracket of these functions we get

$$\{\boldsymbol{J},\boldsymbol{J}\} = \boldsymbol{J},\ \{\boldsymbol{J},\boldsymbol{P}\} = \boldsymbol{P},\ \{\boldsymbol{J},\boldsymbol{K}\} = \boldsymbol{K},\ \{\boldsymbol{J},H\} = \{\boldsymbol{P},H\} = 0,$$
$$\{H,\boldsymbol{K}\} = -\boldsymbol{P},\ \{\boldsymbol{P},\boldsymbol{P}\} = 0,\ \{\boldsymbol{K},\boldsymbol{K}\} = -\boldsymbol{J}/c^2,\ \{\boldsymbol{K},\boldsymbol{P}\} = H/c^2.$$

These are the commutation relations (3.11) and (3.12) of the Poincaré group, up to a global sign. Now our conserved quantity $\boldsymbol{K}$ has the same dimensions as the corresponding Galilei kinematical momentum and must be put in correspondence with the Poincaré generator $\boldsymbol{K}/c$ because it has dimensions of angular momentum divided by velocity. In the low velocity limit of the particle ($c \to \infty$), these Poisson brackets reduce to those of the extended Galilei group, given in (2.36) and (2.37), because $\boldsymbol{J}/c^2 \to 0$ and $H/c^2 \to m$.

# 3. RELATIVISTIC SPINNING PARTICLES

There are three maximal homogeneous spaces of $\mathcal{P}$, all of them at first parameterized by the variables $(t, \boldsymbol{r}, \boldsymbol{u}, \boldsymbol{\rho})$, where the velocity variable $\boldsymbol{u}$ can be either $u < c$, $u = c$ or $u > c$. We shall call these kinds of particles by the following names: The first one, since the motion of the position of the charge $\boldsymbol{r}$ satisfies $u < c$, we call a **Bradyon**, from the Greek term $\beta\rho\alpha\delta\upsilon\varsigma \equiv$ slow. Bradyons are thus particles for which point $\boldsymbol{r}$ never reaches the speed of light. The second class of particles $(u = c)$ will be called **Luxons** because point $\boldsymbol{r}$ is always moving at the speed of light for every observer, and finally those of the third group, because $u > c$, are called **Tachyons**, from the Greek $\tau\alpha\chi\upsilon\varsigma \equiv$ fast.

For the second class we use the Latin denomination Luxons in spite of the Greek one of photons, because this class of particles will supply the description not only of classical photons but also a classical model of the electron. This class of models is very important and it has no nonrelativistic limit. Therefore the models this manifold produce have no nonrelativistic equivalent.

The first class corresponds to a kinematical space that is the Poincaré group itself and will be analyzed in what follows. The analysis of the other two will be left to subsequent sections.

## 3.1 BRADYONS

A spinning Bradyon is defined as a dynamical system whose kinematical space $X$ is the Poincaré group manifold. [3]

Similarly as in the Galilei case, it is characterized by the ten real kinematical variables $x(\tau) \in X$, $x \equiv (t(\tau), \boldsymbol{r}(\tau), \boldsymbol{u}(\tau), \boldsymbol{\rho}(\tau))$ with domains $t \in \mathbb{R}$, $\boldsymbol{r} \in \mathbb{R}^3$, $\boldsymbol{u} \in \mathbb{R}^3$ but now $u < c$, and $\boldsymbol{\rho} \in \mathbb{R}^3_c$, like the corresponding group parameters, in such a way that under the action of a group element $g \equiv (b, \boldsymbol{a}, \boldsymbol{v}, \boldsymbol{\mu})$ of $\mathcal{P}$ they transform as $x' = gx$:

$$t'(\tau) = \gamma t(\tau) + \gamma(\boldsymbol{v} \cdot R(\boldsymbol{\mu})\boldsymbol{r}(\tau))/c^2 + b, \tag{3.49}$$

$$\boldsymbol{r}'(\tau) = R(\boldsymbol{\mu})\boldsymbol{r}(\tau) + \gamma \boldsymbol{v} t(\tau) + \frac{\gamma^2}{(1+\gamma)c^2}(\boldsymbol{v} \cdot R(\boldsymbol{\mu})\boldsymbol{r}(\tau))\boldsymbol{v} + \boldsymbol{a}, \tag{3.50}$$

$$\boldsymbol{u}'(\tau) = \frac{R(\boldsymbol{\mu})\boldsymbol{u}(\tau) + \gamma \boldsymbol{v} + (\boldsymbol{v} \cdot R(\boldsymbol{\mu})\boldsymbol{u}(\tau))\boldsymbol{v}\gamma^2/(1+\gamma)c^2}{\gamma(1 + \boldsymbol{v} \cdot R(\boldsymbol{\mu})\boldsymbol{u}(\tau)/c^2)}, \tag{3.51}$$

$$\boldsymbol{\rho}'(\tau) = \frac{\boldsymbol{\mu} + \boldsymbol{\rho}(\tau) + \boldsymbol{\mu} \times \boldsymbol{\rho}(\tau) + \boldsymbol{F}(\boldsymbol{v}, \boldsymbol{\mu}; \boldsymbol{u}(\tau), \boldsymbol{\rho}(\tau))}{1 - \boldsymbol{\mu} \cdot \boldsymbol{\rho}(\tau) + G(\boldsymbol{v}, \boldsymbol{\mu}; \boldsymbol{u}(\tau), \boldsymbol{\rho}(\tau))}. \tag{3.52}$$

Functions $\boldsymbol{F}$ and $G$ are given in (3.9-3.10), and the parametrization of rotations is the one given in (2.8).

The way these variables transform allow us to interpret them respectively as the time, position, velocity and orientation of the particle. As generalized coordinates they are not independent; there exist among them three constraints $\boldsymbol{u}(\tau) = \dot{\boldsymbol{r}}(\tau)/\dot{t}(\tau)$ that together with the homogeneity condition of the Lagrangian will reduce to six the number of essential degrees of freedom of the system.

The six independent variables are $\boldsymbol{r}(t)$ and $\boldsymbol{\rho}(t)$, such that the Lagrangian will be a function of $\boldsymbol{r}(t)$ up to its second derivative and up to the first derivative of the orientation $\boldsymbol{\rho}(t)$.

Because the possible Lagrangians are explicit functions of the kinematical variables and their first derivatives, and must be Poincaré invariant, we need to form invariant expressions in terms of these variables, such that they become homogeneous functions of first degree in terms of the derivatives. The best way to do that will be to find first some tensor functions of these variables and afterwards to obtain scalar magnitudes by saturation of the tensor indexes.

If we look at expressions (3.51) and (3.52), they come from the composition law of the Lorentz group $\Lambda(\boldsymbol{u}',\boldsymbol{\rho}') = \Lambda(\boldsymbol{v},\boldsymbol{\mu})\Lambda(\boldsymbol{u},\boldsymbol{\rho})$ that can be considered, by looking at the different columns of matrix $\Lambda(\boldsymbol{u},\boldsymbol{\rho})$, as

$$\left(\boldsymbol{e}'_{(0)}, \boldsymbol{e}'_{(1)}, \boldsymbol{e}'_{(2)}, \boldsymbol{e}'_{(3)}\right) = \Lambda(\boldsymbol{v},\boldsymbol{\mu}) \, \left(\boldsymbol{e}_{(0)}, \boldsymbol{e}_{(1)}, \boldsymbol{e}_{(2)}, \boldsymbol{e}_{(3)}\right),$$

where the $\boldsymbol{e}_{(\alpha)}$ are the $\alpha$-th column of the corresponding Lorentz matrix $\Lambda(\boldsymbol{u},\boldsymbol{\rho})$. But this expression amounts for each $\alpha$ to

$$\boldsymbol{e}'_{(\alpha)} = \Lambda(\boldsymbol{v},\boldsymbol{\mu})\, \boldsymbol{e}_{(\alpha)}, \tag{3.53}$$

so that each column is transformed into the corresponding one by means of a Lorentz transformation. But each column is in fact a normalized four-vector, that, according to the metric we have chosen, and by (3.1), satisfy

$$\boldsymbol{e}_{(\alpha)} \cdot \boldsymbol{e}_{(\beta)} = \eta_{\alpha\beta},$$

*i.e.*, $\boldsymbol{e}_{(0)}$ is time-like and the three $\boldsymbol{e}_{(i)}$ are space-like orthogonal four-vectors, and where a $\cdot$ between two four-vectors represents their corresponding scalar product in Minkowski space-time. Their contravariant components are $e^{\mu}_{(\alpha)} \equiv \Lambda(\boldsymbol{u},\boldsymbol{\rho})^{\mu}{}_{\alpha}$, when expressed in terms of a set of basis four-vectors $\boldsymbol{f}_{(\mu)}$, $\mu = 0,1,2,3$ in the laboratory frame. These vectors have the contravariant components $f^{\nu}_{(\mu)} = \delta^{\nu}_{\mu}$, where we use a subindex between brackets to denote some vector of a set, while subindexes or superindexes without brackets refer to the corresponding covariant or contravariant components. Thus $\boldsymbol{e}_{(\alpha)} = e^{\mu}_{(\alpha)}\boldsymbol{f}_{(\mu)} = \Lambda(\boldsymbol{u},\boldsymbol{\rho})^{\mu}{}_{\alpha}\boldsymbol{f}_{(\mu)}$.

The four-vectors $\boldsymbol{e}_{(\alpha)}$ are invariant in form, because each $\boldsymbol{e}_{(\alpha)}$ is transformed into the corresponding $\boldsymbol{e}'_{(\alpha)}$, such that its contravariant components $e'^{\mu}_{(\alpha)}$ have the same analytical expression in terms of $\boldsymbol{u}'$ and $\boldsymbol{\rho}'$ as the components of $e^{\mu}_{(\alpha)}$ in terms of $\boldsymbol{u}$ and $\boldsymbol{\rho}$. The same thing happens to the derivative with respect to the invariant evolution parameter $\tau$ of both terms of expression (3.53)

$$\dot{e}'_{(\alpha)} = \Lambda(\boldsymbol{v}, \boldsymbol{\mu})\, \dot{e}_{(\alpha)}, \tag{3.54}$$

but with the additional condition that the components of $\dot{e}_{(\alpha)}$ are invariant in form and are linear functions of $\dot{\boldsymbol{u}}$ and $\dot{\boldsymbol{\rho}}$.

We have now an alternative viewpoint to describe a general Bradyon. It is described by the knowledge of the time $t$ and the position $\boldsymbol{r}$, thus by a four-vector of contravariant components $x^{\mu} \equiv (ct, \boldsymbol{r})$, and an associated set of four orthonormal four-vectors $e_{(\alpha)}$ moving along the trajectory of point $\boldsymbol{r}$. The kinematics of this kind of particles is what is known as a transport of tetrads or transport of frames. [4]

We write the Lorentz matrix $\Lambda$ at any instant $\tau$ as the product $\Lambda(\boldsymbol{u}, \boldsymbol{\rho}) \equiv L(\boldsymbol{u}(\tau))R(\boldsymbol{\rho}(\tau))$, in this order, where $L(\boldsymbol{u})$ is a boost matrix and $R(\boldsymbol{\rho})$ an orthogonal rotation matrix, both expressed in terms of the above mentioned parameters. Then, four-vector $e_{(0)}$ is just the dimensionless four-velocity with contravariant components $(\gamma(u), \gamma(u)\boldsymbol{u}/c)$ and the other three four-vectors $e_{(i)}$, $i = 1, 2, 3$ are the orthogonal spacelike unit vectors linked to point $\boldsymbol{r}$, that correspond to the three column vectors of the orthogonal matrix $R(\boldsymbol{\rho}(\tau))$, and finally boosted with the pure Lorentz transformation $L(\boldsymbol{u})$.

Equations (3.49)-(3.52) can be written, in a covariant form, as

$$x'^{\mu}(\tau) = \Lambda^{\mu}{}_{\nu}(\boldsymbol{v}, \boldsymbol{\mu})x^{\nu}(\tau) + a^{\mu}, \quad e'^{\mu}_{(\alpha)}(\tau) = \Lambda^{\mu}{}_{\nu}(\boldsymbol{v}, \boldsymbol{\mu})e^{\nu}_{(\alpha)}(\tau), \tag{3.55}$$

and taking the $\tau$-derivative

$$\dot{x}'^{\mu}(\tau) = \Lambda^{\mu}{}_{\nu}(\boldsymbol{v}, \boldsymbol{\mu})\dot{x}^{\nu}(\tau), \quad \dot{e}'^{\mu}_{(\alpha)}(\tau) = \Lambda^{\mu}{}_{\nu}(\boldsymbol{v}, \boldsymbol{\mu})\dot{e}^{\nu}_{(\alpha)}(\tau). \tag{3.56}$$

We get for $\dot{t}$ and $\dot{\boldsymbol{r}}$ the transformation equations (3.35) and (3.36), and very complicated expressions for $\dot{\boldsymbol{u}}(\tau)$ and $\dot{\boldsymbol{\rho}}(\tau)$. For instance, for $\dot{\boldsymbol{u}}(\tau)$ we get:

$$\dot{\boldsymbol{u}}' = \frac{\gamma(1 + \dfrac{\boldsymbol{v} \cdot R(\boldsymbol{\mu})\boldsymbol{u}}{c^2})R(\boldsymbol{\mu})\dot{\boldsymbol{u}} - \gamma\dfrac{\boldsymbol{v} \cdot R(\boldsymbol{\mu})\dot{\boldsymbol{u}}}{c^2}R(\boldsymbol{\mu})\boldsymbol{u} - \dfrac{\gamma^2}{1+\gamma}\dfrac{\boldsymbol{v} \cdot R(\boldsymbol{\mu})\dot{\boldsymbol{u}}}{c^2}\boldsymbol{v}}{\gamma^2\left(1 + \dfrac{\boldsymbol{v} \cdot R(\boldsymbol{\mu})\boldsymbol{u}}{c^2}\right)^2}, \tag{3.57}$$

and therefore the acceleration $\boldsymbol{a}(\tau) = \dot{\boldsymbol{u}}(\tau)/\dot{t}(\tau)$ transforms as:

$$\boldsymbol{a}' = \frac{\gamma(1 + \frac{\boldsymbol{v}\cdot R(\boldsymbol{\mu})\boldsymbol{u}}{c^2})\, R(\boldsymbol{\mu})\boldsymbol{a} - \gamma\frac{\boldsymbol{v}\cdot R(\boldsymbol{\mu})\boldsymbol{a}}{c^2}\, R(\boldsymbol{\mu})\boldsymbol{u} - \frac{\gamma^2}{1+\gamma}\frac{\boldsymbol{v}\cdot R(\boldsymbol{\mu})\boldsymbol{a}}{c^2}\boldsymbol{v}}{\gamma^3\left(1 + \frac{\boldsymbol{v}\cdot R(\boldsymbol{\mu})\boldsymbol{u}}{c^2}\right)^3}. \tag{3.58}$$

Since $\boldsymbol{e}_{(\alpha)}(\tau)\cdot\boldsymbol{e}_{(\beta)}(\tau) = \eta_{\alpha\beta}$ at any $\tau$, if we take the $\tau$ derivative of this expression we get

$$\dot{\boldsymbol{e}}_{(\alpha)}\cdot\boldsymbol{e}_{(\beta)} + \boldsymbol{e}_{(\alpha)}\cdot\dot{\boldsymbol{e}}_{(\beta)} = \dot{\boldsymbol{e}}_{(\alpha)}\cdot\boldsymbol{e}_{(\beta)} + \dot{\boldsymbol{e}}_{(\beta)}\cdot\boldsymbol{e}_{(\alpha)} = \Psi_{\alpha\beta} + \Psi_{\beta\alpha} = 0,$$

and the tensor-looking antisymmetric magnitudes $\Psi_{\alpha\beta} = -\Psi_{\beta\alpha}$, are six invariant functions of the kinematical variables $\boldsymbol{u}$ and $\boldsymbol{\rho}$ and their first derivatives. They are not tensors because under an arbitrary Poincaré transformation they transform as $\Psi'_{\alpha\beta} = \Psi_{\alpha\beta}$, as it corresponds to the invariance of the scalar product of two four-vectors $\dot{\boldsymbol{e}}_{(\alpha)}$ and $\boldsymbol{e}_{(\beta)}$ for any arbitrary inertial observer.

If we express the laboratory tetrad $\boldsymbol{f}_{(\mu)}$ in terms of the body tetrad $\boldsymbol{e}_{(\alpha)}$,

$$\boldsymbol{f}_{(\mu)} = f^{\alpha}_{(\mu)}\boldsymbol{e}_{(\alpha)}, \quad f^{\alpha}_{(\mu)} = \Lambda_{\mu}{}^{\alpha}(\boldsymbol{u},\boldsymbol{\rho}) = \eta_{\mu\nu}\eta^{\alpha\beta}\Lambda^{\nu}{}_{\beta}(\boldsymbol{u},\boldsymbol{\rho}).$$

A change of inertial reference frame corresponds to a change of basic tetrads $\boldsymbol{f}_{(\mu)} \to \boldsymbol{f}'_{(\mu)}$ given by

$$\boldsymbol{f}'_{(\mu)} = \Lambda_{\mu}{}^{\nu}(\boldsymbol{v},\boldsymbol{\mu})\boldsymbol{f}'_{(\nu)}, \quad \text{where} \quad \Lambda_{\mu}{}^{\nu}(\boldsymbol{v},\boldsymbol{\mu}) = \eta_{\mu\sigma}\eta^{\nu\lambda}\Lambda^{\sigma}{}_{\lambda}(\boldsymbol{v},\boldsymbol{\mu}),$$

in terms of the components of the Lorentz matrix $\Lambda(\boldsymbol{v},\boldsymbol{\mu})$. But if $\Lambda^{\mu}{}_{\nu}$ is the set of coefficients for transforming the contravariant components of a tensor, $\Lambda_{\mu}{}^{\nu}$ are the coefficients for the transformation of the covariant components, because

$$v'^{\mu} = \Lambda^{\mu}{}_{\nu}v^{\nu}, \;\Rightarrow\quad v'_{\sigma} = \eta_{\sigma\mu}v'^{\mu} = \eta_{\sigma\mu}\Lambda^{\mu}{}_{\nu}v^{\nu} = \eta_{\sigma\mu}\Lambda^{\mu}{}_{\nu}\eta^{\nu\lambda}v_{\lambda} = \Lambda_{\sigma}{}^{\lambda}v_{\lambda}.$$

From the viewpoint of the particle frame the components of the laboratory frame $f^{\alpha}_{(\mu)}$ are no longer constant magnitudes, but they depend on $\tau$. Therefore, taking the $\tau$-derivative of $\eta_{\alpha\beta}f^{\alpha}_{(\mu)}(\tau)f^{\beta}_{(\nu)}(\tau) \equiv \boldsymbol{f}_{(\mu)}\cdot\boldsymbol{f}_{(\nu)} = \eta_{\mu\nu}$, we get

$$\eta_{\alpha\beta}\dot{f}^{\alpha}_{(\mu)}\, f^{\beta}_{(\nu)} + \eta_{\alpha\beta}f^{\alpha}_{(\mu)}\, \dot{f}^{\beta}_{(\nu)} = \Omega_{\mu\nu} + \Omega_{\nu\mu} = 0,$$

but the $\tau$-dependent magnitudes $\Omega_{\mu\nu}$ transform like the components of an antisymmetric rank 2, covariant tensor. Explicitly,

$$\Omega_{\mu\nu}(\boldsymbol{u}, \boldsymbol{\rho}, \dot{\boldsymbol{u}}, \dot{\boldsymbol{\rho}}) = \eta_{\alpha\beta}\dot{\Lambda}_{\mu}{}^{\alpha}(\boldsymbol{u}, \boldsymbol{\rho})\Lambda_{\nu}{}^{\beta}(\boldsymbol{u}, \boldsymbol{\rho}).$$

Then the different magnitudes we find may have indexes of the type $\mu$ with respect to the laboratory frame and indexes of the type $\alpha$ in the body frame. It is only with respect to the laboratory frame that they transform like tensor magnitudes. This is why $\Psi_{\alpha\beta}$ which is a magnitude referred to the particle frame is in fact an invariant magnitude.

It is not difficult to see that

$$\Psi_{\alpha\beta} = \Omega_{\mu\nu}e^{\mu}_{(\alpha)}e^{\nu}_{(\beta)}, \quad \Omega_{\mu\nu} = \Psi_{\alpha\beta}f^{\alpha}_{(\mu)}f^{\beta}_{(\nu)}.$$

**We can reach the same result with an alternative method using matrices. If $\Lambda(\tau) = L(\boldsymbol{u})R(\boldsymbol{\rho})$ is any Lorentz matrix, it satisfies at any instant $\tau$, $\Lambda(\tau)\eta\Lambda^T(\tau) = \eta$, where $\eta = \mathrm{diag}(1, -1, -1, -1)$ is Minkowski's metric tensor written in matrix form, and $T$ means the transpose matrix. Then, taking the $\tau$-derivative, we get**

$$\dot{\Lambda}\eta\,\Lambda^T + \Lambda\eta\,\dot{\Lambda}^T = \Omega + \Omega^T = 0.$$

**Similarly, $\Lambda(\tau)$ also satisfies $\Lambda^T(\tau)\eta\,\Lambda(\tau) = \eta$, and thus**

$$\dot{\Lambda}^T\eta\,\Lambda + \Lambda^T\eta\,\dot{\Lambda} = \Psi + \Psi^T = 0.$$

**Since $\Lambda'(\tau) = \Lambda(\boldsymbol{v}, \boldsymbol{\mu})\Lambda(\tau)$, then $\dot{\Lambda}'(\tau) = \Lambda(\boldsymbol{v}, \boldsymbol{\mu})\dot{\Lambda}(\tau)$, and therefore**

$$\Omega'(\tau) = \dot{\Lambda}'(\tau)\eta\Lambda'^T(\tau) = \Lambda(\boldsymbol{v}, \boldsymbol{\mu})\dot{\Lambda}(\tau)\eta\,\Lambda^T(\tau)\Lambda^T(\boldsymbol{v}, \boldsymbol{\mu}) = \Lambda(\boldsymbol{v}, \boldsymbol{\mu})\Omega(\tau)\Lambda^T(\boldsymbol{v}, \boldsymbol{\mu}),$$

**and**

$$\Psi'(\tau) = \dot{\Lambda}'^T(\tau)\eta\,\Lambda'(\tau) = \dot{\Lambda}^T(\tau)\Lambda^T(\boldsymbol{v}, \boldsymbol{\mu})\eta\,\Lambda(\boldsymbol{v}, \boldsymbol{\mu})\Lambda(\tau) = \dot{\Lambda}^T(\tau)\eta\,\Lambda(\tau) = \Psi(\tau).$$

**This implies, using matrix indices, that**

$$\Omega'^{\mu\nu} = \Lambda^{\mu}{}_{\sigma}\Omega^{\sigma\lambda}(\Lambda^T)^{\sigma}{}_{\nu} = \Lambda^{\mu}{}_{\sigma}\Omega^{\sigma\lambda}\Lambda^{\nu}{}_{\lambda},$$

**but this is just $\Omega'^{\mu\nu} = \Lambda^{\mu}{}_{\sigma}\Lambda^{\nu}{}_{\lambda}\Omega'^{\sigma\lambda}$, by considering that the first index of each object, irrespective of up or down, represents a row index and the second a column index. Then this is the transformation of the contravariant components of a tensor. The other $\Psi'_{\alpha\beta} = \Psi_{\alpha\beta}$ are invariant magnitudes.**

If we write in matrix form the magnitudes $\Psi_{\alpha\beta}$ and $\Omega_{\mu\nu}$, with $\alpha$ and $\mu$ as row indexes and $\beta$ and $\nu$ column indexes respectively, then

$$\Psi_{\alpha\beta} = \begin{pmatrix} 0 & \pi_1 & \pi_2 & \pi_3 \\ -\pi_1 & 0 & \lambda_3 & -\lambda_2 \\ -\pi_2 & -\lambda_3 & 0 & \lambda_1 \\ -\pi_3 & \lambda_2 & -\lambda_1 & 0 \end{pmatrix}, \tag{3.59}$$

$$\Omega_{\mu\nu} = \begin{pmatrix} 0 & \alpha_1 & \alpha_2 & \alpha_3 \\ -\alpha_1 & 0 & \omega_3 & -\omega_2 \\ -\alpha_2 & -\omega_3 & 0 & \omega_1 \\ -\alpha_3 & \omega_2 & -\omega_1 & 0 \end{pmatrix}. \tag{3.60}$$

The invariant vector $\boldsymbol{\pi}$ and pseudovector $\boldsymbol{\lambda}$ and non-invariant vector $\boldsymbol{\alpha}$ and pseudovector $\boldsymbol{\omega}$ are computed through rather tedious calculations and they are expressed as

$$\boldsymbol{\pi} = \frac{\gamma(u)}{c} R(-\boldsymbol{\rho}) \left( \dot{\boldsymbol{u}} + \frac{\gamma(u)^2}{(1+\gamma(u))c^2} (\dot{\boldsymbol{u}} \cdot \boldsymbol{u})\boldsymbol{u} \right), \tag{3.61}$$

$$\boldsymbol{\lambda} = R(-\boldsymbol{\rho})\,(\boldsymbol{\omega}_0 - \boldsymbol{\omega}_T), \tag{3.62}$$

$$\boldsymbol{\alpha} = \frac{\gamma(u)}{c} \left[ \dot{\boldsymbol{u}} + \frac{\gamma(u)^2}{1+\gamma(u)} \frac{\boldsymbol{u} \cdot \dot{\boldsymbol{u}}}{c^2} \boldsymbol{u} + \boldsymbol{u} \times \boldsymbol{\omega}_0 \right], \tag{3.63}$$

$$\boldsymbol{\omega} = \gamma(u)\boldsymbol{\omega}_0 - \frac{\gamma(u)^2}{1+\gamma(u)} \frac{\boldsymbol{u} \cdot \boldsymbol{\omega}_0}{c^2} \boldsymbol{u} + \boldsymbol{\omega}_T. \tag{3.64}$$

The variables $\boldsymbol{\omega}_0$ and $\boldsymbol{\omega}_T$ are given by:

$$\boldsymbol{\omega}_0 = \frac{2(\dot{\boldsymbol{\rho}} + \boldsymbol{\rho} \times \dot{\boldsymbol{\rho}})}{1+\rho^2}, \tag{3.65}$$

$$\boldsymbol{\omega}_T = \frac{\gamma(u)^2}{1+\gamma(u)} \frac{\dot{\boldsymbol{u}} \times \boldsymbol{u}}{c^2}. \tag{3.66}$$

In (3.65) if $\tau$ would be the time, $\boldsymbol{\omega}_0$ would be the angular velocity expressed in terms of the orientation variables and their derivatives as in the nonrelativistic case (2.50), and $\boldsymbol{\omega}_T$ is the Thomas angular velocity. In the case $\boldsymbol{u} = 0$, the time-space part of $\Omega_{\mu\nu}$, $\boldsymbol{\alpha}$, reduces to $\dot{\boldsymbol{u}}/c$, while the space-space part $\boldsymbol{\omega}$ is the angular velocity $\boldsymbol{\omega}_0$. We shall use throughout this chapter the notation $\boldsymbol{\omega}_0$ for the angular velocity in a $\tau$-evolution description, not to be confused with variable $\boldsymbol{\omega}$ defined in (3.64) which depends on $\boldsymbol{\omega}_0$ and also on $\dot{\boldsymbol{u}}$.

The invariant functions $\boldsymbol{\pi}$ and $\boldsymbol{\lambda}$ are explicitly dependent on the orientation variables $\boldsymbol{\rho}$ by the term $R(-\boldsymbol{\rho})$, while the magnitudes $\boldsymbol{\alpha}$ and $\boldsymbol{\omega}$ are only dependent on $\boldsymbol{\rho}$ through its dependence on $\boldsymbol{\omega}_0$. It is doubtfull whether any Lagrangian should be explicitly dependent on thé orientation variables $\boldsymbol{\rho}$ and therefore we expect to construct the possible invariant terms from the variables $\boldsymbol{\alpha}$ and $\boldsymbol{\omega}$.

The expression of $\dot{\boldsymbol{u}}$ and $\boldsymbol{\omega}_0$ in terms of $\boldsymbol{\alpha}$ and $\boldsymbol{\omega}$ is the inverse of (3.63) and (3.64),

$$\dot{\boldsymbol{u}} = c\boldsymbol{\alpha} - \frac{(\boldsymbol{u} \cdot \boldsymbol{\alpha})\boldsymbol{u}}{c} - \boldsymbol{u} \times \boldsymbol{\omega}, \tag{3.67}$$

$$\boldsymbol{\omega}_0 = \boldsymbol{\omega} + \frac{\gamma(u)}{(1+\gamma(u))c}\boldsymbol{u}\times\boldsymbol{\alpha}. \qquad (3.68)$$

Since $\Omega_{\mu\nu}$ transforms under $\mathcal{P}$ like a covariant tensor, variables $\boldsymbol{\alpha}$ and $\boldsymbol{\omega}$ in (3.63) and (3.64) respectively, transform as:

$$\boldsymbol{\alpha}'(\tau) = \gamma R(\boldsymbol{\mu})\boldsymbol{\alpha}(\tau) - \frac{\gamma^2}{(1+\gamma)c^2}(\boldsymbol{v}\cdot R(\boldsymbol{\mu})\boldsymbol{\alpha}(\tau))\boldsymbol{v} + \frac{\gamma}{c}(\boldsymbol{v}\times R(\boldsymbol{\mu})\boldsymbol{\omega}(\tau)), \qquad (3.69)$$

$$\boldsymbol{\omega}'(\tau) = \gamma R(\boldsymbol{\mu})\boldsymbol{\omega}(\tau) - \frac{\gamma^2}{(1+\gamma)c^2}(\boldsymbol{v}\cdot R(\boldsymbol{\mu})\boldsymbol{\omega}(\tau))\boldsymbol{v} - \frac{\gamma}{c}(\boldsymbol{v}\times R(\boldsymbol{\mu})\boldsymbol{\alpha}(\tau)), \qquad (3.70)$$

*i.e.*, as it corresponds to the components of a completely antisymmetric tensor.

In covariant notation

$$\Omega^{\mu\nu} = a^\mu u^\nu - a^\nu u^\mu + u_\alpha \omega_\beta \epsilon^{\alpha\beta\mu\nu}, \qquad (3.71)$$

with the antisymmetric tensor $\epsilon_{0123} = +1$, and in terms of the following three dimensionless four-vectors $u^\mu \equiv (\gamma(u), \gamma(u)\boldsymbol{u}/c)$, $a^\mu \equiv \dot{u}^\mu = (\dot{\gamma}(u), \dot{\gamma}(u)\boldsymbol{u}/c + \gamma(u)\dot{\boldsymbol{u}}/c)$, and $\omega^\mu$, which is the boosted four-vector of components $(0, \widetilde{\boldsymbol{\omega}}_0)$. Here $\widetilde{\boldsymbol{\omega}}_0$ is the angular velocity of the particle measured by the inertial observer at rest with the particle, *i.e.*, that measures $\boldsymbol{u}=0$.

$$\omega^\mu \equiv \left(\gamma(u)\frac{\boldsymbol{u}\cdot\boldsymbol{\omega}_0}{c}, \boldsymbol{\omega}_0 - \frac{\gamma(u)^2}{(1+\gamma(u))c^2}\dot{\boldsymbol{u}}\times\boldsymbol{u} + \frac{\gamma(u)^2}{(1+\gamma(u))c^2}(\boldsymbol{u}\cdot\boldsymbol{\omega}_0)\boldsymbol{u}\right).$$

These three four-vectors satisfy:

$$u^\mu u_\mu = 1, \quad u^\mu a_\mu = u^\mu \omega_\mu = 0, \quad a^\mu a_\mu = -\frac{\dot{\boldsymbol{u}}_0^2}{c^2}, \quad \omega^\mu\omega_\mu = -\widetilde{\boldsymbol{\omega}}_0^2,$$

$$a^\mu \omega_\mu = -\frac{\dot{\boldsymbol{u}}_0\cdot\widetilde{\boldsymbol{\omega}}_0}{c},$$

where $\dot{\boldsymbol{u}}_0$ and $\widetilde{\boldsymbol{\omega}}_0$ are the acceleration and angular velocity in terms of the dimensionless evolution parameter $\tau$, when measured by the instantaneous rest frame observer ($\boldsymbol{u}=0$), and where $\boldsymbol{\omega}_0$ is given in (3.65).

The homogeneity condition of the Lagrangian, in terms of the variables $(\dot{t}, \dot{\boldsymbol{r}}, \dot{\boldsymbol{u}}, \dot{\boldsymbol{\rho}})$, allows us to write it as:

$$L = T\dot{t} + \boldsymbol{R}\cdot\dot{\boldsymbol{r}} + \boldsymbol{U}\cdot\dot{\boldsymbol{u}} + \boldsymbol{V}\cdot\dot{\boldsymbol{\rho}}, \qquad (3.72)$$

and, in terms of the new defined variables $(\dot{t}, \dot{\boldsymbol{r}}, \boldsymbol{\alpha}, \boldsymbol{\omega})$,

$$L = T\dot{t} + \boldsymbol{R}\cdot\dot{\boldsymbol{r}} + \boldsymbol{D}\cdot\boldsymbol{\alpha} + \boldsymbol{Z}\cdot\boldsymbol{\omega}, \qquad (3.73)$$

or, in terms of the variables $(\dot{t}, \dot{\boldsymbol{r}}, \dot{\boldsymbol{u}}, \boldsymbol{\omega}_0)$,

$$L = T\dot{t} + \boldsymbol{R}\cdot\dot{\boldsymbol{r}} + \boldsymbol{U}\cdot\dot{\boldsymbol{u}} + \boldsymbol{W}\cdot\boldsymbol{\omega}_0. \tag{3.74}$$

Here the different functions are defined by $T = \partial L/\partial\dot{t}$, $R_i = \partial L/\partial\dot{r}^i$, $D_i = \partial L/\partial\alpha^i$, $Z_i = \partial L/\partial\omega^i$, $U_i = \partial L/\partial\dot{u}^i$, $V_i = \partial L/\partial\dot{\rho}^i$ and $W_i = \partial L/\partial{\omega_0}^i$. They will be functions of the kinematical variables and homogeneous functions of zero degree of the derivatives. It must be remarked that the relationship between the different sets of variables involving derivatives of the kinematical variables $(\dot{\boldsymbol{u}}, \dot{\boldsymbol{\rho}})$, $(\boldsymbol{\alpha}, \boldsymbol{\omega})$ and $(\dot{\boldsymbol{u}}, \boldsymbol{\omega}_0)$ is a linear one. It turns out that the same thing happens to the relation between the corresponding sets of functions $(\boldsymbol{U}, \boldsymbol{V})$, $(\boldsymbol{D}, \boldsymbol{Z})$ and $(\boldsymbol{U}, \boldsymbol{W})$. We see that, irrespective of the particular Lagrangian model we choose, while keeping fixed the same kinematical variables, it is in terms of these partial derivatives of the Lagrangian that we can construct Noether's constants of the motion.

For instance, by using expressions (3.63), (3.64), (3.67) and (3.68) we obtain the relationship between functions $\boldsymbol{D}$, $\boldsymbol{Z}$ and functions $\boldsymbol{U}$ and $\boldsymbol{W}$. It can be shown that

$$\boldsymbol{Z} = \boldsymbol{u}\times\boldsymbol{U} + \boldsymbol{W}, \tag{3.75}$$

as in the Galilei case, and

$$\boldsymbol{D}/c = \boldsymbol{U} - \frac{\boldsymbol{u}\cdot\boldsymbol{U}}{c^2}\boldsymbol{u} - \frac{\gamma(u)}{(1+\gamma(u))c^2}\boldsymbol{u}\times\boldsymbol{W}. \tag{3.76}$$

Conversely,

$$\boldsymbol{U} = \frac{\gamma(u)}{c}\left(\boldsymbol{D} + \frac{\gamma(u)^2}{(1+\gamma(u))c^2}(\boldsymbol{u}\cdot\boldsymbol{D})\boldsymbol{u} + \frac{\gamma(u)}{(1+\gamma(u))c}\boldsymbol{u}\times\boldsymbol{Z}\right) \tag{3.77}$$

$$\boldsymbol{W} = \gamma(u)\left(\boldsymbol{Z} - \frac{\gamma(u)}{(1+\gamma(u))c^2}(\boldsymbol{u}\cdot\boldsymbol{Z})\boldsymbol{u} - \frac{1}{c}\boldsymbol{u}\times\boldsymbol{D}\right). \tag{3.78}$$

Since the Poincaré group has no exponents, the Lagrangian $L$ is invariant under $\mathcal{P}$, thus, for instance, the different functions of Lagrangian (3.73) transform in the following way:

$$T'(\tau) = \gamma T(\tau) - \gamma(\boldsymbol{v}.R(\boldsymbol{\mu})\boldsymbol{R}(\tau)), \tag{3.79}$$

$$\boldsymbol{R}'(\tau) = R(\boldsymbol{\mu})\boldsymbol{R}(\tau) - \gamma\boldsymbol{v}T(\tau)/c^2 + \frac{\gamma^2}{(1+\gamma)c^2}(\boldsymbol{v}\cdot R(\boldsymbol{\mu})\boldsymbol{R}(\tau))\boldsymbol{v}, \tag{3.80}$$

$$D'(\tau) = \gamma R(\mu)D(\tau) - \frac{\gamma^2}{(1+\gamma)c^2}(v \cdot R(\mu)D(\tau))v - \frac{\gamma}{c}(v \times R(\mu)Z(\tau)), \tag{3.81}$$

$$Z'(\tau) = \gamma R(\mu)Z(\tau) - \frac{\gamma^2}{(1+\gamma)c^2}(v \cdot R(\mu)Z(\tau))v + \frac{\gamma}{c}(v \times R(\mu)D(\tau)). \tag{3.82}$$

Observables $-T$ and $R$ transform like the contravariant components of a four-vector, and $D$ and $Z$ transform like the strict components of an antisymmetric tensor $Z^{\mu\nu} = -Z^{\nu\mu}$, $D_i = Z^{0i}$, $Z_i = \frac{1}{2}\epsilon_{ijk}Z^{jk}$. We see that they are translation invariant and therefore they will be in general functions of $(u, \rho, \alpha/\dot{t}, \omega/\dot{t})$, although we expect no explicit dependence on $\rho$ variables.

In covariant notation the magnitudes $Z^{\mu\nu} = \partial L/\partial\Omega_{\mu\nu}$, and the invariant part of the Lagrangian involving $\alpha$ and $\omega$, can also be written as

$$D \cdot \alpha + Z \cdot \omega \equiv \frac{1}{2}Z^{\mu\nu}\Omega_{\mu\nu}.$$

The generalized canonical momenta are, as in the Galilei case:

$$p_{r(1)} = R - \frac{dU}{dt}, \tag{3.83}$$

$$p_{r(2)} = U, \tag{3.84}$$

$$p_\rho = V, \tag{3.85}$$

and thus, under time translations, the term $L - p^i_{(s)}q^{(s)}_i$ of (1.52) reduces to

$$T + R\cdot\frac{\dot{r}}{\dot{t}} + U\cdot\frac{\dot{u}}{\dot{t}} + V\cdot\frac{\dot{\rho}}{\dot{t}} - R\cdot\frac{\dot{r}}{\dot{t}} + \frac{dU}{dt}\cdot\frac{\dot{r}}{\dot{t}} - U\cdot\frac{\dot{u}}{\dot{t}} - V\cdot\frac{\dot{\rho}}{\dot{t}} = T + \frac{dU}{dt}\cdot\frac{\dot{r}}{\dot{t}}.$$

We thus obtain the same expressions for $H$ and $P$, in terms of $T$, $R$ and $U$, as in the nonrelativistic case.

Under a pure Lorentz transformation

$$\delta t = \frac{r}{c^2}\cdot\delta v, \quad \delta r = t\delta v, \quad \delta u = \delta v - \frac{1}{c^2}(u\cdot\delta v)u,$$

$$\delta\rho = \frac{1}{2}\Gamma(u \times \delta v + \delta v(u\cdot\rho) - u(\rho\cdot\delta v) - \rho(\delta v\cdot(u\times\rho))),$$

i.e.,

$$M_i = \frac{r_i}{c^2}, \quad M^{(0)}_{ij} = t\delta_{ij}, \quad M^{(1)}_{ij} = \delta_{ij} - \frac{u_i u_j}{c^2},$$

$$M_{ik}^{(\rho)} = \frac{1}{2}\Gamma(\epsilon_{ijk}u_j + \delta_{ik}(\boldsymbol{u}\cdot\boldsymbol{\rho}) - u_i\rho_k - \epsilon_{kjl}u_j\rho_l\rho_i).$$

Here we set $\Gamma \equiv \gamma(u)/(1+\gamma(u))c^2$. The part

$$p_{\rho_i}M_{ik}^{(\rho)} = \frac{\partial L}{\partial \dot{\rho}_i}M_{ik}^{(\rho)} = \left(\frac{\partial L}{\partial \alpha_l}\frac{\partial \alpha_l}{\partial \omega_{0r}} + \frac{\partial L}{\partial \omega_l}\frac{\partial \omega_l}{\partial \omega_{0r}}\right)\frac{\partial \omega_{0r}}{\partial \dot{\rho}_i}M_{ik}^{(\rho)},$$

with the term

$$p_{r(2)_i}M_{rik}^{(1)} = \frac{\partial L}{\partial \dot{u}_i}M_{rik}^{(1)} = \left(\frac{\partial L}{\partial \alpha_l}\frac{\partial \alpha_l}{\partial \dot{u}_i} + \frac{\partial L}{\partial \omega_l}\frac{\partial \omega_l}{\partial \dot{u}_i}\right)M_{rik}^{(1)},$$

together with the relations

$$\frac{\partial \omega_{0r}}{\partial \dot{\rho}_i} = \frac{2}{1+\rho^2}(\delta_{ri} + \epsilon_{rji}\rho_j),$$

$$\frac{\partial \alpha_l}{\partial \omega_{0r}} = \frac{\gamma(u)}{c}\epsilon_{ljr}u_j, \qquad \frac{\partial \omega_l}{\partial \omega_{0r}} = \gamma(u)\delta_{lr} - \frac{\gamma(u)^2}{1+\gamma(u)}\frac{u_l u_r}{c^2},$$

$$\frac{\partial \alpha_l}{\partial \dot{u}_i} = \frac{\gamma(u)}{c}\delta_{li} + \frac{\gamma(u)^3}{1+\gamma(u)}\frac{u_l u_i}{c^3}, \qquad \frac{\partial \omega_l}{\partial \dot{u}_i} = -\frac{\gamma(u)^2}{(1+\gamma(u))c^2}\epsilon_{lji}u_j,$$

finally reduces to the function $\boldsymbol{D}/c$ while the remaining terms cancel out.

Similarly, although $\boldsymbol{\alpha}$ and $\boldsymbol{\omega}$ transform under rotations in the form $\boldsymbol{\alpha}' = R(\boldsymbol{\mu})\boldsymbol{\alpha}$ and $\boldsymbol{\omega}' = R(\boldsymbol{\mu})\boldsymbol{\omega}$, only the terms $\boldsymbol{r}\times\boldsymbol{P}$ and $\boldsymbol{Z}$ survive in the final expression of the angular momentum.

Collecting all results of applying Noether's theorem to the whole Poincaré group, we find the following constants of the motion:

$$\text{Energy} \quad H = -T - (d\boldsymbol{U}/dt)\cdot\boldsymbol{u}, \tag{3.86}$$

$$\text{linear momentum} \quad \boldsymbol{P} = \boldsymbol{R} - d\boldsymbol{U}/dt, \tag{3.87}$$

$$\text{kinematical momentum} \quad \boldsymbol{K} = H\boldsymbol{r}/c^2 - \boldsymbol{P}t - \boldsymbol{D}/c, \tag{3.88}$$

$$\text{angular momentum} \quad \boldsymbol{J} = \boldsymbol{r}\times\boldsymbol{P} + \boldsymbol{Z}, \tag{3.89}$$

where $\boldsymbol{Z} = \boldsymbol{u}\times\boldsymbol{U} + \boldsymbol{W}$, as given in (3.75), plays the role in the relativistic case of the observable $\boldsymbol{Z}$ in the Galilei one (2.67).

The only difference with the Galilei case is the expression of the kinematical momentum. In spite of the term $H\boldsymbol{r}/c^2$ there, we have $m\boldsymbol{r}$ and instead of $\boldsymbol{D}/c$ we found the term $\boldsymbol{U}$. In fact, the function $\boldsymbol{D}/c$, given in (3.76), reduces to $\boldsymbol{U}$ in the nonrelativistic limit when $u/c \to 0$ and both terms transform into the Galilei ones.

This difference arises because the relativistic Lagrangian depends on the same variables as the nonrelativistic one. All these variables transform analogously under translations and rotations, respectively, and therefore $H$, $\boldsymbol{P}$ and $\boldsymbol{J}$ will have the same expression in terms of the partial derivatives $\partial L/\partial \dot{x}_i$, of the Lagrangian. The only difference comes from the different way the kinematical variables transform under boosts in the Galilei and Poincaré approach and the non-vanishing gauge function associated to the pure Galilei transformations. This is important because the spin operator will arise from the general expression of $\boldsymbol{J}-\boldsymbol{q}\times\boldsymbol{P}$. But this general expression is the same in both formalisms with the only exception of the definition of the center of mass position in terms of the other functions. Therefore, when quantizing these systems, we will show that the spin operator in the center of mass frame takes exactly the same form in the relativistic and nonrelativistic case.

Total energy and linear momentum transform like the components of a four-vector $P^\mu \equiv (H/c, \boldsymbol{P})$, and the six magnitudes $\boldsymbol{K}$ and $\boldsymbol{J}$ can be written in a covariant way as the essential components of the antisymmetric tensor

$$J_{\mu\nu} = x_\mu P_\nu - x_\nu P_\mu + Z_{\mu\nu}. \tag{3.90}$$

It has two parts, one $Z_{\mu\nu}$ that is translation invariant and another with the form of a generalized orbital angular momentum $x_\mu P_\nu - x_\nu P_\mu$, similarly as in the point particle case. One is tempted to consider that the part $\boldsymbol{Z}$ of the total angular momentum is to be considered as the spin of the system, as is usually done in quantum mechanics. However we shall delay the definition of spin until the definition of the center of mass of the system, because this magnitude is not a constant of the motion for a free particle, but satisfies the dynamical equation $d\boldsymbol{Z}/dt = \boldsymbol{P}\times\boldsymbol{u}$.

If we consider material systems for which $H > 0$, we can define the magnitude $\boldsymbol{k}$ with dimensions of length $\boldsymbol{k} = c\boldsymbol{D}/H$, in such a way that taking the $\tau$-derivative of both sides of the kinematical momentum (3.88) we get $\dot{\boldsymbol{K}} = (\dot{\boldsymbol{r}} - \dot{\boldsymbol{k}})H/c^2 - \boldsymbol{P}\dot{t} = 0$. Therefore, the linear momentum takes the form $\boldsymbol{P} = (H/c^2)d(\boldsymbol{r}-\boldsymbol{k})/dt$, thus, defining the position of the center of mass of the system $\boldsymbol{q} = \boldsymbol{r} - \boldsymbol{k}$, that moves at constant velocity, and the observable $\boldsymbol{k} = \boldsymbol{r} - \boldsymbol{q}$ is the relative position of the system with respect to the center of mass.

We see again the similarity with the nonrelativistic case, because $\boldsymbol{D}/c$ is playing the role of $\boldsymbol{U}$ in the nonrelativistic approach, and the relative position vector $\boldsymbol{k} = (c^2/H)\boldsymbol{D}/c$ is the equivalent to $\boldsymbol{k} = \boldsymbol{U}/m$ whenever we can replace the energy $H$ by $mc^2$. However, there are some differences. In the nonrelativistic case it is only the function $\boldsymbol{U}$, and therefore the dependence of the Lagrangian on $\dot{\boldsymbol{u}}$, that produces the zitterbewegung. In the relativistic case, as we see from (3.76), both functions $\boldsymbol{U}$ and $\boldsymbol{W}$ contribute to $\boldsymbol{D}$ and therefore to the existence of the non-vanishing vector $\boldsymbol{k}$, which measures the separation between the center of mass from the center of charge. Both the functions $\boldsymbol{U}$ and $\boldsymbol{W}$, which contribute to the spin structure, are related to the zitterbewegung. This is because the Lorentz boosts do not form a subgroup of the Lorentz group, and velocity and orientation variables become intertwined.

In terms of the center of mass position, the kinematical momentum (3.88) can be written as:

$$\boldsymbol{K} = H\boldsymbol{q}/c^2 - \boldsymbol{P}\,t. \tag{3.91}$$

The center of mass observer is defined, as in the nonrelativistic case, by the conditions $\boldsymbol{P} = \boldsymbol{K} = 0$. It defines the class of observers, up to an arbitrary rotation and time translation, for which the center of mass is at rest and placed at the origin of the reference system. The six conditions $\boldsymbol{P} = \boldsymbol{K} = 0$, lead to $d\boldsymbol{q}/dt = 0$, and $\boldsymbol{q} = 0$ respectively, thus justifying this definition. We see again that $\boldsymbol{P}$ is not pointing along the direction of the velocity $\boldsymbol{u}$. This is because the point $\boldsymbol{r}$ is not the center of mass of the system. Using the same arguments as in the nonrelativistic case, for instance a minimal coupling with the external potentials, independent of the acceleration, $-e\phi(t,\boldsymbol{r})\dot{t} + e\boldsymbol{A}(t,\boldsymbol{r})\dot{\boldsymbol{r}}$, dynamical equations become

$$\frac{d\boldsymbol{P}}{dt} = \frac{d}{dt}\left(\frac{H}{c^2}\frac{d\boldsymbol{q}}{dt}\right) = e(\boldsymbol{E}(t,\boldsymbol{r}) + \boldsymbol{u} \times \boldsymbol{B}(t,\boldsymbol{r})).$$

We reach the conclusion that $\boldsymbol{r}$ is in fact the point where the external fields are defined, and therefore it represents the center of charge of the particle.

The spin of the system is defined as

$$\boldsymbol{S} \equiv \boldsymbol{J} - \boldsymbol{q} \times \boldsymbol{P} = \boldsymbol{J} - \frac{c^2}{H}\boldsymbol{K} \times \boldsymbol{P} = \boldsymbol{Z} + \boldsymbol{k} \times \boldsymbol{P}. \tag{3.92}$$

It is expressed in terms of the ten constants of the motion $H$, $\boldsymbol{J}$, $\boldsymbol{P}$ and $\boldsymbol{K}$, and thus it is also another constant of the motion for the free particle, similarly as in the Galilei case, where it takes the expression (2.69).

From the constants of the motion (3.86 - 3.89) we can define new constants:

$$w^0 = \boldsymbol{P} \cdot \boldsymbol{J} = \boldsymbol{P} \cdot \boldsymbol{S}, \tag{3.93}$$

$$\boldsymbol{w} = H\boldsymbol{J}/c - c\boldsymbol{K} \times \boldsymbol{P} = H(\boldsymbol{Z} + \boldsymbol{k} \times \boldsymbol{P})/c = H\boldsymbol{S}/c, \tag{3.94}$$

that can be expressed in terms of the four-vector $P^\mu$ and the antisymmetric tensor $Z^{\mu\nu}$ in the form $w_\sigma = \frac{1}{2}\epsilon_{\sigma\mu\nu\lambda}Z^{\mu\nu}P^\lambda$. The observables (3.93, 3.94) are the components of the Pauli-Lubanski four-vector.

It turns out that the four-vectors $P^\mu$ and $w^\mu$ are orthogonal to each other, $P^\mu w_\mu = 0$, and, $P_\mu P^\mu = m^2c^2$ and $-w_\mu w^\mu = m^2c^2S^2$ are two functionally independent invariants, constants of the motion that define two invariant properties of the system $m$ and $S$, the mass and the absolute value of the spin, respectively. These two properties are the only invariant properties that characterize the elementary Poincaré particle.

Since the first part of the Lagrangian $T\dot{t} + \boldsymbol{R}\cdot\dot{\boldsymbol{r}} = -H\dot{t} + \boldsymbol{P}\cdot\dot{\boldsymbol{r}} = -P_\mu \dot{x^\mu}$, is itself Poincaré invariant, then the second part that reduces to $Z_{\mu\nu}\Omega^{\mu\nu}/2$, will also be invariant. In terms of $\Omega^{\mu\nu}$ we can form different invariants, namely the contractions $(1/2)\Omega_{\mu\nu}\Omega^{\mu\nu} = \boldsymbol{\alpha}^2 - \boldsymbol{\omega}^2$ and in terms of the dual tensor $(1/2)\epsilon_{\mu\nu\sigma\rho}\Omega^{\mu\nu}\Omega^{\sigma\rho} = \boldsymbol{\alpha}\cdot\boldsymbol{\omega}$. Then we have two possibilities for $Z_{\mu\nu}$; it might be either proportional to $\Omega_{\mu\nu}$ or to $\epsilon_{\mu\nu\sigma\rho}\Omega^{\sigma\rho}$. This amounts to the choice of functions $\boldsymbol{D}$ and $\boldsymbol{Z}$, respectively, either proportional to $\boldsymbol{\alpha}$ and $\boldsymbol{\omega}$ or vice versa, $\boldsymbol{D}$ proportional to $\boldsymbol{\omega}$ and $\boldsymbol{Z}$ to $\boldsymbol{\alpha}$, with the same proportionality coefficients, as we found in the nonrelativistic approach in Section 6. of Chapter 2.

These invariants take the explicit form:

$$\boldsymbol{\alpha}^2 - \boldsymbol{\omega}^2 = \frac{2\gamma(u)^2\left(\dot{\boldsymbol{u}}^2 + \dot{\boldsymbol{u}}\cdot(\boldsymbol{u}\times\boldsymbol{\omega}_0)\right)}{(1+\gamma(u))c^2} + \frac{(2+2\gamma(u)+\gamma(u)^2)\gamma(u)^4}{(1+\gamma(u))^2c^4}(\boldsymbol{u}\cdot\dot{\boldsymbol{u}})^2 - \boldsymbol{\omega}_0^2, \quad (3.95)$$

$$\boldsymbol{\alpha}\cdot\boldsymbol{\omega} = \frac{\gamma(u)}{c}\left[\dot{\boldsymbol{u}}\cdot\boldsymbol{\omega}_0 + \frac{\gamma(u)^2}{1+\gamma(u)}\frac{(\boldsymbol{u}\cdot\dot{\boldsymbol{u}})(\boldsymbol{u}\cdot\boldsymbol{\omega}_0)}{c^2}\right]. \quad (3.96)$$

These expressions are homogeneous functions of second degree in terms of the derivatives of the kinematical variables. If we divide these terms by the homogeneous invariant function of first degree $c\dot{t}/\gamma(u) = \sqrt{c^2\dot{t}^2 - \dot{\boldsymbol{r}}^2}$ we get first order invariants that can be used to construct possible relativistic invariant Lagrangians.

These two possible selections lead to the following Lagrangians:

$$L_B = -mc\sqrt{c^2\dot{t}^2 - \dot{\boldsymbol{r}}^2} + \frac{S^2}{mc}\frac{\boldsymbol{\alpha}^2 - \boldsymbol{\omega}^2}{\sqrt{c^2\dot{t}^2 - \dot{\boldsymbol{r}}^2}}, \quad (3.97)$$

$$L_F = -mc\sqrt{c^2\dot{t}^2 - \dot{\boldsymbol{r}}^2} + \frac{S^2}{mc}\frac{\boldsymbol{\alpha}\cdot\boldsymbol{\omega}}{\sqrt{c^2\dot{t}^2 - \dot{\boldsymbol{r}}^2}}, \quad (3.98)$$

where $\boldsymbol{\alpha}^2 - \boldsymbol{\omega}^2$ and $\boldsymbol{\alpha}\cdot\boldsymbol{\omega}$ are given in (3.95) and (3.96) respectively. The coefficients $S^2/mc$ have been introduced by dimensional considerations and when $S = 0$ we recover the point particle case. Other possibilities could be the use of the first order terms of the form $\sqrt{\boldsymbol{\alpha}^2 - \boldsymbol{\omega}^2}$ or

$\sqrt{\boldsymbol{\alpha} \cdot \boldsymbol{\omega}}$, instead. Particular cases of these will be considered later. In the absence of rotation $\boldsymbol{\omega}_0 \equiv 0$, the invariant $\boldsymbol{\alpha} \cdot \boldsymbol{\omega}$ vanishes but the other $\boldsymbol{\alpha}^2 - \boldsymbol{\omega}^2$ is different from zero. In this case the spin is only related to the zitterbewegung and a particular example will be analyzed in Section 3.2 of this Chapter.

Lagrangians of the kind (3.97) can be found in the literature. In fact, Constantelos [5] mentions a Lagrangian which is a particular case of this one with $\boldsymbol{\omega}_0 = 0$, and thus the particle has zitterbewegung but no rotation. The Lagrangian depends on the velocity and acceleration but not on the angular variables. In a different context, the work by Hanson and Regge [6] assumes that $\dot{\boldsymbol{u}} = 0$, and the invariant $\boldsymbol{\alpha}^2 - \boldsymbol{\omega}^2$ reduces to $-\boldsymbol{\omega}_0^2$. The particle has no internal motion but it is a spinning top that rotates with a certain angular velocity, and the spin is pointing along the direction of the angular velocity. This model, which is not devoid of problems, will be analyzed later in Chapter 5.

However, to our knowledge, Lagrangians of the type (3.98) have not been described earlier. Both kinds of Lagrangians (3.97, 3.98) lead to nonlinear dynamical equations for the position of the charge.

Lagrangian (3.97) gives rise, for the center of mass observer, to the following dynamical equations:

$$\boldsymbol{r} = -\frac{2S^2}{m^2c^4}\gamma(u)^2\left[\frac{d\boldsymbol{u}}{dt} + \boldsymbol{u}\times\boldsymbol{\Omega} + \frac{\gamma(u)^2}{1+\gamma(u)}\left(\frac{\boldsymbol{u}}{c}\cdot\frac{d\boldsymbol{u}}{dt}\right)\frac{\boldsymbol{u}}{c}\right], \quad (3.99)$$

$$\boldsymbol{S} = \frac{2S^2}{mc^2}\gamma(u)\left[\gamma(u)\boldsymbol{\Omega} - \frac{\gamma(u)^2}{(1+\gamma(u))c^2}\left((\boldsymbol{u}\cdot\boldsymbol{\Omega})\boldsymbol{u} + \boldsymbol{u}\times\frac{d\boldsymbol{u}}{dt}\right)\right]. \quad (3.100)$$

Here $\boldsymbol{\Omega} = \boldsymbol{\omega}_0/t$ is the angular velocity and the spin $\boldsymbol{S}$ is a constant vector. We see that $\boldsymbol{u}\cdot\boldsymbol{r} \sim \boldsymbol{u}\cdot d\boldsymbol{u}/dt$ and $\boldsymbol{u}\cdot\boldsymbol{S} \sim \boldsymbol{u}\cdot\boldsymbol{\Omega}$. Then, there exist solutions of equations (3.99, 3.100) with constant values of the velocity $|\boldsymbol{u}|$ and $|\boldsymbol{\Omega}|$ and also with the condition $\boldsymbol{\Omega}\cdot\boldsymbol{u} = 0$. In this case, the zitterbewegung is circular and takes place in a plane orthogonal to the angular velocity and to the spin, as is shown in the picture. We see that for this model, although the analytical expression of spin in terms of the angular velocity is rather involved, nevertheless for the center of mass observer $\boldsymbol{S}$ and $\boldsymbol{\Omega}$ lie parallel to each other. In general the zitterbewegung, contained in a plane orthogonal to the spin in the center of mass frame, will be more complicated than in the nonrelativistic case. We will show this in the next section when analyzing a simpler model in which the spin depends only on this internal motion and where some alternative trajectories for the center of charge are depicted.

Similarly, Lagrangian (3.98) also describes motions with constant absolute value of the angular velocity, being orthogonal $\boldsymbol{u}$ and $\boldsymbol{\Omega}$. Dynam-

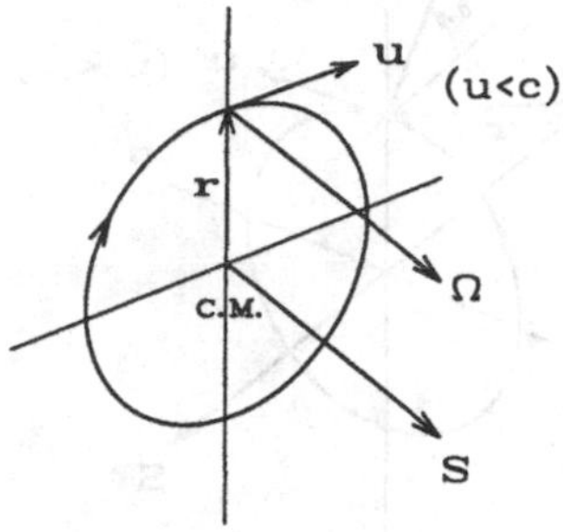


*Figure 3.1.* Motion of the center of charge of the system (3.97) around the C.M.

ical equations for the center of mass observer are:

$$\boldsymbol{r} = \frac{S^2}{m^2c^3}\gamma(u)\left[\gamma(u)\boldsymbol{\Omega} - \frac{\gamma(u)^2}{(1+\gamma(u))c^2}\left((\boldsymbol{u}\cdot\boldsymbol{\Omega})\boldsymbol{u} + \boldsymbol{u}\times\frac{d\boldsymbol{u}}{dt}\right)\right], \tag{3.101}$$

$$\boldsymbol{S} = \frac{S^2}{mc^3}\gamma(u)^2\left[\frac{d\boldsymbol{u}}{dt} + \boldsymbol{u}\times\boldsymbol{\Omega} + \frac{\gamma(u)^2}{1+\gamma(u)}\left(\frac{\boldsymbol{u}}{c}\cdot\frac{d\boldsymbol{u}}{dt}\right)\frac{\boldsymbol{u}}{c}\right]. \tag{3.102}$$

This motion is described in Figure 3.2, where the spin $\boldsymbol{S}$ is a constant vector which is not lying along the angular velocity.

In the examples shown above, the total spin and the zitterbewegung part, which is related to the orbital motion of the charge position around the center of mass, are both orthogonal to the plane of this internal motion. This will have consequences for the definition of the gyromagnetic ratio of the particle because the magnetic moment, according to the classical definition, is related to the current and therefore is orthogonal to the zitterbewegung plane. When expressed in terms of the total spin their relationship will be different than the usual one for a point particle. Another general feature is that in the center of mass frame and for low energy processes, the behaviour of this system is like that of a point charge placed at the time-averaged position, *i.e.*, at the center of mass but with the addition of a constant magnetic moment $\boldsymbol{\mu}$ produced by the current and an oscillating electric dipole $\boldsymbol{d} = e\boldsymbol{k}$ of time average value zero, and thus negligible except in cases of a very close interaction

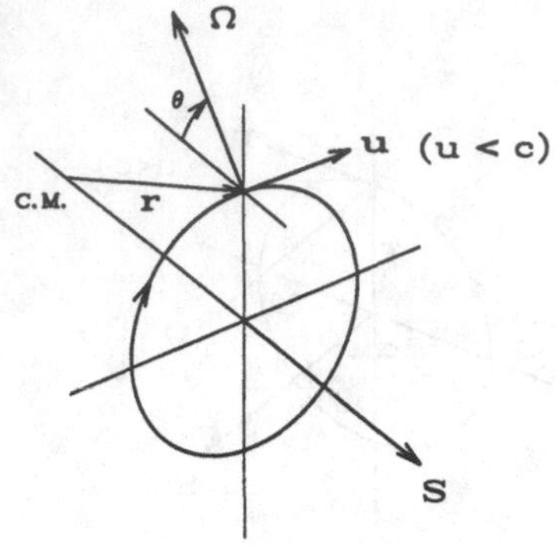


*Figure 3.2.* Motion of the center of charge of the system (3.98) around the C.M.

between particles or in high energy events where the time of flight will be of the same order as the period of this internal motion.

Similarly, as in the Galilei case, if the dependence of the Lagrangian on the orientation variables $\boldsymbol{\rho}$ is only through its dependence on the angular velocity $\boldsymbol{\omega}_0(\boldsymbol{\rho}, \dot{\boldsymbol{\rho}})$, then we also obtain $d\boldsymbol{W}/d\tau = \boldsymbol{\omega}_0 \times \boldsymbol{W}$. Therefore, taking the $\tau-$derivative of the total angular momentum, we reach expression (2.74)

$$\dot{\boldsymbol{r}} \times \boldsymbol{R} + \dot{\boldsymbol{u}} \times \boldsymbol{U} + \boldsymbol{\omega}_0 \times \boldsymbol{W} = 0.$$

Nevertheless, in the relativistic case, $\boldsymbol{R}$ observable is not lying along $\dot{\boldsymbol{r}}$ but by taking the $\tau-$derivative of the kinematical momentum and using (3.88) and (3.76) we arrive at:

$$\boldsymbol{R} = \frac{H}{c^2}\boldsymbol{u} + \frac{d}{dt}\left(\frac{\boldsymbol{u}\cdot\boldsymbol{U}}{c^2}\boldsymbol{u} + \frac{\gamma}{(1+\gamma)c^2}\boldsymbol{W}\times\boldsymbol{u}\right) \tag{3.103}$$

which clearly has as a nonrelativistic limit the expression $\boldsymbol{R} = m\boldsymbol{u}$.

We obtain an alternative expression for the linear momentum

$$\boldsymbol{P} = \frac{H}{c^2}\boldsymbol{u} - \frac{1}{c}\frac{d\boldsymbol{D}}{dt}, \tag{3.104}$$

which shows that $\boldsymbol{P}$ and $\boldsymbol{u}$ are not parallel vectors in general.

Then, even in the case where the Lagrangian is independent of $\boldsymbol{\omega}_0$, there exists a relationship between the velocity and acceleration that produces things like a precession of the orbits of the zitterbewegung, as we shall analyze in the next Section 3.2, in contrast with the nonrelativistic ones where orbits are, in general, closed ellipses.

## 3.2 RELATIVISTIC PARTICLES WITH (ANTI)ORBITAL SPIN

Let us analyze first a class of Bradyons for which the spin structure is simpler than in the general case. In the Galilei approach we got two simpler models by defining the corresponding kinematical spaces as $\mathcal{G}/SO(3)$ and $\mathcal{G}/\{\mathbb{R}^3,+\}$, where $\{\mathbb{R}^3,+\}$ is the subgroup of pure Galilei transformations or boosts. These kinematical spaces gave rise respectively to the free particle with spin related to the zitterbewegung (Section 4., Chapter 2) and the one with spin related only to the angular velocity (Section 5., Chapter 2).

In the present formalism the kinematical space $\mathcal{P}/SO(3)$ can also be defined but the equivalent to the other manifold $\mathcal{G}/\{\mathbb{R}^3,+\}$ is no longer possible because the Lorentz boosts do not form a subgroup of the Poincaré group and it is not possible to define the corresponding homogeneous space. It turns out that in the relativistic case it is not possible to isolate angular variables from acceleration variables and whenever our system has directional properties, and thus orientation, it also necessarily has zitterbewegung. Nevertheless let us analyze first a simple model of spinning particle that the spin is independent of the angular velocity and therefore no orientation variables are involved.

The kinematical space of this system is the seven-dimensional manifold spanned by the variables $(t, \boldsymbol{r}, \boldsymbol{u})$, $u < c$, interpreted as time, position and velocity respectively which is the homogeneous space of $\mathcal{P}$, $X = \mathcal{P}/SO(3)$. Since $u < c$ they will have as a nonrelativistic limit the examples shown in Sections 4. and 4.5 of Chapter 2.

If $\boldsymbol{u}$ is the velocity of the particle, we form in terms of $\boldsymbol{u}$ the dimensionless four-velocity vector $u^\mu \equiv (\gamma(u), \gamma(u)\boldsymbol{u}/c)$ with $u^\mu u_\mu = 1$. If we assume that all kinematical variables are functions of the dimensionless parameter $\tau$, then, taking the derivative with respect to $\tau$ of this four-vector we get another dimensionless four-vector

$$\dot{u}^\mu \equiv (\dot{\gamma}(u), \dot{\gamma}(u)\boldsymbol{u}/c + \gamma(u)\dot{\boldsymbol{u}}/c).$$

We see that

$$\dot{u}^\mu \dot{u}_\mu = \dot{\gamma}(u)^2/\gamma(u)^2 - \gamma(u)^2\dot{\boldsymbol{u}}^2/c^2 - 2\gamma(u)\dot{\gamma}(u)\boldsymbol{u}\cdot\dot{\boldsymbol{u}}/c^2,$$

is a Poincaré invariant expression, homogeneous of second order in the derivatives of $\boldsymbol{u}$, and where $\dot{\gamma}(u) = \gamma(u)^3\boldsymbol{u}\cdot\dot{\boldsymbol{u}}/c^2$.

Since $\dot{t}/\gamma(u)$ is another first order invariant in terms of the derivatives of the kinematical variables, we can form two first order Poincaré invariant terms depending on the acceleration:

$$-\frac{\gamma(u)}{\dot{t}}\dot{u}^\mu\dot{u}_\mu = \frac{\gamma(u)^3}{c^2\dot{t}}\left(\dot{\boldsymbol{u}}^2 + \frac{\gamma(u)^2}{c^2}(\boldsymbol{u}\cdot\dot{\boldsymbol{u}})^2\right), \qquad (3.105)$$

and

$$c\sqrt{-\dot{u}^\mu \dot{u}_\mu} = \sqrt{\gamma(u)^2 \dot{\boldsymbol{u}}^2 + \frac{\gamma(u)^4}{c^2}(\boldsymbol{u}\cdot\dot{\boldsymbol{u}})^2} = \frac{\gamma(u)}{c}\sqrt{c^2\dot{\boldsymbol{u}}^2 + \gamma(u)^2(\boldsymbol{u}\cdot\dot{\boldsymbol{u}})^2}. \tag{3.106}$$

Thus we can have the following Lagrangians:

$$L_1 = -ac\sqrt{c^2\dot{t}^2 - \dot{\boldsymbol{r}}^2} - b\frac{\gamma(u)^3}{\dot{t}}\left(\dot{\boldsymbol{u}}^2 + \frac{\gamma(u)^2}{c^2}(\boldsymbol{u}\cdot\dot{\boldsymbol{u}})^2\right), \tag{3.107}$$

$$L_2 = -ac\sqrt{c^2\dot{t}^2 - \dot{\boldsymbol{r}}^2} - b\gamma(u)\sqrt{c^2\dot{\boldsymbol{u}}^2 + \gamma(u)^2(\boldsymbol{u}\cdot\dot{\boldsymbol{u}})^2}, \tag{3.108}$$

in terms of two arbitrary constant parameters $a$ and $b$.

The Lagrangian $L_1$ in the case of low velocity and in a time evolution description, gives rise to:

$$L_{1NR} = -ac^2\left(1 - \frac{\boldsymbol{u}^2}{2c^2}\right) - b\left(\frac{d\boldsymbol{u}}{dt}\right)^2, \tag{3.109}$$

that has as a nonrelativistic limit the Galilei Lagrangian (2.76) and thus the parameters are identified with $a = m$ and $b = m/2\omega^2$, $m$ being the Galilei mass of the system, and $\omega$ the frequency parameter of the internal zitterbewegung.

Total linear momentum and energy are expressed as usual as

$$\boldsymbol{P} = \boldsymbol{R} - \frac{d\boldsymbol{U}}{dt}, \quad H = -T - \boldsymbol{u}\cdot\frac{d\boldsymbol{U}}{dt},$$

$$H = ac^2\gamma(u) - b\gamma(u)^5\left(\frac{d\boldsymbol{u}}{dt}\right)^2 + 2b\gamma(u)^5\,\boldsymbol{u}\cdot\frac{d^2\boldsymbol{u}}{dt^2} + \frac{5b\gamma(u)^7}{c^2}\left(\boldsymbol{u}\cdot\frac{d\boldsymbol{u}}{dt}\right)^2.$$

The observable $\boldsymbol{U}$ is:

$$\boldsymbol{U} = \frac{\partial L_1}{\partial \dot{\boldsymbol{u}}} = -2b\gamma^3\frac{d\boldsymbol{u}}{dt} - \frac{2b\gamma^5}{c^2}\left(\boldsymbol{u}\cdot\frac{d\boldsymbol{u}}{dt}\right)\boldsymbol{u},$$

and its time derivative

$$\begin{aligned}\frac{d\boldsymbol{U}}{dt} &= -2b\gamma^3\frac{d^2\boldsymbol{u}}{dt^2} - \frac{8b\gamma^5}{c^2}\left(\boldsymbol{u}\cdot\frac{d\boldsymbol{u}}{dt}\right)\frac{d\boldsymbol{u}}{dt} \\ &- \frac{2b\gamma^5}{c^2}\left(\boldsymbol{u}\cdot\frac{d^2\boldsymbol{u}}{dt^2} + \frac{5\gamma^2}{c^2}\left(\boldsymbol{u}\cdot\frac{d\boldsymbol{u}}{dt}\right)^2 + \left(\frac{d\boldsymbol{u}}{dt}\right)^2\right)\boldsymbol{u},\end{aligned}$$

and the observable $\boldsymbol{R} = \partial L/\partial \dot{\boldsymbol{r}}$, becomes

$$\boldsymbol{R} = \left(a\gamma - \frac{3b\gamma^5}{c^2}\left(\frac{d\boldsymbol{u}}{dt}\right)^2 - \frac{5b\gamma^7}{c^4}\left(\boldsymbol{u}\cdot\frac{d\boldsymbol{u}}{dt}\right)^2\right)\boldsymbol{u} - \frac{2b\gamma^5}{c^2}\left(\boldsymbol{u}\cdot\frac{d\boldsymbol{u}}{dt}\right)\frac{d\boldsymbol{u}}{dt}.$$

In this example in which $L$ does not depend on the angular velocity we also have that $\dot{\boldsymbol{r}} \times \boldsymbol{R} + \dot{\boldsymbol{u}} \times \boldsymbol{U} = 0$.

The kinematical momentum is

$$\boldsymbol{K} = \frac{H\boldsymbol{r}}{c^2} - \boldsymbol{P}t - \boldsymbol{U} + \frac{\boldsymbol{u}\cdot\boldsymbol{U}}{c^2}\boldsymbol{u} = \frac{H}{c^2}(\boldsymbol{r} - \boldsymbol{k}) - \boldsymbol{P}t, \tag{3.110}$$

where the relative position vector $\boldsymbol{k}$ is defined by

$$\frac{H}{c^2}\boldsymbol{k} = \boldsymbol{U} - \frac{\boldsymbol{u}\cdot\boldsymbol{U}}{c^2}\boldsymbol{u} = -2b\gamma^3\frac{d\boldsymbol{u}}{dt}. \tag{3.111}$$

We see that it has direction opposite to the acceleration, provided $b > 0$, thus suggesting an internal central motion, as it happens by the corresponding nonrelativistic limit.

The center of mass position is defined as

$$\boldsymbol{q} = \boldsymbol{r} - \boldsymbol{k} = \boldsymbol{r} + \frac{2b\gamma(u)^3c^2}{H}\frac{d\boldsymbol{u}}{dt}. \tag{3.112}$$

The linear momentum can be expressed in terms of the center of mass position $\boldsymbol{q}$ with the usual relation

$$\boldsymbol{P} = \frac{H}{c^2}\frac{d\boldsymbol{q}}{dt},$$

which in terms of the kinematical variables, substituting the expressions of these observables from above, yields

$$\begin{aligned}\boldsymbol{P} &= \frac{H}{c^2}\boldsymbol{u} + \frac{d}{dt}\left(2b\gamma^3\frac{d\boldsymbol{u}}{dt}\right)\\ &= \frac{H}{c^2}\boldsymbol{u} + \frac{6b\gamma^5}{c^2}\left(\boldsymbol{u}\cdot\frac{d\boldsymbol{u}}{dt}\right)\frac{d\boldsymbol{u}}{dt} + 2b\gamma^3\frac{d^2\boldsymbol{u}}{dt^2}.\end{aligned} \tag{3.113}$$

It has components along $\boldsymbol{u}$, $d\boldsymbol{u}/dt$ and $d^2\boldsymbol{u}/dt^2$.

The angular momentum takes the form

$$\boldsymbol{J} = \boldsymbol{r} \times \boldsymbol{P} + \boldsymbol{u} \times \boldsymbol{U},$$

and the spin becomes

$$\boldsymbol{S} = \boldsymbol{J} - \frac{c^2}{H}\boldsymbol{K} \times \boldsymbol{P} = \boldsymbol{u} \times \boldsymbol{U} + \boldsymbol{k} \times \boldsymbol{P}.$$

For the center of mass observer $\boldsymbol{P} = 0$, $\boldsymbol{K} = 0$, $\boldsymbol{k} = \boldsymbol{r}$, $\boldsymbol{u} = d\boldsymbol{k}/dt$ and $H = mc^2$. The spin in this frame reduces to

$$\boldsymbol{S} = \boldsymbol{u} \times \boldsymbol{U} = -2b\gamma^3\boldsymbol{u} \times \frac{d\boldsymbol{u}}{dt} = -m\boldsymbol{k} \times \frac{d\boldsymbol{k}}{dt}, \tag{3.114}$$

which is also of (anti)orbital type, as in the nonrelativistic case. Since $\boldsymbol{S}$ is constant in this frame, the motion takes place in a plane orthogonal to the spin.

If expressed in polar coordinates $r$ and $\phi$, this equation leads to

$$r^2\dot{\phi} = -S/m = \text{const.} \tag{3.115}$$

*i.e.*, a constant areolar velocity as in the usual central motions.

If in (3.111) we make the scalar product of both sides with vector $\boldsymbol{u} = d\boldsymbol{k}/dt$ in this frame and divide the result by $c^2$, we find:

$$m\boldsymbol{k}\cdot\boldsymbol{u}/c^2 = -2b\gamma^3\frac{d\boldsymbol{u}}{dt}\cdot\boldsymbol{u}/c^2,$$

and thus

$$\frac{d}{dt}\left(\frac{m}{4bc^2}r^2 + \gamma(u)\right) = 0.$$

We get another first integral of the form

$$\alpha = \frac{m}{4bc^2}r^2 + \gamma(u), \tag{3.116}$$

which is another constant of the motion, positive, dimensionless, with the meaning of a certain conservation of energy of the internal motion. In fact, if we take $b = m/2\omega^2$, from (3.111), we get

$$\frac{d^2\boldsymbol{k}}{dt^2} + \omega^2\gamma(u)^{-3}\,\boldsymbol{k} = 0, \tag{3.117}$$

which is a kind of generalized relativistic isotropic harmonic motion. Due to the factor $\gamma(u)^{-3}$, the velocity can never increase above the value $c$ because $\gamma(u)$ rises as $u$ does and it turns out that the acceleration decreases with $\gamma(u)^{-3}$. By neglecting terms of order of $u/c$, it is an isotropic harmonic oscillator, whose energy conservation equation is precisely (3.116).

If we take the $XOY$ plane as the plane of the trajectory and define the dimensionless time variable $\theta = \omega t$ and the constant length $a = c/\omega$, then the system of differential equations to be solved is in terms of the dimensionless variables $\hat{x} = x/a$ and $\hat{y} = y/a$, which once we have removed the hats, satisfies the ordinary system:

$$\frac{d^2x}{d\theta^2} + x\left(1 - \left(\frac{dx}{d\theta}\right)^2 - \left(\frac{dy}{d\theta}\right)^2\right)^{3/2} = 0,$$

$$\frac{d^2y}{d\theta^2} + y\left(1 - \left(\frac{dx}{d\theta}\right)^2 - \left(\frac{dy}{d\theta}\right)^2\right)^{3/2} = 0.$$

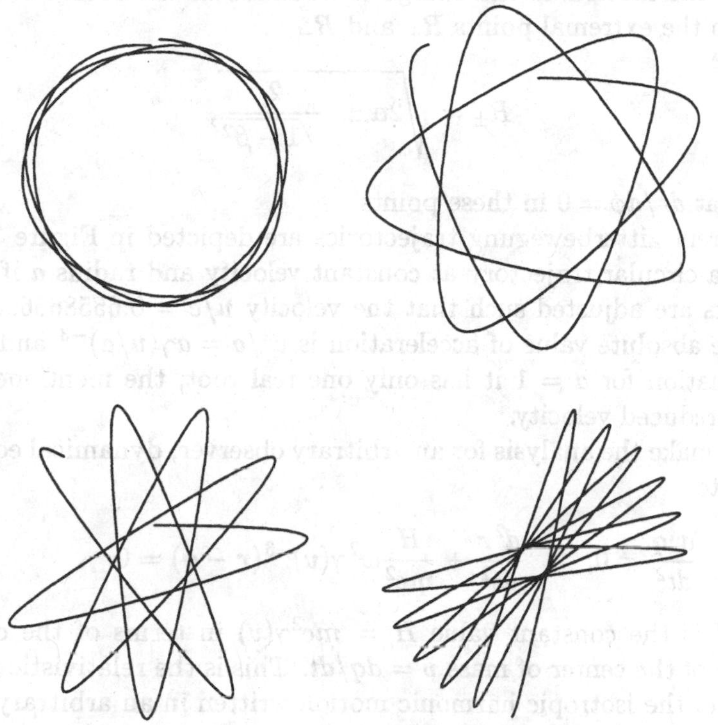

*Figure 3.3.* **Possible zitterbewegung trajectories of spinning relativistic particles.**

These equations can be solved in terms of elliptic integrals. The constant (3.116) takes the form

$$\alpha = \frac{1}{2}\left(x^2 + y^2\right) + \left(1 - \dot{x}^2 - \dot{y}^2\right)^{-1/2},$$

where now the dot means $\theta$–derivative. In these dimensionless units the above constants of the motion (3.115) and (3.116) take the form

$$r^2\dot{\phi} = -\beta, \quad \frac{1}{2}r^2 + \left(1 - \dot{r}^2 - r^2\dot{\phi}^2\right)^{-1/2} = \alpha, \quad \beta = S\omega/mc^2,$$

in polar coordinates, where the dot means $d/d\theta$. From this we get

$$\frac{dr}{d\phi} = \pm\frac{r^3}{\beta}\left(\frac{1}{\beta^2} - 1 - \frac{1}{(\alpha\beta - \beta r^2/2)^2}\right)^{1/2},$$

so that the motion of the charge is bounded in the radial coordinate between the extremal points $R_+$ and $R_-$,

$$R_\pm = \sqrt{2\alpha \pm \frac{2}{\sqrt{1-\beta^2}}},$$

such that $dr/d\phi = 0$ in these points.

Different zitterbewegung trajectories are depicted in Figure 3.3. We obtain a circular trajectory at constant velocity and radius $a$ if the parameters are adjusted such that the velocity $u/c = 0.6558656$. In that case the absolute value of acceleration is $u^2/a = a\gamma(u/a)^{-3}$ and solving this equation for $a = 1$ it has only one real root, the mentioned value for the reduced velocity.

If we make the analysis for an arbitrary observer, dynamical equations reduce to

$$\frac{d^2\boldsymbol{q}}{dt^2} = 0, \qquad \frac{d^2\boldsymbol{r}}{dt^2} + \frac{H}{mc^2}\omega^2\gamma(u)^{-3}(\boldsymbol{r}-\boldsymbol{q}) = 0. \tag{3.118}$$

Here $H$ is the constant value $H = mc^2\gamma(v)$ in terms of the constant velocity of the center of mass $\boldsymbol{v} = d\boldsymbol{q}/dt$. This is the relativistic generalization of the isotropic harmonic motion written in an arbitrary frame.

If there is an interaction with an electromagnetic field, we get again for the relative position vector the same expression as in (3.111) because there are no coupling terms between $\dot{\boldsymbol{u}}$ and the external electromagnetic fields, and the dynamical equations are

$$\frac{d\boldsymbol{P}}{dt} = m\frac{d}{dt}\left(\gamma(v)\frac{d\boldsymbol{q}}{dt}\right) = e\left(\boldsymbol{E} + \boldsymbol{u}\times\boldsymbol{B}\right), \tag{3.119}$$

$$\frac{d^2\boldsymbol{r}}{dt^2} + \gamma(v)\omega^2\gamma(u)^{-3}(\boldsymbol{r}-\boldsymbol{q}) = 0. \tag{3.120}$$

Here, $\gamma(v)$ is the corresponding $\gamma$ factor in terms of the center of mass velocity and where the external fields $\boldsymbol{E}(t,\boldsymbol{r})$ and $\boldsymbol{B}(t,\boldsymbol{r})$ are defined at the center of charge position $\boldsymbol{r}$.

## 3.3 CANONICAL ANALYSIS

Let us consider the generalized Lagrangian (3.107) written in the form

$$L = ac^2\left(1 - \dot{\boldsymbol{r}}^2/c^2\right)^{1/2} - b\gamma(\dot{r})^3\left(\ddot{\boldsymbol{r}}\right)^2 - \frac{b\gamma(\dot{r})^5}{c^2}\left(\dot{\boldsymbol{r}}\cdot\ddot{\boldsymbol{r}}\right)^2,$$

where the dot means now time-derivative. We can obtain the conjugate momenta of the six generalized coordinates, respectively, the three degrees of freedom $\boldsymbol{q}_1 = \boldsymbol{r}$ and their first derivatives $\boldsymbol{q}_2 = \dot{\boldsymbol{r}}$, as

$$\boldsymbol{p}_1 = \frac{\partial L}{\partial \dot{\boldsymbol{r}}} - \frac{d}{dt}\left(\frac{\partial L}{\partial \ddot{\boldsymbol{r}}}\right) = \boldsymbol{R} - \frac{d\boldsymbol{U}}{dt}, \quad \text{and} \quad \boldsymbol{p}_2 = \frac{\partial L}{\partial \ddot{\boldsymbol{r}}} = \boldsymbol{U}.$$

The phase space is a 12-dimensional manifold and the Hamiltonian is in fact the total energy written in terms of the canonical variables

$$H = \boldsymbol{p}_1 \cdot \dot{\boldsymbol{q}}_1 + \boldsymbol{p}_2 \cdot \dot{\boldsymbol{q}}_2 - L = \boldsymbol{p}_1 \cdot \boldsymbol{q}_2 + ac^2 \left(1 - \boldsymbol{q}_2^2/c^2\right)^{1/2}$$

$$-\frac{1}{4b}\left(\boldsymbol{p}_2^2 - \frac{1}{c^2}(\boldsymbol{p}_2 \cdot \boldsymbol{q}_2)^2\right)\left(1 - \boldsymbol{q}_2^2/c^2\right)^{3/2}.$$

Hamilton-Jacobi equations are

$$\dot{\boldsymbol{q}}_1 = \frac{\partial H}{\partial \boldsymbol{p}_1} = \boldsymbol{q}_2, \quad \dot{\boldsymbol{p}}_1 = -\frac{\partial H}{\partial \boldsymbol{q}_1} = 0,$$

$$\dot{\boldsymbol{q}}_2 = \frac{\partial H}{\partial \boldsymbol{p}_2} = -\frac{1}{2b}\left(\boldsymbol{p}_2 - \frac{(\boldsymbol{p}_2 \cdot \boldsymbol{q}_2)}{c^2}\boldsymbol{q}_2\right)\left(1 - \boldsymbol{q}_2^2/c^2\right)^{3/2},$$

$$\dot{\boldsymbol{p}}_2 = -\frac{\partial H}{\partial \boldsymbol{q}_2} = \boldsymbol{p}_1 - a\left(1 - \boldsymbol{q}_2^2/c^2\right)^{-1/2}\boldsymbol{q}_2 + \frac{1}{2bc^2}(\boldsymbol{p}_2 \cdot \boldsymbol{q}_2)\,\boldsymbol{p}_2\left(1 - \boldsymbol{q}_2^2/c^2\right)^{3/2}$$

$$+\frac{1}{2bc^2}\left(\boldsymbol{p}_2^2 - \frac{1}{c^2}(\boldsymbol{p}_2 \cdot \boldsymbol{q}_2)^2\right)\left(1 - \boldsymbol{q}_2^2/c^2\right)^{5/2}\boldsymbol{q}_2.$$

The ten Noether constants of the motion are the generating functions of the corresponding canonical transformations of time and space translations, pure Lorentz transformations and rotations, and are given by

$$\begin{aligned} H &= \boldsymbol{p}_1 \cdot \boldsymbol{q}_2 + ac^2\left(1 - \boldsymbol{q}_2^2/c^2\right)^{1/2} \\ &\quad -\frac{1}{4b}\left(\boldsymbol{p}_2^2 - \frac{1}{c^2}(\boldsymbol{p}_2 \cdot \boldsymbol{q}_2)^2\right)\left(1 - \boldsymbol{q}_2^2/c^2\right)^{3/2}, \\ \boldsymbol{P} &= \boldsymbol{p}_1, \\ \boldsymbol{K} &= \frac{H}{c^2}\boldsymbol{q}_1 - \boldsymbol{p}_1 t - \boldsymbol{p}_2 + \frac{\boldsymbol{p}_2 \cdot \boldsymbol{q}_2}{c^2}\boldsymbol{q}_2, \\ \boldsymbol{J} &= \boldsymbol{q}_1 \times \boldsymbol{p}_1 + \boldsymbol{q}_2 \times \boldsymbol{p}_2. \end{aligned}$$

The Poisson bracket of the above constants of the motion satisfy the commutation relations

$$\{J_i, J_k\} = \epsilon_{ikl}J_l, \quad \{J_i, P_k\} = \epsilon_{ikl}P_l, \quad \{J_i, K_k\} = \epsilon_{ikl}K_l, \quad \{J_i, H\} = 0,$$

$$\{P_i, P_k\} = 0, \quad \{P_i, H\} = 0, \quad \{K_i, K_k\} = -\epsilon_{ikl}J_l/c^2, \quad \{K_i, H\} = P_i,$$

$$\{K_i, P_j\} = H/c^2 \delta_{ij},$$

that are, up to a global sign, the commutation relations of the Poincaré group (3.11-3.12).

Although $K_i$ satisfies $\{K_i, H\} = P_i$, it is a constant of the motion because as we see in (2.85), $\boldsymbol{K}$ is time dependent and its total time derivative in the canonical approach is

$$\frac{d\boldsymbol{K}}{dt} = \{\boldsymbol{K}, H\} + \frac{\partial \boldsymbol{K}}{\partial t} = \boldsymbol{p}_1 - \boldsymbol{p}_1 = 0.$$

The spin observable

$$\boldsymbol{S} = \boldsymbol{J} - \frac{c^2}{H}\boldsymbol{K} \times \boldsymbol{P} = \boldsymbol{q}_2 \times \boldsymbol{p}_2 + \frac{c^2}{H}\boldsymbol{p}_2 \times \boldsymbol{p}_1 - \frac{\boldsymbol{p}_2 \cdot \boldsymbol{q}_2}{H}\boldsymbol{q}_2 \times \boldsymbol{p}_1,$$

satisfies the Poisson bracket commutation relations

$$\{J_i, S_k\} = \epsilon_{ikl} S_l, \quad \{S_i, P_k\} = 0, \quad \{S_i, H\} = 0,$$

showing respectively that transforms like a vector under rotations is invariant under space translations and is a constant of the motion. Also

$$\{S_i, S_k\} = \epsilon_{ikl} S_l - c^2 \frac{\boldsymbol{P} \cdot \boldsymbol{S}}{H^2} \epsilon_{ikl} P_l, \quad \{S_i, K_j\} = \frac{1}{H} S_i P_j - c \frac{\boldsymbol{P} \cdot \boldsymbol{S}}{H} \delta_{ij},$$

i.e., it satisfies the commutation relations of an angular momentum in the center of mass frame and is not invariant under pure Lorentz transformations.

The center of mass position is defined as

$$\boldsymbol{q} = \frac{c^2}{H}(\boldsymbol{K} + t\boldsymbol{P}) = \boldsymbol{q}_1 - \frac{c^2}{H}\boldsymbol{p}_2 + \frac{1}{H}(\boldsymbol{p}_2 \cdot \boldsymbol{q}_2)\boldsymbol{q}_2,$$

and satisfies $\{q_i, q_j\} = -c^2 \epsilon_{ijk} S_k / H^2 \neq 0$, and therefore they cannot be used as canonical variables, as was pointed out by Pryce [7] in the analysis of the center of mass of relativistic particles which is briefly sketched in Chapter 6.

If we had taken as the spin of the system the observable $\boldsymbol{Z} = \boldsymbol{u} \times \boldsymbol{U} = \boldsymbol{q}_2 \times \boldsymbol{p}_2$, then this observable satisfies

$$\begin{aligned}
\{Z_i, Z_k\} &= \epsilon_{ikl} Z_l, \quad \{J_i, Z_k\} = \epsilon_{ikl} Z_l, \quad \{Z_i, P_k\} = 0, \\
\{Z_i, K_j\} &= -\epsilon_{ijk} p_{2k} - \epsilon_{irs} p_{1r} q_{2s} q_{1j} + \epsilon_{ijk} (\boldsymbol{p}_2 \cdot \boldsymbol{q}_2) q_{2k}, \\
\{Z_i, H\} &= (\boldsymbol{p}_1 \times \boldsymbol{q}_2)_i,
\end{aligned}$$

so that it is an angular momentum, that transforms like a vector under rotations, and is invariant under space translations but not under pure

Lorentz transformations and it is not a constant of the motion, but satisfies the dynamical equation $d\boldsymbol{Z}/dt = \boldsymbol{P} \times \boldsymbol{u}$. That is why this observable cannot be considered as the spin of the system.

The Pauli-Lubanski four-vector is defined as usual as the constant of the motion

$$W^0 = \boldsymbol{P} \cdot \boldsymbol{S}, \quad \boldsymbol{W} = H\boldsymbol{S}/c,$$

such that the two constants of the motion

$$P^\mu P_\mu = m^2c^2, \quad \text{and} \quad -W^\mu W_\mu = m^2c^2S^2, \tag{3.121}$$

commute with the above ten generators and they are Poincaré invariant properties of the particle and completely characterize the structure of this particle. They are the two functionally independent Casimir invariants of the Poincaré group.

## 3.4 CIRCULAR ZITTERBEWEGUNG

The second Lagrangian $L_2$ of (3.108) has the following nonrelativistic limit for low velocities $u \ll c$,

$$L_{2NR} = -ac^2\left(1 - \frac{\boldsymbol{u}^2}{2c^2}\right)\dot{t} - bc\sqrt{\dot{\boldsymbol{u}}^2}, \tag{3.122}$$

and we can take for the parameters the values $a = m$ and $b = mR/c$, where $R$ is the average radius of the internal motion. When compared with the nonrelativistic model (2.132), it corresponds to the case of circular zitterbewegung with constant internal radius, a property also shared by this relativistic model.

From Lagrangian $L_2$ we get

$$\boldsymbol{U} = \frac{\partial L_2}{\partial \dot{\boldsymbol{u}}} = -\frac{b\gamma(u)(c^2\dot{\boldsymbol{u}} + \gamma(u)^2(\boldsymbol{u}\cdot\dot{\boldsymbol{u}})\boldsymbol{u})}{\sqrt{c^2\dot{\boldsymbol{u}}^2 + \gamma(u)^2(\boldsymbol{u}\cdot\dot{\boldsymbol{u}})^2}}. \tag{3.123}$$

For the center of mass observer it gives

$$m\boldsymbol{k} = \boldsymbol{U} - \frac{\boldsymbol{u}\cdot\boldsymbol{U}}{c^2}\boldsymbol{u} = \frac{-b\gamma(u)c^2\dot{\boldsymbol{u}}}{\sqrt{c^2\dot{\boldsymbol{u}}^2 + \gamma(u)^2(\boldsymbol{u}\cdot\dot{\boldsymbol{u}})^2}}, \tag{3.124}$$

another internal central motion that can be written in the form

$$\dot{\boldsymbol{u}} = -\frac{\theta\boldsymbol{k}}{\gamma(u)}\sqrt{c^2\dot{\boldsymbol{u}}^2 + \gamma(u)^2(\boldsymbol{u}\cdot\dot{\boldsymbol{u}})^2}, \tag{3.125}$$

with $\theta = m/bc^2$. From here we get the relation

$$\dot{\boldsymbol{u}}^2 = \frac{\theta^2}{\gamma(u)^2}(c^2\dot{\boldsymbol{u}}^2 + \gamma(u)^2(\boldsymbol{u}\cdot\dot{\boldsymbol{u}})^2)\,\boldsymbol{k}^2,$$

and it yields

$$(\boldsymbol{u}\cdot\dot{\boldsymbol{u}})^2 = \frac{\gamma^2 - c^2k^2\theta^2}{k^2\theta^2\gamma^2}\,\dot{\boldsymbol{u}}^2. \qquad (3.126)$$

Taking the scalar product of (3.125) with $\boldsymbol{u}$ and squaring the result we reach

$$(\boldsymbol{u}\cdot\dot{\boldsymbol{u}})^2 = \frac{\theta^2}{\gamma(u)^2}\left(c^2\dot{\boldsymbol{u}}^2 + \gamma(u)^2(\boldsymbol{u}\cdot\dot{\boldsymbol{u}})^2\right)(\boldsymbol{u}\cdot\boldsymbol{k})^2,$$

and therefore

$$(\boldsymbol{u}\cdot\dot{\boldsymbol{u}})^2 = \frac{\theta^2c^2(\boldsymbol{u}\cdot\boldsymbol{k})^2}{\gamma(u)^2(1-\theta^2(\boldsymbol{u}\cdot\boldsymbol{k})^2)}\,\dot{\boldsymbol{u}}^2, \qquad (3.127)$$

from which by identification with (3.126) and after the cancellation of terms in $\dot{\boldsymbol{u}}^2$ in both sides, we get a single equation for the variables $\boldsymbol{u} \equiv d\boldsymbol{k}/dt$ and $\boldsymbol{k}$,

$$1 - k^2c^2\theta^2\gamma(u)^{-2} - \theta^2(\boldsymbol{u}\cdot\boldsymbol{k})^2 = 0.$$

If we take the derivative with respect to $t$, the constant coefficients $\theta^2$ cancel out and the above relation reduces to

$$-(\boldsymbol{u}\cdot\boldsymbol{k})c^2 + k^2(\boldsymbol{u}\cdot\dot{\boldsymbol{u}}) - (\boldsymbol{u}\cdot\boldsymbol{k})(\dot{\boldsymbol{u}}\cdot\boldsymbol{k}) = 0. \qquad (3.128)$$

The last two terms can be grouped together into a single term

$$-(\boldsymbol{u}\cdot\boldsymbol{k})c^2 + (\boldsymbol{k}\times\boldsymbol{u})\cdot(\boldsymbol{k}\times\dot{\boldsymbol{u}}) = 0,$$

that also vanishes because $\boldsymbol{k}$ and $\dot{\boldsymbol{u}}$ are parallel, as can be deduced from (3.125). Therefore, this leads to $\boldsymbol{u}\cdot\boldsymbol{k} = 0$ and from (3.127) $\boldsymbol{u}\cdot\dot{\boldsymbol{u}} = 0$ and it turns out that the internal motion for the center of mass observer is a circle with constant velocity. From (3.124), if we introduce the value of parameter $b = mR/c$, it gives:

$$\boldsymbol{k} = -R\gamma(u)\frac{\dot{\boldsymbol{u}}}{|\dot{\boldsymbol{u}}|}. \qquad (3.129)$$

The motion, similarly as the nonrelativistic case, has a constant radius that in this case is given by

$$R_0 = \gamma(u)R, \qquad (3.130)$$

that reduces to $R$ for the low velocity limit.

The spin is $\boldsymbol{S} = -m\boldsymbol{k}\times d\boldsymbol{k}/dt$ and has, in this frame, the constant absolute value $S = muR_0$, where $u$ is the constant internal velocity for the center of mass observer.

## 4. LUXONS

Let us consider those mechanical systems whose kinematical space is the manifold $X$ generated by the variables $(t, \boldsymbol{r}, \boldsymbol{u}, \boldsymbol{\rho})$ with domains $t \in \mathbb{R}$, $\boldsymbol{r} \in \mathbb{R}^3$, $\boldsymbol{\rho} \in \mathbb{R}^3_c$ as in the previous case, and $\boldsymbol{u} \in \mathbb{R}^3$ but now with $u = c$. Since $u = c$ we shall call this kind of particles **Luxons.** This manifold is in fact a homogeneous space of the Poincaré group $\mathcal{P}$, and therefore, according to our definition of elementary particle has to be considered as a possible candidate for describing the kinematical space of an elementary system. In fact, if we consider the point in this manifold $x \equiv (0, 0, \boldsymbol{u}, 0)$, the little group that leaves $x$ invariant is the one-parameter subgroup $\mathcal{V}_u$ of pure Lorentz transformations in the direction of the vector $\boldsymbol{u}$. Then $X \sim \mathcal{P}/\mathcal{V}_u$, is a nine-dimensional homogeneous space.

For this kind of systems the variables $t$, $\boldsymbol{r}$ and $\boldsymbol{u}$ transform under $\mathcal{P}$ as in the previous case (3.49, 3.50, 3.51) while the transformation of the variable $\boldsymbol{\rho}$, that in general involves $\gamma(u)$ factors which become infinite, can be obtained from (3.52), *i.e.*, from the equivalent to

$$R(\boldsymbol{\rho}') = R(\boldsymbol{w})R(\boldsymbol{\mu})R(\boldsymbol{\rho}) \equiv R(\boldsymbol{\phi})R(\boldsymbol{\rho}).$$

Here, parameter $\boldsymbol{w}$ is given in (3.21) and $\boldsymbol{\phi}$ in (3.22) and by taking the limit $u \to c$ on the r.h.s., we get:

$$\boldsymbol{\rho}'(\tau) = \frac{\boldsymbol{\mu} + \boldsymbol{\rho}(\tau) + \boldsymbol{\mu} \times \boldsymbol{\rho}(\tau) + \boldsymbol{F}_c(\boldsymbol{v}, \boldsymbol{\mu}; \boldsymbol{u}(\tau), \boldsymbol{\rho}(\tau))}{1 - \boldsymbol{\mu} \cdot \boldsymbol{\rho}(\tau) + G_c(\boldsymbol{v}, \boldsymbol{\mu}; \boldsymbol{u}(\tau), \boldsymbol{\rho}(\tau))}, \tag{3.131}$$

where the functions $\boldsymbol{F}_c$ and $G_c$ are given now by:

$$\begin{aligned} \boldsymbol{F}_c(\boldsymbol{v}, \boldsymbol{\mu}; \boldsymbol{u}, \boldsymbol{\rho}) &= \frac{\gamma(v)}{(1+\gamma(v))c^2}\left[\boldsymbol{u} \times \boldsymbol{v} + \boldsymbol{u}(\boldsymbol{v} \cdot \boldsymbol{\mu}) + \boldsymbol{v}(\boldsymbol{u} \cdot \boldsymbol{\rho})\right. \\ &+ \boldsymbol{u} \times (\boldsymbol{v} \times \boldsymbol{\mu}) + (\boldsymbol{u} \times \boldsymbol{\rho}) \times \boldsymbol{v} + (\boldsymbol{u} \cdot \boldsymbol{\rho})(\boldsymbol{v} \times \boldsymbol{\mu}) \\ &+ \left.(\boldsymbol{u} \times \boldsymbol{\rho})(\boldsymbol{v} \cdot \boldsymbol{\mu}) + (\boldsymbol{u} \times \boldsymbol{\rho}) \times (\boldsymbol{v} \times \boldsymbol{\mu})\right], \end{aligned} \tag{3.132}$$

$$\begin{aligned} G_c(\boldsymbol{v}, \boldsymbol{\mu}; \boldsymbol{u}, \boldsymbol{\rho}) &= \frac{\gamma(v)}{(1+\gamma(v))c^2}\left[\boldsymbol{u} \cdot \boldsymbol{v} + \boldsymbol{u} \cdot (\boldsymbol{v} \times \boldsymbol{\mu}) + \boldsymbol{v} \cdot (\boldsymbol{u} \times \boldsymbol{\rho})\right. \\ &- \left.(\boldsymbol{u} \cdot \boldsymbol{\rho})(\boldsymbol{v} \cdot \boldsymbol{\mu}) + (\boldsymbol{u} \times \boldsymbol{\rho}) \cdot (\boldsymbol{v} \times \boldsymbol{\mu})\right]. \end{aligned} \tag{3.133}$$

The kinematical rotation $R(\boldsymbol{\phi})$ is characterized by the three-vector $\boldsymbol{\phi}$, which is given by:

$$\boldsymbol{\phi} = \frac{\boldsymbol{\mu} + \boldsymbol{F}_c(\boldsymbol{v}, \boldsymbol{\mu}; \boldsymbol{u}(\tau), 0)}{1 + G_c(\boldsymbol{v}, \boldsymbol{\mu}; \boldsymbol{u}(\tau), 0)}. \tag{3.134}$$

If no rotation is involved ($\mu = 0$) this kinematical rotation reduces to

$$\phi = \frac{\gamma(v)\boldsymbol{u}\times\boldsymbol{v}}{(1+\gamma(v))c^2+\gamma(v)\boldsymbol{u}\cdot\boldsymbol{v}}. \tag{3.135}$$

In this case there also exist among the kinematical variables the constraints $\boldsymbol{u} = \dot{\boldsymbol{r}}/\dot{t}$. If we analyze equation (3.51) since $u' = u = c$, the absolute value of the velocity vector is conserved and it means that $\boldsymbol{u}'$ can be obtained from $\boldsymbol{u}$ by an orthogonal transformation, so that the transformation equations of the velocity under $\mathcal{P}$ can be expressed as:

$$\boldsymbol{u}' = R(\phi)\boldsymbol{u}. \tag{3.136}$$

Equation (3.131) also corresponds to

$$R(\rho') = R(\phi)R(\rho), \tag{3.137}$$

with the same $\phi$ in both cases, as in (3.134).

Since the variable $u(\tau) = c$, during the whole evolution, we can distinguish two different kinds of systems, because, by taking the derivative with respect to $\tau$ of this expression we get $\dot{\boldsymbol{u}}(\tau)\cdot\boldsymbol{u}(\tau) = 0$, *i.e.*, systems for which $\dot{\boldsymbol{u}} = 0$ or massless systems as we shall see, and systems where $\dot{\boldsymbol{u}} \neq 0$ but always orthogonal to $\boldsymbol{u}$. These systems will correspond to massive particles whose charge internal motion occurs at the constant velocity $c$, although their center of mass moves with velocity below $c$.

## 4.1 MASSLESS PARTICLES. (THE PHOTON)

If $\dot{\boldsymbol{u}} = 0$, $\boldsymbol{u}$ is constant and the system follows a straight trajectory with constant velocity, and therefore the kinematical variables reduce simply to $(t, \boldsymbol{r}, \boldsymbol{\rho})$ with domains and physical meaning as usual as, time, position and orientation, respectively. The derivatives $\dot{t}$ and $\dot{\boldsymbol{r}}$ transform like (3.35) and (3.36) and instead of the variable $\dot{\boldsymbol{\rho}}$ we shall consider the linear function $\boldsymbol{\omega}_0$ defined in (3.65) that transforms under $\mathcal{P}$:

$$\boldsymbol{\omega}_0'(\tau) = R(\phi)\boldsymbol{\omega}_0(\tau), \tag{3.138}$$

where, again, $\phi$ is given by (3.134).

In fact, from (3.137), since $\dot{\boldsymbol{u}} = 0$, taking the $\tau$-derivative,

$$\dot{R}(\rho') = R(\phi)\dot{R}(\rho),$$

and the antisymmetric matrix $\Omega = \dot{R}(\rho)R^T(\rho)$ has as essential components the angular velocity $\boldsymbol{\omega}_0$,

$$\Omega = \begin{pmatrix} 0 & -\omega_{0z} & \omega_{0y} \\ \omega_{0z} & 0 & -\omega_{0x} \\ -\omega_{0y} & \omega_{0x} & 0 \end{pmatrix}. \tag{3.139}$$

It transforms as

$$\Omega' = \dot{R}(\rho')R^T(\rho') = R(\phi)\dot{R}(\rho)R^T(\rho)R^T(\phi) = R(\phi)\Omega R^T(\phi),$$

and this matrix transformation leads for its essential components to (3.138).

For this system there are no constraints among the kinematical variables, and, since $\dot{\boldsymbol{u}} = 0$, the general form of its Lagrangian is

$$L = T\dot{t} + \boldsymbol{R}\cdot\dot{\boldsymbol{r}} + \boldsymbol{W}\cdot\boldsymbol{\omega}_0. \tag{3.140}$$

Funtions $T = \partial L/\partial\dot{t}$, $R_i = \partial L/\partial\dot{r}^i$, $W_i = \partial L/\partial\omega_0^i$, will depend on the variables $(t, \boldsymbol{r}, \boldsymbol{\rho})$ and are homogeneous functions of zero degree in terms of the derivatives of the kinematical variables $(\dot{t}, \dot{\boldsymbol{r}}, \boldsymbol{\omega}_0)$. Since $\dot{t} \neq 0$ they will be expressed in terms of $\boldsymbol{u} = \dot{\boldsymbol{r}}/\dot{t}$ and $\boldsymbol{\Omega} = \boldsymbol{\omega}_0/\dot{t}$, which are the true velocity and angular velocity of the particle respectively.

Invariance of the Lagrangian under $\mathcal{P}$ leads to the following transformation form of these functions under the group $\mathcal{P}$:

$$T' = \gamma T - \gamma(\boldsymbol{v}\cdot R(\boldsymbol{\mu})\boldsymbol{R}), \tag{3.141}$$

$$\boldsymbol{R}' = R(\boldsymbol{\mu})\boldsymbol{R} - \gamma\boldsymbol{v}T/c^2 + \frac{\gamma^2}{(1+\gamma)c^2}(\boldsymbol{v}\cdot R(\boldsymbol{\mu})\boldsymbol{R})\boldsymbol{v}, \tag{3.142}$$

$$\boldsymbol{W}' = R(\boldsymbol{\phi})\boldsymbol{W}. \tag{3.143}$$

They are translation invariant and therefore independent of $t$ and $\boldsymbol{r}$. They will be functions of only $(\boldsymbol{\rho}, \boldsymbol{u}, \boldsymbol{\Omega})$, with the constraint $u = c$. Invariance under rotations forbids the explicit dependence on $\boldsymbol{\rho}$, so that the dependence of these functions on $\boldsymbol{\rho}$ and $\dot{\boldsymbol{\rho}}$ variables is only through the angular velocity $\boldsymbol{\omega}_0$.

Noether's theorem gives rise, as before, to the following constants of the motion:

$$\begin{aligned}
\text{Energy}\quad H &= -T, && (3.144)\\
\text{linear momentum}\quad \boldsymbol{P} &= \boldsymbol{R}, && (3.145)\\
\text{kinematical momentum}\quad \boldsymbol{K} &= H\boldsymbol{r}/c^2 - \boldsymbol{P}t - \boldsymbol{W}\times\boldsymbol{u}/c^2, && (3.146)\\
\text{angular momentum}\quad \boldsymbol{J} &= \boldsymbol{r}\times\boldsymbol{P} + \boldsymbol{W}. && (3.147)
\end{aligned}$$

In this case the system has no zitterbewegung because the Lagrangian does not depend on $\dot{\boldsymbol{u}}$ which vanishes. The particle, located at point $\boldsymbol{r}$, is moving in a straight trajectory at the speed of light and therefore it is not possible to find an inertial rest frame observer. Although we have no center of mass observer, we define the spin by $\boldsymbol{S} = \boldsymbol{J} - \boldsymbol{r}\times\boldsymbol{P} = \boldsymbol{W}$.

If we take in (3.147) the $\tau$-derivative we get $d\boldsymbol{S}/dt = \boldsymbol{P} \times \boldsymbol{u}$. Since $\boldsymbol{P}$ and $\boldsymbol{u}$ are two non-vanishing constant vectors, then the spin has a constant time derivative. It represents a system with a continuously increasing angular momentum. This is not what we understand by an elementary particle except if this constant $d\boldsymbol{S}/dt = 0$. Therefore for this system the spin is a constant of the motion and $\boldsymbol{P}$ and $\boldsymbol{u}$ are collinear vectors.

Energy and linear momentum are in fact the components of a four-vector and with the spin they transform as

$$H' = \gamma H + \gamma(\boldsymbol{v} \cdot R(\boldsymbol{\mu})\boldsymbol{P}), \tag{3.148}$$

$$\boldsymbol{P}' = R(\boldsymbol{\mu})\boldsymbol{P} + \gamma \boldsymbol{v} H/c^2 + \frac{\gamma^2}{(1+\gamma)c^2}(\boldsymbol{v} \cdot R(\boldsymbol{\mu})\boldsymbol{P})\boldsymbol{v}, \tag{3.149}$$

$$\boldsymbol{S}' = R(\boldsymbol{\phi})\boldsymbol{S}. \tag{3.150}$$

The relation between $\boldsymbol{P}$ and $\boldsymbol{u}$ can be obtained from (3.146), taking the $\tau$-derivative and the condition that the spin $\boldsymbol{W}$ is constant, $\dot{\boldsymbol{K}} = 0 = -H\dot{\boldsymbol{r}}/c^2 + \boldsymbol{P}\dot{t}$, *i.e.*, $\boldsymbol{P} = H\boldsymbol{u}/c^2$. If we take the scalar product of this expression with $\boldsymbol{u}$ we also get $H = \boldsymbol{P} \cdot \boldsymbol{u}$.

Then, from (3.148) and (3.149), an invariant and constant of the motion, which vanishes, is $(H/c)^2 - \boldsymbol{P}^2$. The mass of this system is zero. It turns out that for this particle both $H$ and $\boldsymbol{P}$ are non-vanishing for every inertial observer. Otherwise, if one of them vanishes for a single observer they vanish for all of them. By (3.150), $S^2$ is another Poincaré invariant property of the system that is also a constant of the motion.

The first part of the Lagrangian $T\dot{t} + \boldsymbol{R} \cdot \dot{\boldsymbol{r}} = -H\dot{t} + \boldsymbol{P} \cdot \dot{\boldsymbol{r}}$, which can be written as $-(H - \boldsymbol{P} \cdot \boldsymbol{u})\dot{t} = 0$, also vanishes. Then the Lagrangian is reduced to the third term $\boldsymbol{S} \cdot \boldsymbol{\omega}_0$.

We see from (3.136) and (3.150) that the dimensionless magnitude $\epsilon = \boldsymbol{S} \cdot \boldsymbol{u}/Sc$ is another invariant and constant of the motion, and we thus expect that the Lagrangian will be explicitly dependent on both constant parameters $S$ and $\epsilon$. Taking into account the transformation properties under $\mathcal{P}$ of $\boldsymbol{u}$, $\boldsymbol{\omega}_0$ and $\boldsymbol{S}$, given in (3.136), (3.138) and (3.150) respectively, it turns out that the spin must necessarily be a vector function of $\boldsymbol{u}$ and $\boldsymbol{\omega}_0$.

If the spin is not transversal, as it happens for real photons, then $\boldsymbol{S} = \epsilon S\boldsymbol{u}/c$ where $\epsilon = \pm 1$, and thus the Lagrangian finally becomes:

$$L = \left(\frac{\epsilon S}{c}\right) \frac{\dot{\boldsymbol{r}} \cdot \boldsymbol{\omega}_0}{\dot{t}}. \tag{3.151}$$

From this Lagrangian we get that the energy is $H = -\partial L/\partial \dot{t} = \boldsymbol{S}\cdot\boldsymbol{\Omega}$, where $\boldsymbol{\Omega} = \boldsymbol{\omega}_0/\dot{t}$ is the angular velocity of the particle. The linear momentum is $\boldsymbol{P} = \partial L/\partial \dot{\boldsymbol{r}} = \epsilon\, S\, \boldsymbol{\Omega}/c$, and, since $\boldsymbol{P}$ and $\boldsymbol{u}$ are parallel vectors, $\boldsymbol{\Omega}$ and $\boldsymbol{u}$ must also be parallel, and if the energy is definite positive, then $\boldsymbol{\Omega} = \epsilon\Omega\boldsymbol{u}/c$.

This means that the energy $H = S\Omega$. For photons we know that $S = \hbar$, and thus $H = \hbar\Omega = h\nu$. In this way the frequency of a photon is the frequency of its rotational motion around the direction of its trajectory. We thus see that the spin and angular velocity have the same direction, although they are not analytically related, because $S$ is invariant under $\mathcal{P}$ while $\Omega$ is not.

We say that the Lagrangian (3.151) represents a photon of spin $S$ and polarization $\epsilon$. A set of photons of this kind, all with the same polarization, corresponds to circularly polarized light, as has been shown by direct measurement of the angular momentum carried by these photons. [8] Left and right polarized photons correspond to $\epsilon = 1$ and $\epsilon = -1$, respectively. Energy is related to the angular frequency $H = \hbar\Omega$, and linear momentum to the wave number $\boldsymbol{P} = \hbar\boldsymbol{k}$, that therefore is related to the angular velocity vector by $\boldsymbol{k} = \epsilon\boldsymbol{\Omega}/c$. If it is possible to talk about the 'wave-length' of a single photon this will be the distance run by the particle during a complete turn.

## 4.2 MASSIVE PARTICLES. (THE ELECTRON)

If we consider now the other possibility, $\dot{\boldsymbol{u}} \neq 0$ but orthogonal to $\boldsymbol{u}$, then variables $\dot{t}$ and $\dot{\boldsymbol{r}}$ transform as in the previous case (3.35) and (3.36), but for $\dot{\boldsymbol{u}}$ and $\boldsymbol{\omega}_0$ we have:

$$\dot{\boldsymbol{u}}' = R(\boldsymbol{\phi})\dot{\boldsymbol{u}} + \dot{R}(\boldsymbol{\phi})\boldsymbol{u}, \tag{3.152}$$

$$\boldsymbol{\omega}_0' = R(\boldsymbol{\phi})\boldsymbol{\omega}_0 + \boldsymbol{\omega}_{\phi}, \tag{3.153}$$

where the rotation of parameter $\boldsymbol{\phi}$ is again given by (3.134) and vector $\boldsymbol{\omega}_{\phi}$ is:

$$\boldsymbol{\omega}_{\phi} = \frac{\gamma R\boldsymbol{u}\times\boldsymbol{v} - (\gamma-1)R(\boldsymbol{u}\times\dot{\boldsymbol{u}}) + 2\gamma^2(\boldsymbol{v}\cdot R(\boldsymbol{u}\times\dot{\boldsymbol{u}}))\boldsymbol{v}/(1+\gamma)c^2}{\gamma(c^2 + \boldsymbol{v}\cdot R\boldsymbol{u})}. \tag{3.154}$$

Expression (3.152) is the $\tau$-derivative of (3.136) and can also be written in the form:

$$\dot{\boldsymbol{u}}' = \frac{R(\boldsymbol{\phi})\dot{\boldsymbol{u}}}{\gamma(1 + \boldsymbol{v}\cdot R(\boldsymbol{\mu})\boldsymbol{u}/c^2)}. \tag{3.155}$$

Expression (3.153) comes from $R(\boldsymbol{\rho}') = R(\boldsymbol{\phi})R(\boldsymbol{\rho})$ and taking the $\tau$-derivative of this expression $\dot{R}(\boldsymbol{\rho}') = \dot{R}(\boldsymbol{\phi})R(\boldsymbol{\rho}) + R(\boldsymbol{\phi})\dot{R}(\boldsymbol{\rho})$, because parameter $\phi$ depends on $\tau$ through the velocity $\boldsymbol{u}(\tau)$, and therefore

$$\Omega' = \dot{R}(\rho')R^T(\rho') = R(\phi)\Omega R^T(\phi) + \dot{R}(\phi)R^T(\phi).$$

$R(\phi)\Omega R^T(\phi)$ corresponds to $R(\phi)\omega_0$ and the antisymmetric matrix $\Omega_\phi = \dot{R}(\phi)R^T(\phi)$ has as essential components the $\boldsymbol{\omega}_\phi$ vector, *i.e.*, equation (3.154).

The homogeneity condition of the Lagrangian leads to the general form

$$L = T\dot{t} + \boldsymbol{R}\cdot\dot{\boldsymbol{r}} + \boldsymbol{U}\cdot\dot{\boldsymbol{u}} + \boldsymbol{W}\cdot\boldsymbol{\omega}_0, \qquad (3.156)$$

where $T = \partial L/\partial\dot{t}$, $R_i = \partial L/\partial\dot{r}^i$, $U_i = \partial L/\partial\dot{u}^i$ and $W_i = \partial L/\partial\omega_0^i$, and Noether's theorem provides the following constants of the motion:

$$\begin{aligned}
\text{Energy}\quad H &= -T - (d\boldsymbol{U}/dt)\cdot\boldsymbol{u}, & (3.157)\\
\text{linear momentum}\quad \boldsymbol{P} &= \boldsymbol{R} - (d\boldsymbol{U}/dt), & (3.158)\\
\text{kinematical momentum}\quad \boldsymbol{K} &= H\boldsymbol{r}/c^2 - \boldsymbol{P}t - \boldsymbol{Z}\times\boldsymbol{u}/c^2, & (3.159)\\
\text{angular momentum}\quad \boldsymbol{J} &= \boldsymbol{r}\times\boldsymbol{P} + \boldsymbol{Z}. & (3.160)
\end{aligned}$$

In this case the function $\boldsymbol{Z}$ is defined as in the previous example, and also in the Galilei case, by

$$\boldsymbol{Z} = \boldsymbol{u}\times\boldsymbol{U} + \boldsymbol{W}. \qquad (3.161)$$

Expressions (3.157, 3.158) imply that $H/c$ and $\boldsymbol{P}$ transform like the components of a four-vector, similarly as in (3.44-3.45), thus defining the invariant and constant of the motion $(H/c)^2 - \boldsymbol{P}^2 = m^2c^2$, in terms of the parameter $m$ that is interpreted as the mass of the particle.

Observable $\boldsymbol{Z}$ transforms as:

$$\boldsymbol{Z}'(\tau) = \gamma R(\boldsymbol{\mu})\boldsymbol{Z}(\tau) - \frac{\gamma^2}{(1+\gamma)c^2}(\boldsymbol{v}\cdot R(\boldsymbol{\mu})\boldsymbol{Z}(\tau))\boldsymbol{v} + \frac{\gamma}{c^2}(\boldsymbol{v}\times R(\boldsymbol{\mu})(\boldsymbol{Z}(\tau)\times\boldsymbol{u})), \qquad (3.162)$$

an expression that corresponds to the transformation of an antisymmetric tensor $Z^{\mu\nu}$ with strict components $Z^{0i} = (\boldsymbol{Z}\times\boldsymbol{u})^i/c$, and $Z^{ij} = \epsilon^{ijk}Z_k$.

By defining the relative position vector $\boldsymbol{k} = \boldsymbol{Z}\times\boldsymbol{u}/H$, the kinematical momentum (3.159) can be cast into the form

$$\boldsymbol{K} = H\boldsymbol{q}/c^2 - \boldsymbol{P}t,$$

where $\boldsymbol{q} = \boldsymbol{r} - \boldsymbol{k}$, represents the position of the center of mass of the particle.

The spin is defined as usual

$$\boldsymbol{S} = \boldsymbol{J} - \boldsymbol{q} \times \boldsymbol{P} = \boldsymbol{J} - \frac{c^2}{H}\boldsymbol{K} \times \boldsymbol{P}, \tag{3.163}$$

and is a constant of the motion. It takes the form

$$\boldsymbol{S} = \boldsymbol{Z} + \boldsymbol{k} \times \boldsymbol{P} = \boldsymbol{Z} + \frac{1}{H}(\boldsymbol{Z} \times \boldsymbol{u}) \times \boldsymbol{P}. \tag{3.164}$$

The helicity $\boldsymbol{S} \cdot \boldsymbol{P} = \boldsymbol{Z} \cdot \boldsymbol{P} = \boldsymbol{J} \cdot \boldsymbol{P}$, is also a constant of the motion. We can construct the constant Pauli-Lubanski four-vector

$$w^{\mu} \equiv (\boldsymbol{P} \cdot \boldsymbol{S}, H\boldsymbol{S}/c), \tag{3.165}$$

with $-w^{\mu}w_{\mu} = m^2c^2S^2$, in terms of the invariant properties $m$ and $S$ of the particle.

If we take in (3.159) the $\tau$-derivative and the scalar product with the velocity $\boldsymbol{u}$ we get the Poincaré invariant relation:

$$H = \boldsymbol{P} \cdot \boldsymbol{u} + \boldsymbol{Z} \cdot (\frac{d\boldsymbol{u}}{dt} \times \boldsymbol{u})/c^2. \tag{3.166}$$

This looks like Dirac's Hamiltonian, $H = c\boldsymbol{P} \cdot \boldsymbol{\alpha} + \beta mc^2$ when expressed in the quantum case, in terms of the $\boldsymbol{\alpha}$ and $\beta$ Dirac matrices. Since $c\boldsymbol{\alpha}$ is usually interpreted as the local velocity operator $\boldsymbol{u}$ of the electron, [9] we have $H = \boldsymbol{P} \cdot \boldsymbol{u} + \beta mc^2$ and this relation suggests the identification

$$\beta = \frac{1}{mc^4}\boldsymbol{Z} \cdot \left(\frac{d\boldsymbol{u}}{dt} \times \boldsymbol{u}\right).$$

Here all magnitudes on the right-hand side are measured in the center of mass frame. We shall come back to this relation after quantization of this system.

The center of mass observer is defined by the conditions $\boldsymbol{P} = \boldsymbol{K} = 0$. For this observer $\boldsymbol{Z} = \boldsymbol{S}$ is constant, $H = mc^2$ and thus from (3.159) we get

$$\boldsymbol{k} = \frac{1}{mc^2}\boldsymbol{S} \times \boldsymbol{u}, \tag{3.167}$$

and the internal motion takes place in a plane orthogonal to the constant spin $\boldsymbol{S}$. The scalar product with $\boldsymbol{u} \equiv d\boldsymbol{k}/dt$ leads to $\boldsymbol{k} \cdot d\boldsymbol{k}/dt = 0$, and thus the zitterbewegung radius $k$ is a constant. Taking the time derivative of both sides of (3.167), we obtain $mc^2\boldsymbol{u} = (\boldsymbol{S} \times d\boldsymbol{u}/dt)$, because the spin is constant in this frame, we get that $\boldsymbol{u}$ and $\boldsymbol{S}$ are orthogonal and therefore

$$\boldsymbol{S} = m\boldsymbol{u} \times \boldsymbol{k}. \tag{3.168}$$

Since $S$ and $u = c$ are constant, the motion is a circle of radius $R_0 = S/mc$. For the electron we take $S = \hbar/2$, and the radius is $\hbar/2m_ec = 1.93 \times 10^{-13}$ m., half the Compton wave length of the electron. The frequency of this motion in the C.M. frame is $\nu = 2m_ec^2/h = 2.47 \times 10^{20}$ $\mathrm{s}^{-1}$, and $\omega = 2\pi\nu = 1.55 \times 10^{21}$ rad $\mathrm{s}^{-1}$. The ratio of this radius to the so-called classical radius $R_{cl} = e^2/8\pi\varepsilon_0 m_e c^2 = 1.409 \times 10^{-15}$ m, is precisely $R_{cl}/R_0 = e^2/2\varepsilon_0 hc = 1/136.97 = \alpha$, the fine structure constant. Both radii are larger than today's estimated lower bound of electron radius, based on a model that integrates electromagnetism with gravitation, giving a value of the order $10^{-36}$ m. [10]

Motions of this sort, in which the particle is moving at the speed of light, can be found in early literature, but the distinction between the motion of center of charge and center of mass is not sufficiently clarified. [11, 12]

Nevertheless, in the model we are analyzing, the idea that the electron has a size of the order of the zitterbewegung radius is a plausible macroscopic vision but is not necessary to maintain any longer, because the only important point from the dynamical point of view is the center of charge position, whose motion completely determines the dynamics of the system. In this form, elementary particles, the kind of objects we are describing, look like extended objects. Nevertheless, although some kind of related length can be defined, they are dealt with as point particles with orientation because the physical attributes are all located at the single point $\boldsymbol{r}$.

The transformation equation for the function $\boldsymbol{Z}$, (3.162) can also be written as

$$\boldsymbol{Z}' = \gamma(1 + \boldsymbol{v} \cdot R(\boldsymbol{\mu})\boldsymbol{u}/c^2)R(\boldsymbol{\phi})\boldsymbol{Z}, \qquad (3.169)$$

and therefore $\boldsymbol{Z} \cdot \dot{\boldsymbol{u}} = \boldsymbol{Z}' \cdot \dot{\boldsymbol{u}}'$ and $\boldsymbol{Z}' \cdot \boldsymbol{u}' = \gamma(1 + \boldsymbol{v} \cdot R(\boldsymbol{\mu})\boldsymbol{u}/c^2)\boldsymbol{Z} \cdot \boldsymbol{u}$. Since it is orthogonal to $\boldsymbol{u}$ and $\dot{\boldsymbol{u}}$, for the center of mass observer, it is also orthogonal to $\boldsymbol{u}$ and $\dot{\boldsymbol{u}}$ for any other inertial observer.

**An alternative method of verifying this is to take the time derivative in (3.159) and (3.160), and thus**

$$H\boldsymbol{u} - c^2\boldsymbol{P} - \frac{d\boldsymbol{Z}}{dt} \times \boldsymbol{u} - \boldsymbol{Z} \times \frac{d\boldsymbol{u}}{dt} = 0,$$

$$\frac{d\boldsymbol{Z}}{dt} = \boldsymbol{P} \times \boldsymbol{u},$$

***i.e.,***

$$\boldsymbol{Z} \times \frac{d\boldsymbol{u}}{dt} = (H - \boldsymbol{u} \cdot \boldsymbol{P})\boldsymbol{u}.$$

**and a final scalar product with $\boldsymbol{Z}$, leads to $(H - \boldsymbol{u}\cdot\boldsymbol{P})\boldsymbol{u}\cdot\boldsymbol{Z} = 0$. The first factor does not vanish since the invariant $H^2/c^2 - P^2 = m^2c^2$ is positive definite and if $H = \boldsymbol{u}\cdot\boldsymbol{P}$, then $(\boldsymbol{u}\cdot\boldsymbol{P})^2/c^2 - P^2$ with $u \leq c$ is always negative, then $\boldsymbol{Z}\cdot\boldsymbol{u} = 0$.**

**If we take the time derivative of this last expression, with the condition that $dZ/dt$ is orthogonal to $\boldsymbol{u}$, we obtain $\boldsymbol{Z}\cdot\dot{\boldsymbol{u}}=0$. The observable $\boldsymbol{Z}$ has always the direction of the non-vanishing vector $\dot{\boldsymbol{u}}\times\boldsymbol{u}$ for positive energy particles and the opposite direction for particles of negative energy.**

Equation (3.166) can be recast into the form

$$\frac{H}{c}c\dot{t}-\boldsymbol{P}\cdot\dot{\boldsymbol{r}}-\frac{1}{c^2}\boldsymbol{Z}\cdot(\dot{\boldsymbol{u}}\times\boldsymbol{u})=0,$$

where the first two terms give rise to the invariant term $P_\mu\dot{x}^\mu=mc^2\dot{t}_{cm}$, and the third to the invariant relation

$$\boldsymbol{Z}\cdot(\dot{\boldsymbol{u}}\times\boldsymbol{u})=mc^4\dot{t}_{cm}. \tag{3.170}$$

Here $t_{cm}$ is the time observable measured in the center of mass frame, and the right-hand side, which is positive definite for particles, implies that $\boldsymbol{Z}$ has precisely the direction of $\dot{\boldsymbol{u}}\times\boldsymbol{u}$. In the case of antiparticles it has the opposite direction.

We see that the particle has mass and spin, and the center of charge moves in circles at the speed of light in a plane orthogonal to the spin, for the center of mass observer. All these features are independent of the particular Lagrangian of the type (3.156) we can consider. All that remains is to describe the evolution of the orientation and therefore its angular velocity. The analysis developed until now is compatible with many different possibilities for the angular velocity. The behaviour of the angular velocity depends on the particular model we work with.

For the task of analyzing invariant Lagrangians, the term $T\dot{t}+\boldsymbol{R}\cdot\dot{\boldsymbol{r}}=-H\dot{t}+\boldsymbol{P}\cdot\dot{\boldsymbol{r}}$ is Poincaré invariant and also the term $\boldsymbol{U}\cdot\dot{\boldsymbol{u}}+\boldsymbol{W}\cdot\boldsymbol{\omega}_0$. The Lagrangian can be written as the sum of two invariant terms, depending on the two scalars $m$ and $S$.

To construct invariant terms depending on the kinematical variables $\boldsymbol{u}$ and $\boldsymbol{\rho}$ and their derivatives we start with the transformation equations of these variables

$$\boldsymbol{u}'(\tau)=R(\phi)\boldsymbol{u}(\tau),\quad R(\boldsymbol{\rho}'(\tau))=R(\phi)R(\boldsymbol{\rho}(\tau)),$$

and because $\phi$ also depends on $\tau$, taking the $\tau$-derivative

$$\dot{\boldsymbol{u}}'=\dot{R}(\phi)\boldsymbol{u}+R(\phi)\dot{\boldsymbol{u}},\quad \dot{R}(\boldsymbol{\rho}')=\dot{R}(\phi)R(\boldsymbol{\rho})+R(\phi)\dot{R}(\boldsymbol{\rho}). \tag{3.171}$$

In terms of this last equation we know that $\Omega=\dot{R}(\boldsymbol{\rho})R^T(\boldsymbol{\rho})$ transforms as

$$\Omega'=(\dot{R}(\phi)R(\boldsymbol{\rho})+R(\phi)\dot{R}(\boldsymbol{\rho}))R^T(\boldsymbol{\rho})R^T(\phi)=R(\phi)\Omega R^T(\phi)+\dot{R}(\phi)R^T(\phi),$$

which corresponds to equation (3.153). Because $\Omega$ is the antisymmetric matrix (3.139), its action on a three-dimensional vector $\boldsymbol{u}$, gives the vector $\boldsymbol{w} = \Omega\boldsymbol{u} \equiv \boldsymbol{\omega}_0 \times \boldsymbol{u}$, and thus this magnitude transforms as

$$\boldsymbol{w}' = \Omega'\boldsymbol{u}' = \left(R(\phi)\Omega R^T(\phi) + \dot{R}(\phi)R^T(\phi)\right) R(\phi)\boldsymbol{u} = R(\phi)\boldsymbol{w} + \dot{R}(\phi)\boldsymbol{u}.$$

If we compare this with the transformation of $\dot{\boldsymbol{u}}$ in (3.171), we see that the dimensionless variables $\boldsymbol{\alpha}_e = (\dot{\boldsymbol{u}} - \boldsymbol{w})/c$, linear in the derivatives $\dot{\boldsymbol{u}}$ and $\dot{\boldsymbol{\rho}}$ or in $\dot{\boldsymbol{u}}$ and $\boldsymbol{\omega}_0$, explicitly given by

$$\boldsymbol{\alpha}_e = \frac{1}{c}\left(\dot{\boldsymbol{u}} + \boldsymbol{u} \times \boldsymbol{\omega}_0\right), \tag{3.172}$$

and

$$\boldsymbol{\omega}_e = \boldsymbol{\alpha}_e \times \frac{\boldsymbol{u}}{c} = \boldsymbol{\omega}_0 - \frac{1}{c^2}(\boldsymbol{\omega}_0 \cdot \boldsymbol{u})\boldsymbol{u} - \frac{1}{c^2}\boldsymbol{u} \times \dot{\boldsymbol{u}}, \tag{3.173}$$

are both orthogonal to the velocity $\boldsymbol{u}$ and they transform under $\mathcal{P}$ in the form:

$$\boldsymbol{\alpha}_e' = R(\phi)\boldsymbol{\alpha}_e, \quad \boldsymbol{\omega}_e' = R(\phi)\boldsymbol{\omega}_e. \tag{3.174}$$

At any instant $\tau$ the particle has associated with it three orthogonal dimensionless vectors, $\boldsymbol{u}/c$, $\boldsymbol{\alpha}_e$ and $\boldsymbol{\omega}_e$, (see Figure 3.4) the first of absolute value 1 and the other two of the same invariant size $\alpha_e = \omega_e \equiv y$. In this case the velocity $\boldsymbol{u}$ never vanishes but vectors $\boldsymbol{\alpha}_e$ and $\boldsymbol{\omega}_e$ are the equivalent for this system to vectors $\boldsymbol{\alpha}$ and $\boldsymbol{\omega}$ given in (3.63) and (3.64) respectively, in the case of Bradyons.

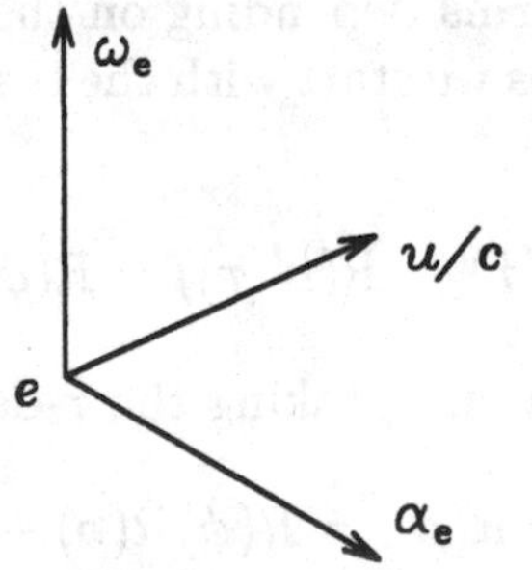


*Figure 3.4.* System of three orthogonal vectors linked to the motion of the charge of the electron with $\alpha_e = \omega_e = y$.

We can form different Poincaré invariant terms as functions of the kinematical variables and their derivatives. The above $y$ is a first order invariant in terms of the derivatives $\dot{u}$ and $\omega_0$. Similarly, $y^2$ is a second order invariant term and expressions $(\omega_0 \cdot \dot{u})\dot{t}$ and $\dot{u}^2\dot{t}^2$ are invariant terms of third and fourth order, respectively. Since

$$\dot{u}' \times u' = \frac{R(\phi)\,(\dot{u} \times u)}{\gamma(v)\,(1 + v \cdot R(\mu)u/c^2)},$$

this implies that

$$\dot{u}' \times \dot{r}' = R(\phi)\,(\dot{u} \times \dot{r}), \tag{3.175}$$

and the term $(\dot{u} \times \dot{r})^2$ is also a fourth order invariant in the derivatives of the kinematical variables. We just have to construct among them, first order invariant terms. For instance,

$$y \equiv \alpha_e = \omega_e = \frac{1}{c}|\dot{u} + u \times \omega_0|, \qquad \frac{\omega_0 \cdot \dot{u}}{\sqrt{\dot{u}^2}}, \tag{3.176}$$

$$\frac{(\omega_0 \cdot \dot{u})\dot{t}}{y^2}, \quad \dot{t}\sqrt{\frac{\dot{u}^2}{y^2}}, \quad \frac{\dot{u}^2\dot{t}^2}{y^3}, \quad \frac{\dot{u}^2\dot{t}}{(\omega_0 \cdot \dot{u})}, \tag{3.177}$$

are possible terms to be included in invariant Lagrangians.

Let us consider the Poincaré invariant Lagrangian expressed in terms of the first of the invariants in (3.177)

$$L = a\frac{(\omega_0 \cdot \dot{u})\,\dot{t}}{(\dot{u} + u \times \omega_0)^2} \equiv a\frac{\Omega \cdot \dot{u}}{(du/dt + u \times \Omega)^2}. \tag{3.178}$$

If we set $F = du/dt + u \times \Omega = (\dot{u} + u \times \omega_0)/\dot{t}$, then,

$$U = \frac{a}{F^2}\Omega - \frac{2a}{F^4}\left(\Omega \cdot \frac{du}{dt}\right)F, \tag{3.179}$$

$$W = \frac{a}{F^2}\frac{du}{dt} - \frac{2a}{F^4}\left(\Omega \cdot \frac{du}{dt}\right)F \times u, \tag{3.180}$$

and therefore

$$Z = u \times U + W = \frac{a}{F^2}F. \tag{3.181}$$

When analyzed in the center of mass frame, (3.181) is the constant spin of the particle. If we choose for the parameter $a = mc^3/2$, we get for the angular velocity that its three components are $\Omega_u = 0$, $\Omega_s = -\Omega_l = -c/R_0$ and $\Omega_\perp = a\Omega_s/mc^3 = -c/2R_0$, as shown in the Figure 3.5.

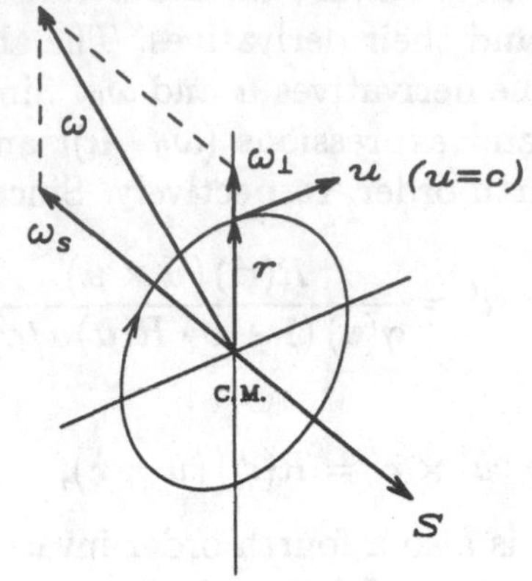


*Figure 3.5.* Center of charge motion of system (3.178) in the C.M. frame.

The time evolution of the body, corresponds to the precession of $\Omega_\perp$ around the spin direction with the constant angular velocity $\Omega_l$ which is twice the other. This means that the motion is periodic and that the body frame comes back to its initial position after two turns of the charge.

If $a$ is negative, then the spin has the opposite direction to the one depicted in the figure.

To end this section and with the above model of the electron in mind, it is convenient to remember some of the features that Dirac [13] obtained for the motion of a free electron. Let point $\boldsymbol{r}$ be the position vector on which Dirac's spinor $\psi(t, \boldsymbol{r})$ is defined. When computing the velocity of point $\boldsymbol{r}$, Dirac arrives at:

a) The velocity $\boldsymbol{u} = i/\hbar[H, \boldsymbol{r}] = c\boldsymbol{\alpha}$, is expressed in terms of $\boldsymbol{\alpha}$ matrices and writes, '... *a measurement of a component of the velocity of a free electron is certain to lead to the result* $\pm c$'.

b) The linear momentum does not have the direction of this velocity $\boldsymbol{u}$, but must be related to some average value of it: ... '*the* $x_1$ *component of the velocity,* $c\alpha_1$*, consists of two parts, a constant part* $c^2p_1H^{-1}$*, connected with the momentum by the classical relativistic formula, and an oscillatory part, whose frequency is at least* $2mc^2/h$*, ...*'.

c) About the position $\boldsymbol{r}$: '*The oscillatory part of* $x_1$ *is small, ... , which is of order of magnitude* $\hbar/mc$*, ...*'.

And when analyzing, in his original 1928 paper, [14] the interaction of the electron with an external electromagnetic field, after performing the square of Dirac's operator, he obtains two new interaction terms:

$$\frac{e\hbar}{2mc}\boldsymbol{\Sigma}\cdot\boldsymbol{B}+\frac{ie\hbar}{2mc}\boldsymbol{\alpha}\cdot\boldsymbol{E}, \tag{3.182}$$

where the electron spin is written as $\boldsymbol{S}=\hbar\boldsymbol{\Sigma}/2$ and

$$\boldsymbol{\Sigma}=\begin{pmatrix}\boldsymbol{\sigma} & 0\\ 0 & \boldsymbol{\sigma}\end{pmatrix},$$

in terms of $\sigma$-Pauli matrices and $\boldsymbol{E}$ and $\boldsymbol{B}$ are the external electric and magnetic fields, respectively. He says, *'The electron will therefore behave as though it has a magnetic moment $(e\hbar/2mc)\,\boldsymbol{\Sigma}$ and an electric moment $(ie\hbar/2mc)\,\boldsymbol{\alpha}$. The magnetic moment is just that assumed in the spinning electron model'* (Pauli model). *'The electric moment, being a pure imaginary, we should not expect to appear in the model.'*

However, if we look at our classical model, we see that for the center of mass observer, there is a non-vanishing electric and magnetic dipole moment

$$\boldsymbol{d}=e\boldsymbol{k}=\frac{e}{mc^2}\boldsymbol{S}\times\boldsymbol{u},\quad \boldsymbol{\mu}=\frac{e}{2}\boldsymbol{k}\times\frac{d\boldsymbol{k}}{dt}=-\frac{e}{2m}\boldsymbol{Y}, \tag{3.183}$$

where $\boldsymbol{S}$ is the total spin and $\boldsymbol{Y}=-m\boldsymbol{k}\times d\boldsymbol{k}/dt$ is the zitterbewegung part of spin. The time average value of $\boldsymbol{d}$ is zero, and the average value of $\boldsymbol{\mu}$ is the constant vector $\boldsymbol{\mu}$.

This classical model gives rise to the same kinematical prediction as the nonrelativistic model described in Sec.4.1. If the charge of the particle is negative, the current of Fig.3.5 produces a magnetic moment that necessarily has the same direction as the spin. If the electron spin and magnetic moments are antiparallel, then we need another contribution to the total spin, different from the zitterbewegung. All real experiments to determine very accurately the gyromagnetic ratio are based on the determination of precession frequencies, but these precession frequencies are independent of the spin orientation. However, the difficulty to separate electrons in a Stern-Gerlach type experiment, suggests to perform polarization experiments in order to determine in a direct way whether spin and magnetic moment for elementary particles are either parallel or antiparallel.

Another consequence of the classical model is that it enhances the role of the so-called minimal coupling interaction $j_\mu A^\mu$. The magnetic properties of the electron are produced by the current of its internal motion and not by some possible distribution of magnetic dipoles, so that the only possible interaction of a point charge at $\boldsymbol{r}$ with the external electromagnetic field is that of the current $j^\mu$, associated to the motion of point $\boldsymbol{r}$, with the external potentials.

## 5. TACHYONS

Let us consider the manifold spanned by the variables $(t, \boldsymbol{r}, \boldsymbol{u}, \boldsymbol{\rho})$ with domains $t \in \mathbb{R}$, $\boldsymbol{r} \in \mathbb{R}^3$, $\boldsymbol{\rho} \in \mathbb{R}^3_c$ as in the previous cases, but now $\boldsymbol{u} \in \mathbb{R}^3$, with the lower bound $u > c$. We see that the transformation equations under $\mathcal{P}$ of the variables $(t, \boldsymbol{r}, \boldsymbol{u})$ are the same as the ones described in (3.49 - 3.51), while for the orientation variables (3.52), in the case $u > c$, there is no real limit because the $\gamma(u)$ factors that appear in the functions $\boldsymbol{F}$ and $G$, defined in (3.9-3.10), become imaginary, thus producing a complex result for the transformed variables $\boldsymbol{\rho}'$. Then, with the constraint $u > c$, the maximal homogeneous space of $\mathcal{P}$ is spanned by the variables $(t, \boldsymbol{r}, \boldsymbol{u})$.

Because $u > c$, the particle is called a **Tachyon**, and it is a mechanical system whose kinematical space $X$ is the manifold spanned by the mentioned variables $(t, \boldsymbol{r}, \boldsymbol{u})$, which are interpreted as the time, position and velocity of the particle. This manifold is isomorphic to the homogeneous space $\mathcal{P}/SO(3)$. Because there are no orientation variables, these systems can never give rise, when quantized, to spin 1/2 particles, as we shall see in Chapter 4, so that fermions can never have a tachyonic zitterbewegung from the classical point of view.

There are three differential constraints among the kinematical variables, $\boldsymbol{u} = \dot{\boldsymbol{r}}/\dot{t}$, and with the homogeneity condition, this reduces to three the number of independent degrees of freedom. The Lagrangian can be written as

$$L = T\dot{t} + \boldsymbol{R}\cdot\dot{\boldsymbol{r}} + \boldsymbol{U}\cdot\dot{\boldsymbol{u}}, \tag{3.184}$$

where, as usual, $T = \partial L/\partial \dot{t}$, $R_i = \partial L/\partial \dot{r}^i$, $U_i = \partial L/\partial \dot{u}^i$ and they transform under $\mathcal{P}$ as:

$$T'(\tau) = \gamma T(\tau) - \gamma(\boldsymbol{v}\cdot R(\boldsymbol{\mu})\boldsymbol{R}(\tau)), \tag{3.185}$$

$$\boldsymbol{R}'(\tau) = R(\boldsymbol{\mu})\boldsymbol{R}(\tau) - \gamma\boldsymbol{v}T(\tau)/c^2 + \frac{\gamma^2}{(1+\gamma)c^2}(\boldsymbol{v}\cdot R(\boldsymbol{\mu})\boldsymbol{R}(\tau))\boldsymbol{v}, \tag{3.186}$$

$$\boldsymbol{U}'(\tau) = \gamma(1 + \boldsymbol{v}\cdot R(\boldsymbol{\mu})\boldsymbol{u}/c^2)\Bigg[R(\boldsymbol{\mu})\boldsymbol{U}(\tau) + \frac{\gamma^2}{(1+\gamma)c^2}(\boldsymbol{v}\cdot R(\boldsymbol{\mu})\boldsymbol{U}(\tau))\boldsymbol{v}$$
$$+\frac{\gamma^2}{c^2}(\boldsymbol{u}\cdot\boldsymbol{U}(\tau))\boldsymbol{v}\Bigg]. \tag{3.187}$$

Noether's theorem defines the following constants of the motion

$$\text{energy } H = -T - (d\boldsymbol{U}/dt)\cdot\boldsymbol{u}, \tag{3.188}$$

$$\text{linear momentum } \boldsymbol{P} = \boldsymbol{R} - (d\boldsymbol{U}/dt), \tag{3.189}$$

$$\begin{aligned} \text{kinematical momentum } \boldsymbol{K} &= H\boldsymbol{r}/c^2 - \boldsymbol{P}t - \boldsymbol{U} + (\boldsymbol{u}\cdot\boldsymbol{U})\frac{\boldsymbol{u}}{c^2}, && (3.190)\\ \text{angular momentum } \boldsymbol{J} &= \boldsymbol{r}\times\boldsymbol{P} + \boldsymbol{u}\times\boldsymbol{U}. && (3.191)\end{aligned}$$

The function $\boldsymbol{Z} = \boldsymbol{u}\times\boldsymbol{U}$ is always orthogonal to the velocity, and the energy $H$ and linear momentum $\boldsymbol{P}$ transform like the components of a four-vector.

If we define the relative position vector $\boldsymbol{k}$ by

$$H\boldsymbol{k}/c^2 = \boldsymbol{U} - (\boldsymbol{u}\cdot\boldsymbol{U})\boldsymbol{u}/c^2, \qquad (3.192)$$

then, from (3.190) we get:

$$\boldsymbol{P} = \frac{H}{c^2}\frac{d(\boldsymbol{r}-\boldsymbol{k})}{dt} = \frac{H}{c^2}\frac{d\boldsymbol{q}}{dt}, \qquad (3.193)$$

and point $\boldsymbol{q}$ moves with constant velocity. Function $\boldsymbol{Z}$ and vector $H\boldsymbol{k}/c$, transform as

$$\boldsymbol{Z}'(\tau) = \gamma R(\boldsymbol{\mu})\boldsymbol{Z}(\tau) - \frac{\gamma^2}{(1+\gamma)c^2}(\boldsymbol{v}\cdot R(\boldsymbol{\mu})\boldsymbol{Z}(\tau))\boldsymbol{v} + \frac{\gamma H}{c^2}(\boldsymbol{v}\times R(\boldsymbol{\mu})\boldsymbol{k}(\tau)), \qquad (3.194)$$

$$H'\boldsymbol{k}'(\tau) = \gamma R(\boldsymbol{\mu})H\boldsymbol{k}(\tau) - \frac{\gamma^2 H}{(1+\gamma)c^2}(\boldsymbol{v}\cdot R(\boldsymbol{\mu})\boldsymbol{k}(\tau))\boldsymbol{v} - \gamma(\boldsymbol{v}\times R(\boldsymbol{\mu})\boldsymbol{Z}(\tau)), \qquad (3.195)$$

like the strict components of an antisymmetric tensor $Z^{\mu\nu}$.

It must be observed that the expression that defines $H\boldsymbol{k}/c^2$ is a vector equation that is always different from zero. This is true if $\boldsymbol{U}$ and $\boldsymbol{u}$ have different directions. Otherwise, if $\boldsymbol{U}$ and $\boldsymbol{u}$ have the same direction it can never vanish because $u > c$. This suggests that the energy $H$ and vector $\boldsymbol{k}$ are always non-vanishing magnitudes for every observer. If the energy is positive definite, the center of mass will have a motion with velocity below $c$.

For Tachyons the spin is only of (anti)orbital type. It takes the general expression

$$\boldsymbol{S} = \boldsymbol{J} - \frac{c^2}{H}\boldsymbol{K}\times\boldsymbol{P} = \boldsymbol{u}\times\boldsymbol{U} + \boldsymbol{k}\times\boldsymbol{P},$$

where the relative position of the charge $\boldsymbol{k}$, given in (3.192), is related only to the function $\boldsymbol{U}$, similarly as the particle described in Section 3.2.

The following invariant Lagrangian for tachyonic particles:

$$L = a\dot{t}(u^2-c^2)^{1/2} - \frac{b\dot{t}}{(u^2-c^2)^{3/2}}\left[(d\boldsymbol{u}/dt)^2 - \frac{(\boldsymbol{u}\cdot d\boldsymbol{u}/dt)^2}{u^2-c^2}\right], \qquad (3.196)$$

leads in the center of mass frame to the dynamical equations:

$$m\boldsymbol{k} = -\frac{2b}{(u^2 - c^2)^{3/2}} \frac{d^2\boldsymbol{k}}{dt^2}, \tag{3.197}$$

where $\boldsymbol{u} \equiv d\boldsymbol{k}/dt$.

Similarly to the example shown in Section 3.2, if parameter $b$ is written as $b = m/2\omega^2$, then this is again a generalization of an isotropic harmonic oscillator

$$\frac{d^2\boldsymbol{k}}{dt^2} + \omega^2(u^2 - c^2)^{3/2}\,\boldsymbol{k} = 0.$$

Due to the factor $(u^2 - c^2)^{3/2}$ in front of $\boldsymbol{k}$ and because $u > c$, the velocity can never decrease to the value $c$.

Charge trajectories of this equations are similar to the ones depicted in Figure 3.3, but now with velocity $u > c$.

The internal motion of the charge is a central motion that, being the spin constant, gives rise to a first integral $\boldsymbol{S} = -\boldsymbol{k} \times m\boldsymbol{u}$, and the motion takes place in a plane orthogonal to the constant vector $\boldsymbol{S}$. In polar coordinates $(r, \theta)$ equations (3.197) give:

$$\frac{d^2r}{dt^2} - r\left(\frac{d\theta}{dt}\right)^2 + \frac{m}{2b}\left[\left(\frac{dr}{dt}\right)^2 + r^2\left(\frac{d\theta}{dt}\right)^2 - c^2\right]^{3/2} r = 0, \tag{3.198}$$

$$2\frac{dr}{dt}\frac{d\theta}{dt} + r\frac{d^2\theta}{dt^2} = 0, \tag{3.199}$$

and the first integral leads to $d\theta/dt = S/mr^2$. The radial equation (3.198) becomes

$$\frac{d^2r}{dt^2} - \frac{S^2}{m^2r^3} + \frac{m}{2b}\left[\left(\frac{dr}{dt}\right)^2 + \frac{S^2}{m^2r^2} - c^2\right]^{3/2} r = 0, \tag{3.200}$$

that has solutions with $r =$ constant $\neq 0$, and thus circular motions with constant velocity $u > c$ are also allowed.

## 6. INVERSIONS

The space and time inversions are the automorphisms of $\mathcal{P}$, given by

$$\mathbb{P} : (b, \boldsymbol{a}, \boldsymbol{v}, \boldsymbol{\mu}) \longrightarrow (b, -\boldsymbol{a}, -\boldsymbol{v}, \boldsymbol{\mu}), \tag{3.201}$$

$$\mathbb{T} : (b, \boldsymbol{a}, \boldsymbol{v}, \boldsymbol{\mu}) \longrightarrow (-b, \boldsymbol{a}, -\boldsymbol{v}, \boldsymbol{\mu}), \tag{3.202}$$

such that the Poincaré group composition law (3.5-3.8) remains invariant under these transformations. We can extend this action on the kinematical space $X = \mathcal{P}$ in the form:

$$\mathbb{P} : (t(\tau), \boldsymbol{r}(\tau), \boldsymbol{u}(\tau), \boldsymbol{\rho}(\tau)) \longrightarrow (t(\tau), -\boldsymbol{r}(\tau), -\boldsymbol{u}(\tau), \boldsymbol{\rho}(\tau)), \quad (3.203)$$

$$\mathbb{T} : (t(\tau), \boldsymbol{r}(\tau), \boldsymbol{u}(\tau), \boldsymbol{\rho}(\tau)) \longrightarrow (-t(\tau), \boldsymbol{r}(\tau), -\boldsymbol{u}(\tau), \boldsymbol{\rho}(\tau)). \quad (3.204)$$

If we assume that the evolution parameter $\tau$ remains invariant under inversions, we can define the action of $\mathbb{P}$ and $\mathbb{T}$ on the derivatives by:

$$\mathbb{P} : (\dot{t}(\tau), \dot{\boldsymbol{r}}(\tau), \dot{\boldsymbol{u}}(\tau), \dot{\boldsymbol{\rho}}(\tau)) \longrightarrow (\dot{t}(\tau), -\dot{\boldsymbol{r}}(\tau), -\dot{\boldsymbol{u}}(\tau), \dot{\boldsymbol{\rho}}(\tau)), \quad (3.205)$$

$$\mathbb{T} : (\dot{t}(\tau), \dot{\boldsymbol{r}}(\tau), \dot{\boldsymbol{u}}(\tau), \dot{\boldsymbol{\rho}}(\tau)) \longrightarrow (-\dot{t}(\tau), \dot{\boldsymbol{r}}(\tau), -\dot{\boldsymbol{u}}(\tau), \dot{\boldsymbol{\rho}}(\tau)), \quad (3.206)$$

and therefore:

$$\mathbb{P} : (\boldsymbol{\alpha}(\tau), \boldsymbol{\omega}(\tau)) \longrightarrow (-\boldsymbol{\alpha}(\tau), \boldsymbol{\omega}(\tau)), \quad (3.207)$$

$$\mathbb{T} : (\boldsymbol{\alpha}(\tau), \boldsymbol{\omega}(\tau)) \longrightarrow (-\boldsymbol{\alpha}(\tau), \boldsymbol{\omega}(\tau)), \quad (3.208)$$

hence $\alpha^2 - \omega^2$ is invariant under both inversions, while the expression $\boldsymbol{\alpha} \cdot \boldsymbol{\omega}$ changes its sign.

Lagrangians $L_B$, (3.97) and $L_T$, (3.196) are invariant under $\mathbb{P}$ and change of sign under $\mathbb{T}$, while the photon Lagrangian (3.151) changes the sign under both inversions.

However, Lagrangian $L_F$ (3.98) changes the sign of the second term under $P$ while the first remains the same. Under time inversion we have the opposite situation.

Lagrangian (3.178) changes the sign under inversions and can be interpreted as representing a system of mass $-m$.

# 7. INTERACTION WITH AN EXTERNAL FIELD

Let us assume that the total Lagrangian of a relativistic spinning particle, interacting with some external source, is $L = L_0 + L_I$, where $L_0$ is a general Poincaré invariant Lagrangian for a free system. For the interaction term $L_I$ we shall consider the minimal coupling of electromagnetism

$$L_I = -e\phi(t, \boldsymbol{r})\dot{t} + e\boldsymbol{A}(t, \boldsymbol{r}) \cdot \dot{\boldsymbol{r}}, \quad (3.209)$$

linear in the derivatives and where the external potentials are functions only of $t$ and $\boldsymbol{r}$. Most general interactions will not be analyzed in this work.

The total Lagrangian can be written as

$$L = T\dot{t} + \boldsymbol{R}\cdot\dot{\boldsymbol{r}} + \boldsymbol{U}\cdot\dot{\boldsymbol{u}} + \boldsymbol{W}\cdot\boldsymbol{\omega}_0.$$

The difference with the free particle case is contained in the terms $T = T_0 + T_I$ and $\boldsymbol{R} = \boldsymbol{R}_0 + \boldsymbol{R}_I$, with $T_0 = \partial L_0/\partial\dot{t}$, $\boldsymbol{R}_0 = \partial L_0/\partial\dot{\boldsymbol{r}}$, $T_I = -e\phi$ and $\boldsymbol{R}_I = e\boldsymbol{A}$, respectively. Because the interaction term $L_I$ is not an explicit function of $\dot{\boldsymbol{u}}$ and $\boldsymbol{\omega}_0$, the functions $T_0$, $\boldsymbol{R}_0$, $\boldsymbol{U}$ and $\boldsymbol{W}$ have the same expressions as in the free case. Similarly, $\boldsymbol{Z} = \boldsymbol{u}\times\boldsymbol{U} + \boldsymbol{W}$ also has the same expression in terms of the degrees of freedom and their derivatives as in the free case.

If the interaction term $L_I$ is Poincaré invariant, the Noether constants of the motion will have a part coming from $L_0$, which we call the mechanical part and represent it with a subindex $m$, and another part coming from $L_I$. They take the following form in the case of Luxons

$$\begin{aligned}
H &= H_m + e\phi,\\
\boldsymbol{P} &= \boldsymbol{P}_m + e\boldsymbol{A},\\
\boldsymbol{K} &= \frac{H}{c^2}\boldsymbol{r} - \boldsymbol{P}t - \frac{1}{c^2}\boldsymbol{Z}\times\boldsymbol{u} = \boldsymbol{K}_m - e\boldsymbol{A}t,\\
\boldsymbol{J} &= \boldsymbol{r}\times\boldsymbol{P} + \boldsymbol{u}\times\boldsymbol{U} + \boldsymbol{W} = \boldsymbol{J}_m + \boldsymbol{r}\times e\boldsymbol{A}.
\end{aligned}$$

The above mechanical observables have the same analytical expressions in terms of the kinematical variables and their time derivatives as in the free particle case. In particular, the mechanical energy and linear momentum are

$$H_m = -T_0 - \boldsymbol{u}\cdot\frac{d\boldsymbol{U}}{dt},\quad \boldsymbol{P}_m = \boldsymbol{R}_0 - \frac{d\boldsymbol{U}}{dt}.$$

This implies that the invariant

$$\frac{1}{c^2}H_m^2 - \boldsymbol{P}_m^2 = m^2c^2, \tag{3.210}$$

defines the same constant parameter $m$, the rest mass of the system, as in the free case. This yields the usual replacement $P_\mu \to P_\mu - eA_\mu$ when dealing with interacting systems

$$\frac{1}{c^2}\left(H - e\phi\right)^2 - \left(\boldsymbol{P} - e\boldsymbol{A}\right)^2 = m^2c^2,$$

where $H$ and $\boldsymbol{P}$ are the generating functions of translations of the Poincaré group.

In the general case the interaction term $L_I$ is not translation invariant because of the explicit dependence of the potentials on $t$ and $\boldsymbol{r}$, and therefore the above observables are not constants of the motion. In terms of the kinematical variables we have to use the constraints $L_c = \boldsymbol{\lambda}(\tau)\cdot(\boldsymbol{u}\dot{t}-\dot{\boldsymbol{r}})$, so that the dynamical equations for $t$, $\boldsymbol{r}$ and $\boldsymbol{u}$ variables become

$$\frac{\partial L_I}{\partial t} - \frac{d}{d\tau}\left(\frac{\partial(L_0+L_I+L_c)}{\partial \dot{t}}\right) = 0, \tag{3.211}$$

$$\frac{\partial L_I}{\partial \boldsymbol{r}} - \frac{d}{d\tau}\left(\frac{\partial(L_0+L_I+L_c)}{\partial \dot{\boldsymbol{r}}}\right) = 0, \tag{3.212}$$

$$\frac{\partial L_c}{\partial \boldsymbol{u}} - \frac{d}{d\tau}\left(\frac{\partial L_0}{\partial \dot{\boldsymbol{u}}}\right) = 0. \tag{3.213}$$

We have assumed that $L_0$ does not depend on $\boldsymbol{u}$ variables, which have been replaced by $\dot{\boldsymbol{r}}/\dot{t}$ whenever they appear. For the orientation variables we shall assume that the only dependence of $L_0$ of $\boldsymbol{\rho}$ is through the angular velocity $\boldsymbol{\omega}_0$ and therefore dynamical equations are the same as in the free case

$$\frac{d\boldsymbol{W}}{dt} = \boldsymbol{\omega}_0 \times \boldsymbol{W}.$$

Equation (3.213) implies that

$$\boldsymbol{\lambda} = \frac{d\boldsymbol{U}}{dt}.$$

Using this solution for Lagrange's multipliers, the first equation (3.211) can be rewritten as

$$\frac{dH}{dt} = \frac{dH_m}{dt} + \frac{d(e\phi(t,\boldsymbol{r}))}{dt} = -\frac{1}{\dot{t}}\frac{\partial L_I}{\partial t}.$$

Total energy will be conserved only if the potentials are not explicit functions of $t$. In general, by taking the above derivatives, we get

$$\frac{dH_m}{dt} = e\boldsymbol{u}\cdot\left(-\nabla\phi(t,\boldsymbol{r}) - \frac{\partial \boldsymbol{A}(t,\boldsymbol{r})}{\partial t}\right) = e\boldsymbol{u}\cdot\boldsymbol{E}(t,\boldsymbol{r}), \tag{3.214}$$

and therefore the variation of the mechanical energy of the particle is the work of the external electric field along the charge trajectory.

The second equation (3.212) gives rise to

$$\frac{d\boldsymbol{P}}{dt} = \frac{d\boldsymbol{P}_m}{dt} + e\frac{d\boldsymbol{A}}{dt} = \frac{1}{\dot{t}}\frac{\partial L_I}{\partial \boldsymbol{r}},$$

thus showing that total linear momentum is conserved if the potentials are not explicit functions of the position of the charge. Rearranging terms we get

$$\frac{d\boldsymbol{P}_m}{dt} = e\left(\boldsymbol{E} + \boldsymbol{u}\times\boldsymbol{B}\right) = \boldsymbol{F}, \tag{3.215}$$

where the Lorentz force $\boldsymbol{F}$ is expressed in terms of the external fields $\boldsymbol{E} = -\nabla\phi - \partial \boldsymbol{A}/\partial t$ and $\boldsymbol{B} = \nabla \times \boldsymbol{A}$ which are defined at the charge position $\boldsymbol{r}$.

Because the interaction terms do not affect the analytic expressions of $\boldsymbol{U}$ and $\boldsymbol{W}$, and therefore of $\boldsymbol{Z}$, the separation $\boldsymbol{k}$ between the center of mass and center of charge is still given by the same expression as in the free case of Luxons as

$$\boldsymbol{k} = \frac{1}{H_m}\boldsymbol{Z} \times \boldsymbol{u}.$$

The definition of the center of mass $\boldsymbol{q}$ remains the same and this allows us to write

$$\boldsymbol{P}_m = \frac{H_m}{c^2}\frac{d\boldsymbol{q}}{dt}.$$

If we call $\boldsymbol{v} = d\boldsymbol{q}/dt$ the center of mass velocity, then we get as for the point particle

$$H_m = \gamma(v)mc^2, \quad \boldsymbol{P}_m = \gamma(v)m\boldsymbol{v}.$$

If the rest mass is not affected by the interaction, taking the time derivative of (3.210), we get

$$\frac{H_m}{c^2}\frac{dH_m}{dt} - \boldsymbol{P}_m \cdot \frac{d\boldsymbol{P}_m}{dt} = 0 \tag{3.216}$$

and using the above expression in terms of the center of mass velocity we obtain

$$\frac{dH_m}{dt} = \boldsymbol{v} \cdot \boldsymbol{F}(t, \boldsymbol{r}). \tag{3.217}$$

The non-relativistic equivalent to equation (3.216), with the assumption that the internal energy of an elementary particle is not affected by the interaction, is

$$m\frac{dH_m}{dt} - \boldsymbol{P}_m \cdot \frac{d\boldsymbol{P}_m}{dt} = 0,$$

and since there $\boldsymbol{P}_m = m\boldsymbol{v}$, we also get Equation (3.217).

The variation of the mechanical energy is the work of the total external force, defined at the charge position $\boldsymbol{r}$, along the center of mass trajectory. But this has to give the same result as in (3.214). Otherwise, the assumption that the mass of an elementary particle is not affected by the interaction will no longer be true. This will mean, for instance, that an electron under the action of an external field will have a different mass when the interaction is switched off. This is clearly unphysical.

Because (3.214) can also be written as $e\boldsymbol{u} \cdot \boldsymbol{E} = \boldsymbol{u} \cdot \boldsymbol{F}$, this implies that the internal dynamics of the charge is adjusted in such a way that if we write $\boldsymbol{u} = \boldsymbol{v} + \boldsymbol{w}$, then necessarily the charge has a component along the center of mass motion while the other component $\boldsymbol{w}$ must be orthogonal to the total external Lorentz force $\boldsymbol{w} \cdot \boldsymbol{F} = 0$. Therefore, the magnetic field does not contribute to the variation of energy of the particle but contributes to the orientation of the zitterbewegung plane, and thus, to the spin dynamics.

## Notes


1 M.Rivas, M.Valle and J.M.Aguirregabiria, *Eur. J. Phys.* **6**, 128 (1986).

2 L.H. Thomas, *Phil. Mag.* **3**, 1 (1927).

3 M. Rivas, *J. Math. Phys.* **30**, 318 (1989).

4 R.G. Beil, *Found. Phys.* **25**, 717 (1995); **25**, 1577 (1995).

5 G.C. Constantelos, *Nuovo Cimento* **B 21**, 279 (1974).

6 A.J. Hanson and T. Regge, *Ann. Phys.* NY **87**, 498 (1974).

7 M.H.L.Pryce, *Proc. Roy. Soc.* **A195**, 62 (1948).

8 R. A. Beth, *Phys. Rev.* **50**, 115 (1936).

9 J.J. Sakurai, *Advanced Quantum Mechanics*, Addison-Wesley Reading, MA (1967).

10 M. Visser, *Phys. Lett.* **A 139**, 99 (1989).

11 M. Mathisson, *Acta Phys. Pol.* **6**, 163 (1937); **6**, 218 (1937)

12 M.H.L. Weyssenhof, *Acta Phys. Pol.* **9**, 46 (1947). M.H.L. Weyssenhof and A. Raabe, *Acta Phys. Pol.* **9**, 7 (1947); **9**, 19 (1947).

13 P.A.M. Dirac, *The Principles of Quantum mechanics*, Oxford Univ. Press, 4th ed. Oxford (1967).

14 P.A.M. Dirac, *Proc. Roy. Soc. Lon.* **A117**, 610 (1928).

## Chapter 4

# QUANTIZATION OF LAGRANGIAN SYSTEMS

Quantization of generalized Lagrangian systems will suggest that wave functions for elementary particles must be squared integrable functions defined on the kinematical space.

We shall use Feynman's quantization method to show the structure of the wave function and the way it transforms under the kinematical or symmetry group of the theory. Once the Hilbert space structure of the state space is determined, this leads to a specific representation of the generators of the group as self-adjoint operators and the remaining analysis is done within the usual quantum mechanical context, *i.e.*, by choosing the complete commuting set of operators to properly determine a set of orthogonal basis vectors of the Hilbert space. Special emphasis is devoted to the analysis of the different angular momentum operators the formalism supplies. They have a similar structure to the classical ones, and this will help us to properly obtain the identification of the spin observable.

The structure of the spin operator depends on the kind of translation invariant kinematical variables we use to describe the particle, and the way these variables transform under the rotation group. Since in the Galilei and Poincaré case, as we have seen in previous chapters, these variables are the velocity $\boldsymbol{u}$ and orientation $\boldsymbol{\alpha}$ and they transform in the same way under rotations in both approaches, then the structure of the spin operator is exactly the same in both relativistic and nonrelativistic formalisms.

As we have seen in the classical description the position of the charge of the particle and its center of mass are different points, and spin is related to the rotation and internal motion (zitterbewegung) of the charge around the center of mass of the particle. The magnetic properties of the particle are connected only with the motion of the charge and therefore to the zitterbewegung part of spin. It is this double spin structure that gives rise to the concept of gyromagnetic ratio when expressing the magnetic moment in terms of the total spin. If the Lagrangian shows no dependence on the acceleration, the spin is only of rotational nature, and the position and center of mass position define the same point. Spin 1/2 particles arise if the corresponding classical model rotates but no half integer spins are obtained for systems with spin of orbital nature related only to the zitterbewegung. On the manifold spanned by non-compact variables $\boldsymbol{u}$ no half-integer spins can be found, because the spin operator has the form of an orbital angular momentum and eigenvectors are but spherical harmonics.

Dirac's equation will be obtained when quantizing the classical relativistic spinning particles whose center of charge is circling around its center of mass at the speed $c$. In that case, the internal orientation of the electron completely characterizes its Dirac algebra.

Tachyons will be quantized in the same way and because the classical model has no orientation variables, they always have integer spin.

# 1. FEYNMAN'S QUANTIZATION OF LAGRANGIAN SYSTEMS

Let us consider a generalized Lagrangian system as described in previous chapters and whose evolution is considered on the kinematical space between points $x_1$ and $x_2$.

For quantizing these generalized Lagrangian systems we shall follow Feynman's path integral method. [1] The Uncertainty Principle is introduced in Feynman's approach by the condition that if no measurement is performed to determine the trajectory followed by the system from $x_1$ to $x_2$, then all paths $x(\tau)$ are allowed with the same probability. Therefore a probability definition $P[x(\tau)]$, must be given for every path.

But instead of defining the probability associated to each possible path $P[x(\tau)]$, this is calculated in terms of a **probability amplitude**, $\phi[x(\tau)]$ for that path such that $P[x(\tau)] = |\phi[x(\tau)]|^2$, where $0 \leq P \leq 1$. But in general $\phi$ does not need to be a positive real number; in fact it is a complex number. Thus, to every possible trajectory followed by the system, $x(\tau)$ in $X$ space, Feynman associates a complex number $\phi[x(\tau)]$ called the probability amplitude of this alternative, given by

$$\phi[x(\tau)] = N \exp\left\{ \frac{i}{\hbar} \int_{\tau_1}^{\tau_2} L(x(\tau), \dot{x}(\tau)) d\tau \right\} = N \exp\left\{ \frac{i}{\hbar} A_{[x]}(x_1, x_2) \right\}, \tag{4.1}$$

where $N$ is a path independent normalization factor, and where the phase of this complex number in units of $\hbar$ is the classical action of the system $A_{[x]}(x_1, x_2)$ along the path $x(\tau)$. Once we perform the integration along the path, this probability amplitude becomes clearly a function of the initial and final points in $X$ space, $x_1$ and $x_2$, respectively.

In this Feynman statistical procedure, the probability amplitude of the occurrence of any alternative of a set of independent alternatives is the sum of the corresponding probability amplitudes of the different independent events. The probability of the whole process is the square of the absolute value of the total probability amplitude. This produces the effect that the probability of the whole process can be less than the probability of any single alternative of the set. This is what Feynman calls **interfering statistics**.

Then, the total probability amplitude that the system arrives at point $x_2$ coming from $x_1$, *i.e.*, Feynman's kernel $K(x_1, x_2)$, is obtained as the sum or integration over all paths, of terms of the form of Eq. (4.1). Feynman's kernel $K(x_1, x_2)$, will be in general a function, or more precisely a distribution, on the $X \times X$ manifold. If information concerning the initial point is lost, and the final point is left arbitrary, say $x$, the kernel reduces to the probability amplitude for finding the system at point $x$, *i.e.*, the usual interpretation of the quantum mechanical wave function $\Phi(x)$. By the above discussion we see that wave functions must be complex functions of the kinematical variables.

We thus see that Feynman's quantization method enhances the role of the kinematical variables to describe the quantum state of an arbitrary system, in spite of the independent degrees of freedom. We consider that this is one of the reasons why the kinematical variables have to play a leading role also in the classical approach.

We are used to consider in quantum mechanics, instead of a single function $\Phi(x)$, multicomponent wave functions, *i.e*, a set of linearly independent functions $\psi_i(t, \boldsymbol{r})$ defined on space-time and labeled with a discrete subindex that runs over a finite range, such that it can be considered as a vector valued function in a finite dimensional complex vector space. In general this finite space carries some irreducible representation of the rotation group and each component $\psi_i$ represents a definite spin state of the system. Nevertheless, our wave function $\Phi(x)$ depends on more variables than space-time variables. Once we define later the complete commuting set of observables to obtain, in terms of their simultaneous eigenvectors, an orthonormal basis for the Hilbert space of states, we shall find that $\Phi(x)$ can be separated in two parts. One part $\phi(t, \boldsymbol{r})$ depending on space-time variables and another part $\chi$ that depends on the remaining translation invariant kinematical variables, that in our case will reduce to the velocity $\boldsymbol{u}$ and orientation $\boldsymbol{\alpha}$. It is this possible separation of our wave function that will produce the emergence of the different components of the usual formalism.

To see how the wave function transforms between inertial observers, and therefore to obtain its transformation equations under the kinematical groups, let us consider that $O$ and $O'$ are two inertial observers related by means of a transformation $g \in G$, such that the kinematical variables transform as:

$$x'^i = f^i(x, g). \tag{4.2}$$

If observer $O$ considers that the system follows the path $\bar{x}(\tau)$, then it follows for $O'$ the path $\bar{x}'(\tau) = f(\bar{x}(\tau), g)$ and because the action along classical paths transforms according to Eq. (1.60), the probability amplitude for observer $O'$ is just

$$\phi'[\bar{x}'(\tau)] = N \exp\left\{ \frac{i}{\hbar} \int_{\tau_1}^{\tau_2} L(\bar{x}'(\tau), \dot{\bar{x}}'(\tau)) d\tau \right\}$$

$$= N \exp\left\{ \frac{i}{\hbar} \int_{\tau_1}^{\tau_2} L(\bar{x}(\tau), \dot{\bar{x}}(\tau)) d\tau \right\} \exp\left\{ \frac{i}{\hbar} \int_{\tau_1}^{\tau_2} \frac{d\alpha(g; \bar{x}(\tau))}{d\tau} d\tau \right\},$$

*i.e.*,

$$\phi'[\bar{x}'(\tau)] = \phi[\bar{x}(\tau)] \exp\left\{ \frac{i}{\hbar} \left( \alpha(g; x_2) - \alpha(g; x_1) \right) \right\},$$

where the last phase factor is independent of the integration path. If we add all probability amplitudes of this form, it turns out that Feynman's kernel transforms as:

$$K'(x_1', x_2') = K(x_1, x_2) \exp\left\{ \frac{i}{\hbar} \left( \alpha(g; x_2) - \alpha(g; x_1) \right) \right\}. \tag{4.3}$$

If information concerning the initial point $x_1$ is lost, the wave function transforms as the part related to the variables $x_2$, up to an arbitrary function on $G$,

$$\Phi'(x'(x)) = \Phi'(gx) = \Phi(x) \exp\left\{ \frac{i}{\hbar} \left( \alpha(g; x) + \theta(g) \right) \right\}, \tag{4.4}$$

or in terms of unprimed $x$ variables

$$\Phi'(x) = \Phi(g^{-1}x) \exp\left\{ \frac{i}{\hbar} \left( \alpha(g; g^{-1}x) + \theta(g) \right) \right\}, \tag{4.5}$$

where $\theta(g)$ is some function defined on $G$ but independent of $x$.

Since our system is somewhere in $X$ space, the probability of finding the system anywhere is 1. Then we have to define the way of adding probabilities at different points $x \in X$. If we define a measure on $X$, $\mu(x)$, such that $d\mu(x)$ is the volume element in $X$ space and $|\Phi(x)|^2 d\mu(x)$ is interpreted as the probability of finding the system inside the volume element $d\mu(x)$ around point $x$, the probability of finding it anywhere in $X$ must be unity, so that

$$\int_X |\Phi(x)|^2 d\mu(x) = 1. \tag{4.6}$$

Since from (4.5)

$$|\Phi'(x')|^2 = |\Phi(x)|^2, \tag{4.7}$$

it is sufficient for the conservation of probability to assume that the measure to be defined $\mu(x)$ is group invariant. In that case, equation (4.7) implies also that inertial observers measure locally the same probability. This will have strong consequences about the possibility of invariance of the formalism under arbitrary changes of phase of the wave function. But the phase can be changed in a different manner at different points $x$. We can use this fact to further impose the local gauge invariance of the theory. It must be remarked that this arbitrary change of phase $\beta(x)$ is not only a phase on space-time, but rather on the whole kinematical space of the system and this enlarges the possibilities of analyzing different transformation groups that can be more general than the original kinematical groups, because they act on a larger manifold.

Consequently, the Hilbert space $\mathcal{H}$ whose unit rays represent the pure states of the system is the space of squared-integrable functions $\mathbb{L}^2(X, \mu)$ defined on the kinematical space $X$, $\mu(x)$ being an invariant measure such that the scalar product on $\mathcal{H}$ is defined as

$$< \Phi|\Psi >= \int_X \Phi^*(x)\Psi(x)d\mu(x), \tag{4.8}$$

$\Phi^*(x)$ being the complex conjugate function of $\Phi(x)$. There is an arbitrariness in the election of the invariant measure $\mu(x)$ but this will be guided by physical arguments. Nevertheless, the invariance condition will restrict the possible measures to be used.

## 1.1 REPRESENTATION OF OBSERVABLES

Wigner's theorem, [2,3] implies that to every symmetry $g \in G$ of a continuous group, there exists a one to one mapping of unit rays into unit rays that is induced on $\mathcal{H}$ by a unitary operator $U(g)$ defined up to a phase that maps a wave function defined on $x$ into an arbitrary wave function of the image unit ray in $x'$. The Relativity Principle is a strong symmetry of physical systems that defines the equivalence between the set of inertial observers whose space-time measurements are related by means of a transformation of a kinematical group G. Now, if we interpret $\Phi(x)$ as the wave function that describes the state of the system for the observer $O$ and $\Phi'(x)$ for $O'$, then we have

$$U(g)\Phi(x) = \Phi'(x) = \Phi(g^{-1}x)\exp\left\{\frac{i}{\hbar}\alpha(g;g^{-1}x) + \theta(g)\right\}. \qquad (4.9)$$

Since the $\theta(g)$ function gives rise to a constant phase we can neglect it and then take as the definition of the unitary representation of the group $G$ on Hilbert space $\mathcal{H}$

$$\Phi'(x) = U(g)\Phi(x) = \Phi(g^{-1}x)\exp\left\{\frac{i}{\hbar}\alpha(g;g^{-1}x)\right\}. \qquad (4.10)$$

Gauge functions satisfy (1.62), and therefore the phase term can be replaced by

$$\alpha(g;g^{-1}x) = -\alpha(g^{-1};x) + \alpha(0;x) + \xi(g,g^{-1}) = -\alpha(g^{-1};x) + \zeta(g), \qquad (4.11)$$

because gauge functions can always be chosen such that $\alpha(0;x) = 0$ and the group function $\zeta(g) = \xi(g,g^{-1})$ giving rise also to a constant phase, can be suppressed. We thus define the transformation of the wave function by

$$\Phi'(x) = U(g)\Phi(x) = \Phi(g^{-1}x)\exp\left\{-\frac{i}{\hbar}\alpha(g^{-1};x)\right\}. \qquad (4.12)$$

If the unitary operator is represented in terms of the corresponding self-adjoint generators of the Lie algebra in the form

$$U(g) = \exp\left\{-\frac{i}{\hbar}g^\sigma X_\sigma\right\}, \qquad (4.13)$$

then, for an infinitesimal transformation of parameters $\delta g^\sigma$ its inverse transformation has infinitesimal parameters $-\delta g^\sigma$, we obtain at first order in $\delta g^\sigma$

$$U(\delta g)\Phi(x) = \left(\mathbb{I} - \frac{i}{\hbar}\delta g^\sigma X_\sigma\right)\Phi(x) = \Phi(x) - \frac{i}{\hbar}\delta g^\sigma X_\sigma\,\Phi(x),$$

while

$$\Phi(\delta g^{-1}x) \equiv \Phi(f(x,\delta g^{-1})) = \Phi(x) - \delta g^\sigma u^i_\sigma(x)\frac{\partial\Phi(x)}{\partial x^i},$$

and

$$\exp\left\{-\frac{i}{\hbar}\alpha(\delta g^{-1};x)\right\} = 1 - \frac{i}{\hbar}\alpha(\delta g^{-1};x).$$

But because $\alpha(0;x) = 0$,

$$\alpha(\delta g^{-1};x) = \left.\frac{\partial\alpha(g;x)}{\partial g^\sigma}\right|_{g=0}(-\delta g^\sigma),$$

and the substitution of the above terms in (4.12) and further identification of the first order terms in $\delta g^\sigma$ imply that the self-adjoint operators $X_\sigma$ when acting on the wave functions have the differential representation

$$X_\sigma = \frac{\hbar}{i}\, u^j_\sigma(x)\,\frac{\partial}{\partial x^j} - v_\sigma(x), \tag{4.14}$$

where

$$u^j_\sigma(x) = \left.\frac{\partial f^j(x,g)}{\partial g^\sigma}\right|_{g=0}, \qquad v_\sigma(x) = \left.\frac{\partial\alpha(g;x)}{\partial g^\sigma}\right|_{g=0}. \tag{4.15}$$

If we restrict ourselves to transformations of the enlarged configuration space $(t,q_i)$ that can be extended to the whole kinematical space $x \equiv (t, q_i, \ldots, q_i^{(k-1)})$, then, using the same notation as in (1.43)-(1.46), if the infinitesimal transformation is of the form

$$t' = t + M_\sigma \delta g^\sigma,\; q'_i = q_i + M_{i\sigma}\delta g^\sigma,\; \ldots,\; q'^{(k-1)}_i = q_i^{(k-1)} + M_{i\sigma}^{(k-1)}\delta g^\sigma,$$

these generators take the form

$$X_\sigma = \frac{\hbar}{i}\left(M_\sigma\frac{\partial}{\partial t} + M_{i\sigma}\frac{\partial}{\partial q_i} + \ldots + M_{i\sigma}^{(k-1)}\frac{\partial}{\partial q_i^{(k-1)}}\right) - v_\sigma(x). \tag{4.16}$$

When compared with the Noether constants of the motion (1.52) written in the form

$$-N_\sigma = -H\,M_\sigma + p^i_{(s+1)}M^{(s)}_{i\sigma} - B_\sigma(x), \tag{4.17}$$

we see a certain kind of **'correspondence recipe'**. When restricted to kinematical groups, the functions $B_\sigma(x)$ of (1.52), are obtained from the Lagrangian gauge functions $\alpha(g;x)$, by (1.61), which is exactly the same derivation as the functions $v_\sigma(x)$ above in (4.15). Now, by identifying the different classical observables and generalized momenta that appear here in (4.17) with the corresponding differential operators of (4.16) that multiply the corresponding $M^{(s)}_{i\sigma}$ function, we get: the generalized Hamiltonian $H = p^i_{(s)}q_i^{(s)} - L$, which is multiplied in (4.17) by the function $M_\sigma$, is identified with the operator $i\hbar\partial/\partial t$ which is also in front of the function $M_\sigma$ in (4.16), and similarly, the generalized momentum $p^i_{(s+1)}$, the factor that multiplies the function $M^{(s)}_{i\sigma}$, with the differential operator $-i\hbar\partial/\partial q^{(s)}_i$, for $s = 0, \ldots, k-1$, because the functions $v_\sigma(x) = B_\sigma(x)$, are the same.

Remember that $p^i_{(s+1)}$ and $q^{(s)}_i$ are canonical conjugate variables. Then, each generalized momentum is replaced by $(\hbar/i)$ times the differential operator that differentiates with respect to its conjugate generalized coordinate and the generalized Hamiltonian by $i\hbar\partial/\partial t$.

The Heisenberg representation is that representation in which the time dependence has been withdrawn from the wave function by means of a time dependent unitary transformation. Then the wave function in this representation depends on the kinematical variables with the time excluded, *i.e.*, it depends only on the generalized coordinates $q^{(r)}_i$. Therefore, when acting on the wave function in the Heisenberg representation $\psi(q_i, q^{(1)}_i, \ldots, q^{(k-1)}_i)$, the observables $q^{(r)}_i$ and $p^j_{(s)}$ satisfy the canonical commutation relations

$$[q^{(r)}_i, p^j_{(s+1)}] = i\hbar\delta^j_i\delta^r_s.$$

If functions $v_\sigma(x)$ in (4.14) vanish, the $X_\sigma$ generators satisfy the commutation relations of the group $G$. But if some $v_\sigma(x) \neq 0$ the $X_\sigma$ generators do not satisfy in general the commutation relations of the initial group $G$ where they come from, but rather the commutation relations of a central extension of $G$. The group representation is not a true representation but a projective representation of $G$ as shown by Bargmann. [4]

In fact, from (4.10) we get

$$U(g_1)\Phi(x) = \Phi(g_1^{-1}x)\exp\{\frac{i}{\hbar}\alpha(g_1; g_1^{-1}x)\},$$

acting now on the left with $U(g_2)$,

$$U(g_2)U(g_1)\Phi(x) = U(g_2)\Phi(g_1^{-1}x)\exp\{\frac{i}{\hbar}\alpha(g_1; g_1^{-1}x)\}$$

$$= \Phi((g_2g_1)^{-1}x)\exp\{\frac{i}{\hbar}\alpha(g_2; g_2^{-1}x)\}\exp\{\frac{i}{\hbar}\alpha(g_1; (g_2g_1)^{-1}x)\}, \quad (4.18)$$

while acting on $\Phi(x)$ with $U(g_2g_1)$,

$$U(g_2g_1)\Phi(x) = \Phi((g_2g_1)^{-1}x)\exp\{\frac{i}{\hbar}\alpha(g_2g_1; (g_2g_1)^{-1}x)\}. \quad (4.19)$$

If we define $(g_2g_1)^{-1}x = g_1^{-1}g_2^{-1}x = z$, then $g_1z = g_2^{-1}x$ and because gauge functions satisfy (1.62), we write

$$\alpha(g_2; g_1z) + \alpha(g_1; z) = \alpha(g_2g_1; z) + \xi(g_2, g_1), \tag{4.20}$$

and by comparing (4.18) with (4.19), taking into account (4.20), we obtain

$$U(g_2)U(g_1)\Phi(x) = U(g_2g_1)\Phi(x)\exp\{\frac{i}{\hbar}\xi(g_2; g_1)\}. \tag{4.21}$$

Since $\Phi(x)$ is arbitrary, we have a projective representation of the group $G$ characterized by the non-trivial exponent $\xi(g, g')$.

For both Galilei and Poincaré particles the kinematical space is the ten-dimensional manifold spanned by the variables $(t, \boldsymbol{r}, \boldsymbol{u}, \boldsymbol{\alpha})$, $t$ being the time, $\boldsymbol{r}$ the charge position, $\boldsymbol{u}$ the velocity and $\boldsymbol{\alpha}$ the orientation of the particle. Thus in the quantum formalism the wave function of the most general elementary particle is a squared-integrable function $\Phi(t, \boldsymbol{r}, \boldsymbol{u}, \boldsymbol{\alpha})$ of these kinematical variables. For point particles, the kinematical space is just the four-dimensional space-time, so that wave functions are only functions of time and position, but spinning particles will have to depend on the additional variables like velocity and orientation. The spin structure will thus be related to these additional variables.

## 2. NONRELATIVISTIC PARTICLES

Let $\mathcal{G}$ be the Galilei group. Although it is a very well-known example, we shall consider first the simplest case of a point-like particle to show the power of the general formalism.

### 2.1 NONRELATIVISTIC POINT PARTICLE

The kinematical space of a point particle is the space-time manifold and therefore the wave function is just a complex function $\psi(t, \boldsymbol{r})$ defined on this manifold. Kinematical variables transform according to

$$t'(\tau) = t(\tau) + b, \tag{4.22}$$
$$\boldsymbol{r}'(\tau) = R(\boldsymbol{\alpha})\boldsymbol{r}(\tau) + \boldsymbol{v}t(\tau) + \boldsymbol{a}, \tag{4.23}$$

and the gauge function for this homogeneous space is

$$\alpha(g; x(\tau)) = m(v^2t(\tau)/2 + \boldsymbol{v} \cdot R(\boldsymbol{\alpha})\boldsymbol{r}(\tau)), \tag{4.24}$$

where $\boldsymbol{v}$ and $\boldsymbol{\alpha}$ are group parameters, $m$ defines the mass of the system and thus the ten selfadjoint generators of the projective unitary representation of the Galilei group $\mathcal{G}$, taking into account (4.22-4.23) and

(4.14) and (4.15), or alternatively the so-called correspondence recipe between the Noether constants of the motion (2.83-2.86), and the group generators by $H \to i\hbar\partial/\partial t$, the generalized momentum $\boldsymbol{p} \equiv \boldsymbol{P} \to -i\hbar\nabla$, they are given by

$$H = i\hbar\frac{\partial}{\partial t}, \quad \boldsymbol{P} = \frac{\hbar}{i}\nabla, \quad \boldsymbol{K} = m\boldsymbol{r} - t\boldsymbol{P} = m\boldsymbol{r} - t\frac{\hbar}{i}\nabla, \tag{4.25}$$

$$\boldsymbol{J} = \boldsymbol{r} \times \boldsymbol{P} = \boldsymbol{r} \times \frac{\hbar}{i}\nabla. \tag{4.26}$$

It is easily checked that these operators satisfy the commutation relations of the extended Galilei group, as described in Section 1. of Chapter 2.

The Casimir operators of the extended Galilei group are the internal energy $H_0 = H - P^2/2m$, the mass operator $M = m\mathbb{I}$ and the absolute value of spin $(\boldsymbol{J} - \boldsymbol{r} \times \boldsymbol{P})^2$ which vanishes in this case. All of them in any irreducible representation are multiples of the unit operator and therefore $H_0$ and $m$ are constant real numbers. When acting with the internal energy operator on any wave function, this satisfies

$$\left(H - \frac{\boldsymbol{P}^2}{2m}\right)\psi(t,\boldsymbol{r}) = H_0\psi(t,\boldsymbol{r}),$$

*i.e.*, Schroedinger's equation

$$\left(i\hbar\frac{\partial}{\partial t} + \frac{\hbar^2}{2m}\nabla^2\right)\psi(t,\boldsymbol{r}) = H_0\psi(t,\boldsymbol{r}). \tag{4.27}$$

If we look at the commutation relations of the extended Galilei group, we can find simultaneous eigenfunctions of the Schroedinger operator, the energy $H$ and the linear momentum $\boldsymbol{P}$, so that these states are described by plane waves.

We can check that Schroedinger's equation is a Galilei invariant wave equation [5] under a proper transformation of the wave function according to the projective representation (4.4). Wave function transforms in the way:

$$\psi'(t',\boldsymbol{r}') = \psi(t,\boldsymbol{r})\exp\left(\frac{im}{\hbar}(\boldsymbol{v}^2t/2 + \boldsymbol{v}\cdot R(\boldsymbol{\alpha})\boldsymbol{r})\right).$$

To see this let us consider a one-dimensional case in which the wave function is $\psi(t,x)$ and the transformation between the unprimed and primed variables is $t = t'$ and $x = x' - vt'$. The gauge function reduces to

$$\alpha(t,x) = m\left(v^2t/2 + vx\right).$$

Then

$$\frac{\partial\psi'}{\partial t'} = \frac{\partial}{\partial t}\left(\psi e^{i\alpha(t,x)/\hbar}\right)\frac{\partial t}{\partial t'} + \frac{\partial}{\partial x}\left(\psi e^{i\alpha(t,x)/\hbar}\right)\frac{\partial x}{\partial t'},$$

the derivatives $\partial t/\partial t' = 1$, $\partial x/\partial t' = -v$, and thus

$$\frac{\partial \psi'}{\partial t'} = \frac{\partial \psi}{\partial t}\, e^{i\alpha(t,x)/\hbar} - \frac{imv^2}{2\hbar}\, \psi e^{i\alpha(t,x)/\hbar} - v\frac{\partial \psi}{\partial x}\, e^{i\alpha(t,x)/\hbar}.$$

Similarly

$$\frac{\partial^2 \psi'}{\partial x'^2} = \frac{\partial^2 \psi}{\partial x^2}\, e^{i\alpha(t,x)/\hbar} + \frac{2imv}{\hbar}\frac{\partial \psi}{\partial x}\, e^{i\alpha(t,x)/\hbar} - \frac{m^2v^2}{\hbar^2}\psi\, e^{i\alpha(t,x)/\hbar},$$

and therefore whenever $\psi$ satisfies equation (4.27), then $\psi'$ satisfies

$$\left(i\hbar\frac{\partial}{\partial t'} + \frac{\hbar^2}{2m}\frac{\partial^2}{\partial x'^2}\right)\psi'(t',x') = H_0\psi'(t',x'),$$

and we see that the projective unitary representations of the Galilei group lead to invariant wave equations. In the three-dimensional case the proof is similar. As was pointed out by Inönü and Wigner [6] the true irreducible unitary representations of the Galilei group are void of physical content, since from the projective representations point of view they correspond to unexistent massless Galilei systems.

This proof of the invariance of Schroedinger's equation is unnecessary and has been done only for pedagogical reasons. We must remark that the internal energy Casimir operator is Galilei invariant, and therefore $H_0 = H' - P'^2/2m = H - P^2/2m$ giving rise also in the primed frame to Schroedinger's equation.

## 2.2 NONRELATIVISTIC SPINNING PARTICLES. BOSONS

Now let us apply the formalism to the most interesting case of spinning particles. Let us consider next Galilei particles with (anti)orbital spin. This corresponds for example to systems for which $X = \mathcal{G}/SO(3)$ and thus the kinematical variables are time, position and velocity. A particular classical example is given in Chapter 2, Section 4. by the free Lagrangian

$$L = \frac{m}{2}\left(\frac{d\boldsymbol{r}}{dt}\right)^2 - \frac{m}{2\omega^2}\left(\frac{d\boldsymbol{u}}{dt}\right)^2, \tag{4.28}$$

with $\boldsymbol{u} = d\boldsymbol{r}/dt$. For the free particle, the center of mass $\boldsymbol{q} = \boldsymbol{r} - \boldsymbol{k}$ has a straight motion while the relative position vector $\boldsymbol{k}$ follows an elliptic trajectory with frequency $\omega$ around its center of mass, being the spin related to this internal motion. It is expressed as $\boldsymbol{S} = -m\boldsymbol{k} \times d\boldsymbol{k}/dt$.

The kinematical variables transform under $\mathcal{G}$ in the form

$$t'(\tau) \;=\; t(\tau) + b, \tag{4.29}$$

$$\begin{aligned} \boldsymbol{r}'(\tau) &= R(\boldsymbol{\alpha})\boldsymbol{r}(\tau) + \boldsymbol{v}t(\tau) + \boldsymbol{a}, & (4.30) \\ \boldsymbol{u}'(\tau) &= R(\boldsymbol{\alpha})\boldsymbol{u}(\tau) + \boldsymbol{v}. & (4.31) \end{aligned}$$

The wave functions are functions on $X$ and thus functions of the variables $(t, \boldsymbol{r}, \boldsymbol{u})$. On this kinematical space the gauge function is the same as in (4.24), where $m$ defines again the mass of the system. Taking into account as in the previous example the correspondence recipe for the Hamiltonian $H \to i\hbar\partial/\partial t$, the first generalized momentum $\boldsymbol{p}_1 \equiv \boldsymbol{P} \to -i\hbar\nabla$ and the other generalized momentum $\boldsymbol{p}_2 \equiv \boldsymbol{U} \to -i\hbar\nabla_u$, the generators of the projective representation are given by

$$H = i\hbar\frac{\partial}{\partial t}, \quad \boldsymbol{P} = \frac{\hbar}{i}\nabla, \quad \boldsymbol{K} = m\boldsymbol{r} - t\frac{\hbar}{i}\nabla - \frac{\hbar}{i}\nabla_u, \tag{4.32}$$

$$\boldsymbol{J} = \boldsymbol{r} \times \frac{\hbar}{i}\nabla + \boldsymbol{u} \times \frac{\hbar}{i}\nabla_u = \boldsymbol{L} + \boldsymbol{Z}, \tag{4.33}$$

where $\nabla$ is the gradient operator with respect to $\boldsymbol{q}_1 \equiv \boldsymbol{r}$ variables and $\nabla_u$ the gradient operator with respect to the $\boldsymbol{q}_2 \equiv \boldsymbol{u}$ variables. It is important to stress that this representation of the generators is independent of the particular Lagrangian that describes the system. It depends only on the kinematical variables $(t, \boldsymbol{r}, \boldsymbol{u})$ we use to describe the system, with the transformation equations (4.29-4.31), and of the gauge function (4.24).

If we define $\boldsymbol{q} = \boldsymbol{r} - \boldsymbol{k} = (\boldsymbol{K} + \boldsymbol{P}t)/m$, it satisfies the commutation relations with $\boldsymbol{P}$,

$$[q_i, P_j] = i\hbar\,\delta_{ij},$$

which are the canonical commutation relations between the linear momentum and position for a point particle and therefore these canonical commutation relations between the total linear momentum and the center of mass position for a spinning particle are already contained in the commutation relations of the Lie algebra of the kinematical group. Therefore the quantum mechanical operator

$$\boldsymbol{q} = \boldsymbol{r} - \frac{\hbar}{im}\nabla_u, \tag{4.34}$$

can be interpreted as the center of mass position operator. Discussion of other possibilities for the center of mass position operator are delayed till Chapter 6.

In this representation, one Casimir operator is the internal energy $H - \boldsymbol{P}^2/2m$. We see that the spin operator is defined as usual

$$\boldsymbol{S} = \boldsymbol{J} - \frac{1}{m}\boldsymbol{K} \times \boldsymbol{P} = \boldsymbol{u} \times \boldsymbol{U} + \boldsymbol{k} \times \boldsymbol{P} = \boldsymbol{u} \times \frac{\hbar}{i}\nabla_u + \frac{\hbar}{im}\nabla_u \times \frac{\hbar}{i}\nabla;$$

written in terms of two non-commuting terms, it satisfies

$$[\boldsymbol{S},\boldsymbol{S}]=i\hbar\boldsymbol{S}, \quad [\boldsymbol{J},\boldsymbol{S}]=i\hbar\boldsymbol{S}, \quad [\boldsymbol{S},\boldsymbol{P}]=[\boldsymbol{S},H]=[\boldsymbol{S},\boldsymbol{K}]=0,$$

*i.e.*, it is an angular momentum operator, transforms like a vector under rotations and is invariant under space and time translations and under Galilei boosts, respectively. The second part of the spin operator is of order $\hbar^2$ so that it produces a very small correction to the first $\boldsymbol{Z}$ part.

Operators $\boldsymbol{Z}$ satisfy the commutation relations

$$[\boldsymbol{Z},\boldsymbol{Z}]=i\hbar\boldsymbol{Z}, \quad [\boldsymbol{J},\boldsymbol{Z}]=i\hbar\boldsymbol{Z}, \quad [\boldsymbol{Z},\boldsymbol{P}]=[\boldsymbol{Z},H]=0,$$

$$[\boldsymbol{Z},\boldsymbol{K}]=-i\hbar\boldsymbol{U}=-\hbar^2\nabla_{\boldsymbol{u}},$$

*i.e.*, $\boldsymbol{Z}$ is an angular momentum operator, transforms like a vector under rotations and is invariant under space and time translations but not under Galilei boosts. It is usually considered as the quantum mechanical spin operator.

We see however, that the angular momentum operator $\boldsymbol{J}$ is split into two commuting terms $\boldsymbol{r}\times\boldsymbol{P}$ and $\boldsymbol{Z}$. They both commute with $H$, but the first one is not invariant under space translations. The $\boldsymbol{Z}$ operators are angular momentum operators that only differentiate the wave function with respect to the velocity variables, and consequently commute with $H$ and $\boldsymbol{P}$, and although it is not the true Galilei invariant spin operator, we can find simultaneous eigenstates of the three commuting operators $H-\boldsymbol{P}^2/2m$, $Z^2$ and $Z_3$. Because the $\boldsymbol{Z}$ operators only affect the wave function in its dependence on $\boldsymbol{u}$ variables, we can choose functions with the variables separated in the form $\Phi(t,\boldsymbol{r},\boldsymbol{u})=\sum_i\psi_i(t,\boldsymbol{r})\chi_i(\boldsymbol{u})$ so that

$$(H-\boldsymbol{P}^2/2m)\psi_i(t,\boldsymbol{r})=E\psi_i(t,\boldsymbol{r}), \tag{4.35}$$

$$Z^2\chi_i(\boldsymbol{u})=z(z+1)\hbar^2\chi_i(\boldsymbol{u}), \tag{4.36}$$

$$Z_3\chi_i(\boldsymbol{u})=m_z\hbar\chi_i(\boldsymbol{u}). \tag{4.37}$$

The space-time dependent wave function $\psi_i(t,\boldsymbol{r})$, satisfies Schroedinger's equation and is uncoupled with the spin part $\chi(\boldsymbol{u})$.

Due to the structure of $Z^2$ in terms of the $\boldsymbol{u}$ variables, which is that of an orbital angular momentum, the spin part of the wave function is of the form

$$\chi(\boldsymbol{u})=f(u)Y_z^{m_z}(\theta,\phi), \tag{4.38}$$

$f(u)$ being an arbitrary function of the modulus of $\boldsymbol{u}$ and $Y_z^{m_z}(\theta,\phi)$ the spherical harmonics on the direction of $\boldsymbol{u}$.

For the center of mass observer, $\boldsymbol{S}=\boldsymbol{Z}$ and both angular momentum operators are the same. But for an arbitrary observer, $\boldsymbol{Z}$ operators do not commute with the boosts generators so that its absolute value is not Galilei invariant, while $\boldsymbol{S}$ is. But the splitting of the wave function into a multiple-component function that reflects its spin structure is an intrinsic property that can be done in any frame.

It turns out that if for an arbitrary observer $\boldsymbol{Z}$ is not the spin of the system, $\boldsymbol{r}\times\boldsymbol{P}$ is not the conserved orbital angular momentum, because $\boldsymbol{r}$ does not represent the position of the center of mass of the particle.

When there is an interaction with an external electromagnetic field, equation (4.35) is satisfied for the mechanical parts $H_m = H - e\phi$ and $\boldsymbol{P}_m = \boldsymbol{P} - e\boldsymbol{A}$ and we thus obtain the usual equation

$$\left(H - e\phi - \frac{(\boldsymbol{P} - e\boldsymbol{A})^2}{2m}\right)\psi_i(t,\boldsymbol{r}) = E\psi_i(t,\boldsymbol{r}). \tag{4.39}$$

This formalism, when the classical spin is of orbital nature, does not lead to half integer spin values, and therefore, from the quantum mechanical point of view these particles can be used only as models for representing bosons.

## 2.3 NONRELATIVISTIC SPINNING PARTICLES. FERMIONS

Other examples of nonrelativistic spinning particles are those which have orientation and thus angular velocity. For instance, if $X = \mathcal{G}/\mathbb{R}^3_v$, $\mathbb{R}^3_v$ being the subgroup $\{\mathbb{R}^3,+\}$ of pure Galilei transformations, then the kinematical space is spanned by the variables $(t,\boldsymbol{r},\boldsymbol{\alpha})$. This corresponds for instance to the Lagrangian system described in Section 4. of Chapter 2,

$$L = \frac{m}{2}\left(\frac{d\boldsymbol{r}}{dt}\right)^2 + \frac{I}{2}\boldsymbol{\omega}^2. \tag{4.40}$$

The particle travels freely at constant velocity while it rotates with constant angular velocity $\boldsymbol{\omega}$. The classical spin is just $\boldsymbol{S} = I\boldsymbol{\omega}$, and the center of charge and center of mass represent the same point.

To describe orientation we can think of the three orthogonal unit vectors $\boldsymbol{e}_i$, $i = 1,2,3$ linked to the body, similarly as in a rigid rotator. If initially they are taken parallel to the spatial Cartesian axis of the laboratory inertial frame, then their nine components considered by columns define an orthogonal rotation matrix $R_{ij}(\boldsymbol{\alpha})$ that describes the triad evolution with the initial condition $R_{ij}(t=0) = \delta_{ij}$.

Now, kinematical variables $t$, $\boldsymbol{r}$ and $\boldsymbol{\rho}$ transform under $\mathcal{G}$ in the form

$$t'(\tau) = t(\tau) + b, \tag{4.41}$$

$$\boldsymbol{r}'(\tau) = R(\boldsymbol{\alpha})\boldsymbol{r}(\tau) + \boldsymbol{v}t(\tau) + \boldsymbol{a}, \tag{4.42}$$

$$\boldsymbol{\rho}'(\tau) = \frac{\boldsymbol{\mu}+\boldsymbol{\rho}(\tau)+\boldsymbol{\mu}\times\boldsymbol{\rho}(\tau)}{1-\boldsymbol{\mu}\cdot\boldsymbol{\rho}(\tau)}. \tag{4.43}$$

On the corresponding Hilbert space, the Galilei generators are given by:

$$H = i\hbar\frac{\partial}{\partial t}, \quad \boldsymbol{P} = \frac{\hbar}{i}\nabla, \quad \boldsymbol{K} = m\boldsymbol{r} - t\frac{\hbar}{i}\nabla, \tag{4.44}$$

$$\boldsymbol{J} = \frac{\hbar}{i}\boldsymbol{r}\times\nabla + \frac{\hbar}{2i}\{\nabla_\rho + \boldsymbol{\rho}\times\nabla_\rho + \boldsymbol{\rho}(\boldsymbol{\rho}\cdot\nabla_\rho)\} = \boldsymbol{L}+\boldsymbol{W}, \tag{4.45}$$

$\nabla_\rho$ being the gradient operator with respect to the $\boldsymbol{\rho}$ variables and in the $\boldsymbol{\rho}$ parameterization of the rotation group.

The $\boldsymbol{W}$ part comes from the general group analysis. The group generators in this parametrization $X_i$ will be obtained from (4.43) and according to (1.33) and (1.35). They are obtained as

$$X_i = \left(\frac{\partial \rho'^k}{\partial \mu^i}\right)\Bigg|_{\mu=0}\frac{\partial}{\partial \rho^k},$$

that can be written in vector notation as

$$\boldsymbol{X} = \nabla_\rho + \boldsymbol{\rho}\times\nabla_\rho + \boldsymbol{\rho}(\boldsymbol{\rho}\cdot\nabla_\rho)$$

They satisfy the commutation relations

$$[X_i, X_k] = -2\epsilon_{ikl}X_l$$

and therefore operators $W_k = \frac{\hbar}{2i}X_k$, or in vector notation

$$\boldsymbol{W} = \frac{\hbar}{2i}\{\nabla_\rho + \boldsymbol{\rho}\times\nabla_\rho + \boldsymbol{\rho}(\boldsymbol{\rho}\cdot\nabla_\rho)\}, \tag{4.46}$$

will satisfy the angular momentum commutation relations

$$[\boldsymbol{W}, \boldsymbol{W}] = i\hbar\boldsymbol{W}. \tag{4.47}$$

In this way since $\boldsymbol{L}$ and $\boldsymbol{W}$ commute among each other, we also get $[\boldsymbol{J}, \boldsymbol{J}] = i\hbar\boldsymbol{J}$.

In this example the center of mass and center of charge are the same point, $\boldsymbol{L} = \boldsymbol{r}\times\boldsymbol{P}$ is the orbital angular momentum associated to the center of mass motion and $\boldsymbol{W} \equiv \boldsymbol{S}$ is the spin operator. The spin operator commutes with $H$, $\boldsymbol{P}$ and $\boldsymbol{K}$ and the wave function can be separated as $\Phi(t, \boldsymbol{r}, \boldsymbol{\rho}) = \sum_i \psi_i(t, \boldsymbol{r})\chi_i(\boldsymbol{\rho})$ leading to the equations

$$(H - \boldsymbol{P}^2/2m)\psi_i(t, \boldsymbol{r}) = E\psi_i(t, \boldsymbol{r}), \tag{4.48}$$

$$S^2\chi_i(\rho) = s(s+1)\hbar^2\chi_i(\rho), \tag{4.49}$$

$$S_3\chi_i(\rho) = m_s\hbar\chi_i(\rho). \tag{4.50}$$

Bopp and Haag [7] succeeded in finding $s = 1/2$ solutions for the system of equations (4.49) and (4.50). They are called Wigner's functions. [8] Solutions of (4.49) for arbitrary spin $s$ are but a linear combination of the matrix elements of a $(2s+1)\times(2s+1)$ irreducible matrix representation of the rotation group as can be derived from the Peter-Weyl theorem on finite representations of compact groups. [9,10,11] We shall deal with the $s = 1/2$ functions in Section 3., where explicit expressions and a short introduction to the Peter-Weyl theorem, will be given.

To describe fermions, the classical particles must necessarily have compact orientation variables as kinematical variables, otherwise no spin 1/2 values can be obtained when the classical spin is related only to the zitterbewegung.

## 2.4 GENERAL NONRELATIVISTIC SPINNING PARTICLE

More general nonrelativistic particles will be described by considering larger kinematical spaces like the whole Galilei group itself, where the particle wave-function will be a function of the ten kinematical variables $\Phi(t, \boldsymbol{r}, \boldsymbol{u}, \boldsymbol{\rho})$ and the spin structure will contain both contributions from the zitterbewegung part and from rotation.

Kinematical variables transform under $\mathcal{G}$ as

$$t'(\tau) = t(\tau) + b, \tag{4.51}$$
$$\boldsymbol{r}'(\tau) = R(\boldsymbol{\alpha})\boldsymbol{r}(\tau) + \boldsymbol{v}t(\tau) + \boldsymbol{a}, \tag{4.52}$$
$$\boldsymbol{u}'(\tau) = R(\boldsymbol{\alpha})\boldsymbol{u}(\tau) + \boldsymbol{v}, \tag{4.53}$$
$$\boldsymbol{\rho}'(\tau) = \frac{\boldsymbol{\mu} + \boldsymbol{\rho}(\tau) + \boldsymbol{\mu}\times\boldsymbol{\rho}(\tau)}{1 - \boldsymbol{\mu}\cdot\boldsymbol{\rho}(\tau)}. \tag{4.54}$$

The differential representation of the ten generators is obtained as

$$H = i\hbar\frac{\partial}{\partial t}, \quad \boldsymbol{P} = \frac{\hbar}{i}\nabla, \quad \boldsymbol{K} = m\boldsymbol{r} - t\frac{\hbar}{i}\nabla - \frac{\hbar}{i}\nabla_u, \tag{4.55}$$

$$\boldsymbol{J} = \boldsymbol{r}\times\frac{\hbar}{i}\nabla + \boldsymbol{u}\times\frac{\hbar}{i}\nabla_u + \frac{\hbar}{2i}\left\{\nabla_\rho + \boldsymbol{\rho}\times\nabla_\rho + \boldsymbol{\rho}(\boldsymbol{\rho}\cdot\nabla_\rho)\right\}, \tag{4.56}$$

where the total angular momentum operator $\boldsymbol{J} = \boldsymbol{L}+\boldsymbol{Y}+\boldsymbol{W}$ is expressed in terms of three commuting angular momentum operators, $\boldsymbol{L} \equiv \boldsymbol{r}\times\boldsymbol{P}$, the zitterbewegung part $\boldsymbol{Y} \equiv \boldsymbol{u} \times \boldsymbol{U}$ and the part related to rotation $\boldsymbol{W}$. All these three operators produce the derivative of the wave function with respect to different kinematical variables.

These ten generators satisfy the commutation relations of the extended Galilei group. We get again that the Casimir operator that describes the internal energy $H_0 = H - P^2/2m$, gives rise to Schroedinger's equation and commutes with the $\boldsymbol{Y}$ and $\boldsymbol{W}$ spin operators. This allows us to consider wave functions of the form

$$\Phi(t, \boldsymbol{r}, \boldsymbol{u}, \boldsymbol{\rho}) = \sum_j \Psi_j(t, \boldsymbol{r}) \chi_j(\boldsymbol{u}, \boldsymbol{\rho}), \tag{4.57}$$

which separates the space-time part from the internal part that describes its spin structure. Index $j$ runs over a finite range for elementary particles. The interest is to obtain quantum mechanical models of the lowest lying spin values, in particular the description of spin 1/2 particles. In this way we shall deal with the analysis of spin 1/2 wave functions related to the $\boldsymbol{u}$ and $\boldsymbol{\rho}$ dependence of the general wave function in the next section, before the analysis of relativistic particles. The structure of the general wave function (4.57) is exactly the same in both formalisms and therefore the spin part $\chi_j(\boldsymbol{u}, \boldsymbol{\rho})$ will be obtained by the same means.

## 3. SPINORS

In this section of mathematical content we shall review the main properties of spinors, in particular those connected with the possible representation of the wave function to describe spin 1/2 particles. We shall describe the representations in terms of eigenfunctions of the different commuting spin operators. But it must be remarked that in addition to the spin operators in the laboratory frame we also have spin operators in the body frame, because our general spinning particle has orientation, and therefore, a local Cartesian frame linked to its motion. This produces the result that for a spin 1/2 particle the wave function necessarily is a four-component object.

The general wave function is a function of the ten kinematical variables, $\Phi(t, \boldsymbol{r}, \boldsymbol{u}, \boldsymbol{\rho})$, and the spin part of the system related to the translation invariant kinematical variables $\boldsymbol{u}$ and $\boldsymbol{\rho}$ is

$$\boldsymbol{S} = \boldsymbol{u} \times \boldsymbol{U} + \boldsymbol{W} = \boldsymbol{Y} + \boldsymbol{W}, \tag{4.58}$$

where $\boldsymbol{Y}$ and $\boldsymbol{W}$ are given by

$$\boldsymbol{Y} = \boldsymbol{u} \times \frac{\hbar}{i} \nabla_u, \quad \boldsymbol{W} = \frac{\hbar}{2i} \{\nabla_\rho + \boldsymbol{\rho} \times \nabla_\rho + \boldsymbol{\rho}(\boldsymbol{\rho} \cdot \nabla_\rho)\}, \tag{4.59}$$

in the $\tan(\alpha/2)$ representation of the rotation group, as has been deduced in previous sections. $\nabla_u$ and $\nabla_\rho$ are respectively the gradient operators with respect to $\boldsymbol{u}$ and $\boldsymbol{\rho}$ variables. These operators always commute with the $H = i\hbar\partial/\partial t$ and $\boldsymbol{P} = -i\hbar\nabla$ operators, and therefore they are translation invariant. This feature allows the separation of the general wave function according to (4.57).

The above spin operators satisfy the commutation relations

$$[\boldsymbol{Y},\boldsymbol{Y}] = i\hbar\boldsymbol{Y}, \quad [\boldsymbol{W},\boldsymbol{W}] = i\hbar\boldsymbol{W}, \quad [\boldsymbol{Y},\boldsymbol{W}] = 0, \tag{4.60}$$

and thus

$$[\boldsymbol{S},\boldsymbol{S}] = i\hbar\boldsymbol{S}.$$

Because we are describing the orientation of the particle by attaching to it a system of three unit vectors $\boldsymbol{e}_i$, whose orientation in space is described by variables $\boldsymbol{\rho}$ or $\boldsymbol{\alpha}$, then, if at initial instant $\tau = 0$ we choose the body axes coincident with the laboratory axes, the components of the unit vectors $\boldsymbol{e}_i$ at any time are

$$(\boldsymbol{e}_i)_j = R_{ji}(\boldsymbol{\alpha}) = \delta_{ji}\cos\alpha + n_j n_i(1-\cos\alpha) - \epsilon_{jik}n_k\sin\alpha, \tag{4.61}$$

in the normal parametrization and also in the $\boldsymbol{\rho}$ parametrization by

$$(\boldsymbol{e}_i)_j = R_{ji}(\boldsymbol{\rho}) = \frac{1}{1+\rho^2}((1-\rho^2)\delta_{ji} + 2\rho_j\rho_i - 2\epsilon_{jik}\rho_k), \tag{4.62}$$

where the Cartesian components of the rotation axis unit vector $\boldsymbol{n}$ are:

$$n_1 = \sin\theta\cos\phi, \qquad n_2 = \sin\theta\sin\phi, \qquad n_3 = \cos\theta, \tag{4.63}$$

where $\theta$ is the polar angle and $\phi$ the usual azimuth angle. Explicitly:

$$\begin{aligned}
e_{11} &= \cos\alpha + \sin^2\theta\cos^2\phi(1-\cos\alpha),\\
e_{12} &= \cos\theta\sin\alpha + \sin^2\theta\sin\phi\cos\phi(1-\cos\alpha),\\
e_{13} &= -\sin\theta\sin\phi\sin\alpha + \sin\theta\cos\theta\cos\phi(1-\cos\alpha),\\
\\
e_{21} &= -\cos\theta\sin\alpha + \sin^2\theta\sin\phi\cos\phi(1-\cos\alpha),\\
e_{22} &= \cos\alpha + \sin^2\theta\sin^2\phi(1-\cos\alpha),\\
e_{23} &= \sin\theta\cos\phi\sin\alpha + \sin\theta\cos\theta\sin\phi(1-\cos\alpha),\\
\\
e_{31} &= \sin\theta\sin\phi\sin\alpha + \sin\theta\cos\theta\cos\phi(1-\cos\alpha),\\
e_{32} &= -\sin\theta\cos\phi\sin\alpha + \sin\theta\cos\theta\sin\phi(1-\cos\alpha),\\
e_{33} &= \cos\alpha + \cos^2\theta(1-\cos\alpha),
\end{aligned}$$

in the $\boldsymbol{\alpha} = \alpha\boldsymbol{n}$, or normal parametrization of the rotation group. In the $\boldsymbol{\rho} = \tan(\alpha/2)\boldsymbol{n}$ parametrization the body frame is

$$e_{11} = (1+\rho_1^2-\rho_2^2-\rho_3^2)/(1+\rho^2),$$

$$
\begin{aligned}
e_{12} &= (2\rho_1\rho_2 + 2\rho_3)/(1+\rho^2),\\
e_{13} &= (2\rho_1\rho_3 - 2\rho_2)/(1+\rho^2),\\
\\
e_{21} &= (2\rho_2\rho_1 - 2\rho_3)/(1+\rho^2),\\
e_{22} &= (1-\rho_1^2+\rho_2^2-\rho_3^2)/(1+\rho^2),\\
e_{23} &= (2\rho_2\rho_3 + 2\rho_1)/(1+\rho^2),\\
\\
e_{31} &= (2\rho_1\rho_3 + 2\rho_2)/(1+\rho^2),\\
e_{32} &= (2\rho_3\rho_2 - 2\rho_1)/(1+\rho^2),\\
e_{33} &= (1-\rho_1^2-\rho_2^2+\rho_3^2)/(1+\rho^2),
\end{aligned}
$$

where $\rho^2 \equiv \rho_1^2+\rho_2^2+\rho_3^2 = \tan^2(\alpha/2)$.

In addition to the different components of the spin operators $S_i$, $Y_i$ and $W_i$ in the laboratory frame, we also have another set of spin operators. They are the spin projections on the body axes $\boldsymbol{e}_i$, *i.e.*, the operators $R_i = \boldsymbol{e}_i \cdot \boldsymbol{S}$, $M_i = \boldsymbol{e}_i \cdot \boldsymbol{Y}$ and $T_i = \boldsymbol{e}_i \cdot \boldsymbol{W}$, respectively. In particular, spin operators $T_i$, collecting terms from (4.62) and (4.59), take the expression

$$
\begin{aligned}
T_i = \sum_{k=1}^{k=3}(e_i)_k W_k &= \frac{\hbar}{2i(1+\rho^2)}\sum_{k=1}^{k=3}\left((1-\rho^2)\delta_{ik} + 2\rho_i\rho_k - 2\epsilon_{kij}\rho_j\right)\\
&\quad\times\left(\frac{\partial}{\partial\rho_k} + \epsilon_{klr}\rho_l\frac{\partial}{\partial\rho_r} + \rho_k(\boldsymbol{\rho}\cdot\nabla_\rho)\right),
\end{aligned}
$$

and after some tedious manipulations we reach the final result, written in vector notation as

$$
\boldsymbol{T} = \frac{\hbar}{2i}\{\nabla_\rho - \boldsymbol{\rho}\times\nabla_\rho + \boldsymbol{\rho}(\boldsymbol{\rho}\cdot\nabla_\rho)\}. \tag{4.64}
$$

We see, by inspection, that this result can also be obtained from the expression of $\boldsymbol{W}$ in (4.59), just by replacing $\boldsymbol{\rho}$ by $-\boldsymbol{\rho}$, followed by a global change of sign. This is because we describe the orientation of the particle by vector $\boldsymbol{\rho}$ in the laboratory frame from the active viewpoint, *i.e.*, with the laboratory reference frame fixed. However, its orientation with respect to the body frame is described by the motion of the laboratory frame, whose orientation for the body is $-\boldsymbol{\rho}$, and the global change of sign comes from the change from the active point of view to the passive one. This is the difference in the spin description in one frame or another.

It satisfies the following commutation relations

$$
[\boldsymbol{T},\boldsymbol{T}] = -i\hbar\boldsymbol{T}, \quad [\boldsymbol{T},\boldsymbol{W}] = 0
$$

and in general all spin projections on the body frame $R_i$, $M_i$ and $T_i$, commute with all the spin projections on the laboratory frame $S_i$, $Y_i$ and $W_i$. This is in agreement with the quantum mechanical uncertainty principle, because spin components with respect to different frames are compatible observables.

To find eigenstates of the spin operator we have to solve equations of the form:

$$S^2\chi(\boldsymbol{u},\boldsymbol{\rho}) = s(s+1)\hbar^2\chi(\boldsymbol{u},\boldsymbol{\rho}), \quad S_3\chi(\boldsymbol{u},\boldsymbol{\rho}) = m\hbar\chi(\boldsymbol{u},\boldsymbol{\rho}).$$

But we also have the orientation of the particle, and therefore the spin projections on the body axes. These projections commute with $S^2$ and $S_3$, and it is possible to choose another commuting spin operator, like the $T_3$ operator, and therefore our wave function can be taken also as an eigenvector of $T_3$,

$$T_3\chi(\boldsymbol{u},\boldsymbol{\rho}) = n\hbar\chi(\boldsymbol{u},\boldsymbol{\rho}),$$

so that the complete commuting set of operators that describe the spin structure must also include spin projections on the body axes.

The spin squared operator is

$$S^2 = \boldsymbol{Y}^2 + \boldsymbol{W}^2 + 2\boldsymbol{Y}\cdot\boldsymbol{W}, \tag{4.65}$$

and we see from (4.60) that is expressed as the sum of three commuting terms and its eigenvectors can be obtained as the simultaneous eigenvectors of the three commuting operators on the right-hand side of (4.65). Operators $\boldsymbol{Y}$ and $\boldsymbol{W}$ produce derivatives of the wave function with respect to $\boldsymbol{u}$ and $\boldsymbol{\rho}$ variables, separately. Thus, each $\chi(\boldsymbol{u},\boldsymbol{\rho})$ can again be separated as

$$\chi(\boldsymbol{u},\boldsymbol{\rho}) = \sum_j U_j(\boldsymbol{u})\,V_j(\boldsymbol{\rho}), \tag{4.66}$$

where the sum runs over a finite range, and where $U_j(\boldsymbol{u})$ will be eigenfunctions of $\boldsymbol{Y}^2$ and $V_j(\boldsymbol{\rho})$ of $\boldsymbol{W}^2$, respectively.

Functions $U_j(\boldsymbol{u})$ are multiples of spherical harmonics defined on the orientation of the velocity vector $\boldsymbol{u}$, because the $\boldsymbol{Y}$ operator has the structure of an orbital angular momentum in terms of the $\boldsymbol{u}$ variables, and thus its eigenvalues are integer numbers. The global factor left out is an arbitrary function depending on the absolute value of the velocity $u$.

It turns out that to find the most general spinor is necessary to seek also solutions of the $V_j(\boldsymbol{\rho})$ part, depending on the orientation variables. This goal will be achieved in the next section, where we consider the action of the rotation group on itself as a transformation group.

## 3.1 SPINOR REPRESENTATION ON SU(2)

We shall describe now in detail the orientation part of the general wave function, $V(\rho)$. If there is no contribution to spin from the zitterbewegung part $\boldsymbol{Y}$, the spin operator (4.58) reduces to the $\boldsymbol{W}$ operator given in (4.59). To solve the corresponding eigenvalue equations we shall first represent the spin operators in spherical coordinates.

If we represent vector $\boldsymbol{\rho} = \tan(\alpha/2)\boldsymbol{n} = r\boldsymbol{n}$ in spherical coordinates $(r, \theta, \phi)$, with $r = |\boldsymbol{\rho}| = \tan(\alpha/2)$ and $\theta$ and $\phi$ the usual polar and azimuth angles, respectively, then unit vector $\boldsymbol{n}$ has the Cartesian components given in (4.63). If from now on we take $\hbar = 1$, the spin operators (4.59) are represented by the differential operators

$$\begin{aligned} W_1 &= \frac{1}{2i}\left[(1+r^2)\sin\theta\cos\phi\frac{\partial}{\partial r} + \left(\frac{1}{r}\cos\theta\cos\phi - \sin\phi\right)\frac{\partial}{\partial\theta}\right. \\ &\left. - \left(\frac{\sin\phi}{r\sin\theta} + \frac{\cos\theta\cos\phi}{\sin\theta}\right)\frac{\partial}{\partial\phi}\right], \end{aligned}$$

$$\begin{aligned} W_2 &= \frac{1}{2i}\left[(1+r^2)\sin\theta\sin\phi\frac{\partial}{\partial r} + \left(\frac{1}{r}\cos\theta\sin\phi + \cos\phi\right)\frac{\partial}{\partial\theta}\right. \\ &\left. - \left(\frac{\cos\theta\sin\phi}{\sin\theta} - \frac{\cos\phi}{r\sin\theta}\right)\frac{\partial}{\partial\phi}\right], \end{aligned}$$

$$W_3 = \frac{1}{2i}\left[(1+r^2)\cos\theta\frac{\partial}{\partial r} - \frac{\sin\theta}{r}\frac{\partial}{\partial\theta} + \frac{\partial}{\partial\phi}\right].$$

The Casimir operator of the rotation group $W^2$ is:

$$\begin{aligned} W^2 = -\frac{1+r^2}{4}&\left[(1+r^2)\frac{\partial^2}{\partial r^2} + \frac{2(1+r^2)}{r}\frac{\partial}{\partial r}\right. \\ &\left. + \frac{1}{r^2}\left\{\frac{\partial^2}{\partial\theta^2} + \frac{\cos\theta}{\sin\theta}\frac{\partial}{\partial\theta} + \frac{1}{\sin^2\theta}\frac{\partial^2}{\partial\phi^2}\right\}\right]. \end{aligned}$$

The up and down spin operators defined as usual by $W_\pm = W_1 \pm iW_2$ are

$$W_+ = \frac{e^{i\phi}}{2i}\left[(1+r^2)\sin\theta\frac{\partial}{\partial r} + \left(\frac{\cos\theta + ir}{r}\right)\frac{\partial}{\partial\theta} - \left(\frac{r\cos\theta - i}{r\sin\theta}\right)\frac{\partial}{\partial\phi}\right],$$

$$W_- = \frac{e^{-i\phi}}{2i}\left[(1+r^2)\sin\theta\frac{\partial}{\partial r} + \left(\frac{\cos\theta - ir}{r}\right)\frac{\partial}{\partial\theta} - \left(\frac{r\cos\theta + i}{r\sin\theta}\right)\frac{\partial}{\partial\phi}\right].$$

They satisfy the commutation relations

$$[W_3, W_+] = W_+, \quad [W_3, W_-] = -W_-, \quad [W_+, W_-] = 2W_3.$$

We can check that $(W_i)^* = -W_i$ and $W_+ = -(W_-)^*$, where $^*$ means to take the complex conjugate of the corresponding operator.

If $F_s^m(r,\theta,\phi)$ is an eigenfunction of $W^2$ and $W_3$, it satisfies the differential equations:

$$W^2 F_s^m(r,\theta,\phi) = s(s+1)F_s^m(r,\theta,\phi), \quad W_3 F_s^m(r,\theta,\phi) = mF_s^m(r,\theta,\phi).$$

To find solutions of the above system we know that we can proceed in the following way. Let us compute first the eigenfunctions of the form $F_s^s$. Then operator $W_+$ annihilates this state $W_+F_s^s = 0$ and by acting on this function with operator $W_-$ we can obtain the remaining eigenstates $F_s^m$ of the same irreducible representation characterized by parameter $s$ and for $-s \leq m \leq s$. Then our task will be to obtain first the $F_s^s$ functions.

Now, let us consider eigenfunctions $F_s^s$ that can be written in separate variables as $F_s^s(r,\theta,\phi) = A(r)B(\theta)C(\phi)$. Then

$$W_3 A(r)B(\theta)C(\phi) = sA(r)B(\theta)C(\phi)$$

gives rise to

$$(1+r^2)\cos\theta A'BC - \frac{\sin\theta}{r}AB'C + ABC' = 2isABC$$

where $A'$ is the derivative of $A$ and so on, and by dividing both sides by $ABC$ we have

$$(1+r^2)\cos\theta\frac{A'(r)}{A(r)} - \frac{\sin\theta}{r}\frac{B'(\theta)}{B(\theta)} + \frac{C'(\phi)}{C(\phi)} = 2is.$$

Now, the third term on the left-hand side must be a constant, because the remaining terms are functions independent of $\phi$. Therefore, this term is written as $C'(\phi)/C(\phi) = ik$ and thus $C(\phi) = e^{ik\phi}$ up to an arbitrary constant factor. Since $C(\phi + 2\pi) = C(\phi)$ this implies that the constant $k$ must be an integer. The other two functions satisfy

$$r(1+r^2)\cos\theta A'B - \sin\theta AB' + ir(k-2s)AB = 0. \tag{4.67}$$

If there exist solutions with real functions $A$ and $B$, then necessarily $k = 2s$ so that the eigenvalue $s$ can be any integer or half integer, and equation (4.67) can be separated in the form:

$$r(1+r^2)\frac{A'(r)}{A(r)} = \frac{\sin\theta}{\cos\theta}\frac{B'(\theta)}{B(\theta)} = p = \text{constant}, \tag{4.68}$$

where, up to constant factors, the general solution is

$$A(r) = \left(\frac{r^2}{1+r^2}\right)^{p/2}, \quad B(\theta) = (\sin\theta)^p.$$

By acting on this solution $F^s_s \equiv A(r)B(\theta)C(\phi)$, with $W_+$, since $W_+F^s_s = 0$, it gives:

$$r(1+r^2)\sin^2\theta A'B + (\sin\theta\cos\theta + ir\sin\theta)AB' - 2s(ir\cos\theta + 1)AB = 0.$$

By dividing all terms by $AB$, taking into account (4.68), we get the condition $(p-2s)(1+ir\cos\theta) = 0$. Then there exist real solutions in separate variables whenever $p = 2s = k$. They are given, up to a constant factor, by

$$F^s_s(r,\theta,\phi) = \left(\frac{r^2}{1+r^2}\right)^s (\sin\theta)^{2s} e^{i2s\phi}. \tag{4.69}$$

For $s = 1/2$ and after the action of $W_-$ we obtain the two orthogonal spinors

$$\Psi^{1/2}_{1/2} = \frac{r}{\sqrt{1+r^2}}\sin\theta\, e^{i\phi}, \qquad W_-\Psi^{1/2}_{1/2} = \Psi^{-1/2}_{1/2} = \frac{r\cos\theta + i}{\sqrt{1+r^2}},$$

that produce a two-dimensional representation of the rotation group. We can similarly check that $W_-\Psi^{-1/2}_{1/2} = 0$.

By inspection of the structure of $W_\pm$ operators, if we take the complex conjugate of expression $W_+F^s_s = 0$ we get $-W_-(F^s_s)^* = 0$ and therefore $(F^s_s)^* \sim G^{-s}_s$ so that taking the complex conjugate spinors of the above representation we obtain another pair of orthogonal $s = 1/2$ spinors,

$$\widetilde{\Psi}^{1/2}_{1/2} = \frac{r\cos\theta - i}{\sqrt{1+r^2}}, \qquad \widetilde{\Psi}^{-1/2}_{1/2} = \frac{r}{\sqrt{1+r^2}}\sin\theta\, e^{-i\phi}.$$

The remaining representations for higher spins can thus be obtained by the same method, or by taking tensor products of the above two-dimensional representations. For instance, for $s = 1$ we can obtain the following three orthogonal representations. From (4.69) with $s = 1$ and acting with the $W_-$ operator we get

$$\begin{aligned}
\Psi^1_1 = (\Psi^{1/2}_{1/2})^2 &= \frac{r^2}{1+r^2}\sin^2\theta\, e^{i2\phi},\\
\Psi^0_1 = (\Psi^{1/2}_{1/2})(\Psi^{-1/2}_{1/2}) &= \frac{r}{1+r^2}\sin\theta\,(i + r\cos\theta)\, e^{i\phi},\\
\Psi^{-1}_1 = (\Psi^{-1/2}_{1/2})^2 &= \frac{(i + r\cos\theta)^2}{1+r^2},
\end{aligned}$$

that can also be obtained as the tensor product $\Psi \otimes \Psi$. The complex conjugate of this comes from $\widetilde{\Psi} \otimes \widetilde{\Psi}$,

$$\Phi^1_1 = (\widetilde{\Psi}^{1/2}_{1/2})^2 = \frac{(r\cos\theta - i)^2}{1+r^2},$$

$$\Phi_1^0 = (\widetilde{\Psi}_{1/2}^{1/2})(\widetilde{\Psi}_{1/2}^{-1/2}) = \frac{r}{1+r^2}\sin\theta\,(r\cos\theta - i)\,e^{-i\phi},$$

$$\Phi_1^{-1} = (\widetilde{\Psi}_{1/2}^{-1/2})^2 = \frac{r^2}{1+r^2}\sin^2\theta\,e^{-i2\phi},$$

and finally from $\Psi\otimes\widetilde{\Psi}$,

$$\Theta_1^1 = (\Psi_{1/2}^{1/2})(\widetilde{\Psi}_{1/2}^{1/2}) = \frac{r\sin\theta(r\cos\theta - i)}{1+r^2}\,e^{i\phi}$$

$$\Theta_1^0 = (\Psi_{1/2}^{1/2})(\widetilde{\Psi}_{1/2}^{-1/2}) - (\Psi_{1/2}^{-1/2})(\widetilde{\Psi}_{1/2}^{1/2}) = \frac{r^2\cos 2\theta + 1}{1+r^2},$$

$$\Theta_1^{-1} = (\Psi_{1/2}^{-1/2})(\widetilde{\Psi}_{1/2}^{-1/2}) = \frac{r\sin\theta(r\cos\theta + i)}{1+r^2}\,e^{-i\phi},$$

and similarly for higher spins. One representation for $s = 3/2$ is

$$\Psi_{3/2}^{3/2} = \frac{r^3}{(1+r^2)^{3/2}}\sin^3\theta\,e^{i3\phi},$$

$$\Psi_{3/2}^{1/2} = \frac{r^2}{(1+r^2)^{3/2}}\sin^2\theta\,(i + r\cos\theta)\,e^{i2\phi},$$

$$\Psi_{3/2}^{-1/2} = \frac{r}{(1+r^2)^{3/2}}\sin\theta\,(r^2\cos^2\theta - 1 + 2ir\cos\theta)\,e^{i\phi},$$

$$\Psi_{3/2}^{-3/2} = \frac{1}{(1+r^2)^{3/2}}[r\cos\theta(3 - r^2\cos^2\theta) + i(1 - 3r^2\cos^2\theta)].$$

If we work in the normal or canonical representation of the rotation group, where the parameters are $\boldsymbol{\alpha} = \alpha\boldsymbol{n}$, this amounts to replacing the variable $r = \tan(\alpha/2)$ in terms of parameter $\alpha$ and expressing the differential operator $\partial/\partial r$ in terms of $\partial/\partial\alpha$, and then the spin operators are given by

$$W_1 = \frac{1}{2i}\left[2\sin\theta\cos\phi\frac{\partial}{\partial\alpha} + \left(\frac{\cos\theta\cos\phi}{\tan(\alpha/2)} - \sin\phi\right)\frac{\partial}{\partial\theta}\right.$$
$$\left. - \left(\frac{\sin\phi}{\tan(\alpha/2)\sin\theta} + \frac{\cos\theta\cos\phi}{\sin\theta}\right)\frac{\partial}{\partial\phi}\right],$$

$$W_2 = \frac{1}{2i}\left[2\sin\theta\sin\phi\frac{\partial}{\partial\alpha} + \left(\frac{\cos\theta\sin\phi}{\tan(\alpha/2)} + \cos\phi\right)\frac{\partial}{\partial\theta}\right.$$
$$\left. - \left(\frac{\cos\theta\sin\phi}{\sin\theta} - \frac{\cos\phi}{\tan(\alpha/2)\sin\theta}\right)\frac{\partial}{\partial\phi}\right],$$

$$W_3 = \frac{1}{2i}\left[2\cos\theta\frac{\partial}{\partial\alpha} - \frac{\sin\theta}{\tan(\alpha/2)}\frac{\partial}{\partial\theta} + \frac{\partial}{\partial\phi}\right],$$

$$W^2 = -\left[\frac{\partial^2}{\partial\alpha^2} + \frac{1}{\tan(\alpha/2)}\frac{\partial}{\partial\alpha} + \frac{1}{4\sin^2(\alpha/2)}\left\{\frac{\partial^2}{\partial\theta^2} + \frac{\cos\theta}{\sin\theta}\frac{\partial}{\partial\theta} + \frac{1}{\sin^2\theta}\frac{\partial^2}{\partial\phi^2}\right\}\right],$$

$$W_+ = \frac{e^{i\phi}}{2i}\left[2\sin\theta\frac{\partial}{\partial\alpha} + \left(\frac{\cos\theta}{\tan(\alpha/2)}+i\right)\frac{\partial}{\partial\theta} - \left(\frac{\cos\theta\tan(\alpha/2)-i}{\tan(\alpha/2)\sin\theta}\right)\frac{\partial}{\partial\phi}\right],$$

$$W_- = \frac{e^{-i\phi}}{2i}\left[2\sin\theta\frac{\partial}{\partial\alpha} + \left(\frac{\cos\theta}{\tan(\alpha/2)}-i\right)\frac{\partial}{\partial\theta} - \left(\frac{\cos\theta\tan(\alpha/2)+i}{\tan(\alpha/2)\sin\theta}\right)\frac{\partial}{\partial\phi}\right]$$

and the orthogonal spinors of the two two-dimensional representations can be written as

$$\Psi^{1/2}_{1/2} = i\sin\frac{\alpha}{2}\sin\theta\, e^{i\phi}, \qquad \Psi^{-1/2}_{1/2} = \cos\frac{\alpha}{2} - i\sin\frac{\alpha}{2}\cos\theta \qquad (4.70)$$

and

$$\widetilde{\Psi}^{1/2}_{1/2} = \cos\frac{\alpha}{2} + i\sin\frac{\alpha}{2}\cos\theta, \qquad \widetilde{\Psi}^{-1/2}_{1/2} = -i\sin\frac{\alpha}{2}\sin\theta\, e^{-i\phi}. \quad (4.71)$$

We have mentioned that the different spinors are orthogonal. To endow the group manifold with a Hilbert space structure it is necessary to define a hermitian, definite positive, scalar product. The Jacobian matrix of variables $\rho'$ in terms of variables $\rho$ given in (4.43), has the determinant

$$\det\left(\frac{\partial\rho'^i}{\partial\rho^j}\right) = \frac{(1+\mu^2)^2}{(1-\boldsymbol{\mu}\cdot\boldsymbol{\rho})^4},$$

and thus the transformation of the volume element

$$d^3\rho' = \frac{(1+\mu^2)^2}{(1-\boldsymbol{\mu}\cdot\boldsymbol{\rho})^4}\, d^3\rho.$$

We also get from (4.43) that

$$1+\rho'^2 = \frac{(1+\mu^2)}{(1-\boldsymbol{\mu}\cdot\boldsymbol{\rho})^2}\,(1+\rho^2)$$

and then the measure

$$\frac{d^3\rho'}{(1+\rho'^2)^2} = \left(\frac{(1-\boldsymbol{\mu}\cdot\boldsymbol{\rho})^2}{(1+\mu^2)(1+\rho^2)}\right)^2 \frac{(1+\mu^2)^2}{(1-\boldsymbol{\mu}\cdot\boldsymbol{\rho})^4}d^3\rho = \frac{d^3\rho}{(1+\rho^2)^2}$$

is in fact an invariant measure.

In spherical coordinates it is written as

$$\frac{r^2 \sin\theta}{(1+r^2)^2}\, dr d\theta d\phi$$

and in the normal representation is

$$\sin^2(\alpha/2) \sin\theta d\alpha d\theta d\phi.$$

Since the rotation group is a double-connected group, the above measure must be defined on a simply connected manifold, *i.e.*, on the universal covering group of $SO(3)$, which is $SU(2)$. The $SU(2)$ group manifold in the normal representation is given by the three-dimensional sphere of radius $2\pi$ and where points on the surface of this sphere represent a unique $SU(2)$ element, namely the $2 \times 2$ unitary matrix $-\mathbb{I}$. The normalized invariant measure becomes

$$d\mu_N(\alpha, \theta, \phi) \equiv \frac{1}{4\pi^2} \sin^2(\alpha/2) \sin\theta\, d\alpha\, d\theta\, d\phi. \tag{4.72}$$

Therefore, the hermitian scalar product will be defined as

$$< f|g >= \frac{1}{4\pi^2} \int_0^{2\pi} d\alpha \int_0^{\pi} d\theta \int_0^{2\pi} d\phi\, f^*(\alpha, \theta, \phi) g(\alpha, \theta, \phi) \sin^2(\alpha/2) \sin\theta, \tag{4.73}$$

where $f^*$ is the complex conjugate function of $f$.

All the previous computed spinors are orthogonal vectors with respect to the group invariant measure (4.72). In particular, the normalized $s = 1/2$ spinors are those given in (4.70)-(4.71), multiplied by $\sqrt{2}$.

The spin projection operators on the body axis $e_i$ linked to the particle, are given in (4.64) in the $\rho$ parametrization, and we have seen that they differ from the spin operators $W$ only in the change of sign of the second term, or alternatively a global change of sign followed by the change $\rho \to -\rho$. In the normal parametrization this corresponds to the change $\alpha \to -\alpha$ and also a global change of sign.

It can be checked as mentioned before, that

$$[T_i, T_k] = -i\epsilon_{ikl} T_l, \tag{4.74}$$

$$[W_i, T_k] = 0. \tag{4.75}$$

Because of the minus sign on the right-hand side of (4.74) spin operators $T$ are often said to satisfy the so-called 'anomalous' commutation relations of spin, while they commute with the $W_j$. We have seen that this is only a matter of active or passive interpretation.

Since $W^2 = T^2$ we can find simultaneous eigenvectors of the operators $W^2$, $W_3$ and $T_3$, which will be denoted by $D^{(s)}_{mn}(\alpha)$ in such a way that

$$\begin{aligned} W^2 D^{(s)}_{mn}(\alpha) &= s(s+1)D^{(s)}_{mn}(\alpha), \\ W_3 D^{(s)}_{mn}(\alpha) &= mD^{(s)}_{mn}(\alpha), \\ T_3 D^{(s)}_{mn}(\alpha) &= nD^{(s)}_{mn}(\alpha). \end{aligned}$$

Since $W_3(\alpha)D^{(s)}_{mn}(\alpha) = mD^{(s)}_{mn}(\alpha)$, by producing the change $\alpha \to -\alpha$ we get $W_3(-\alpha)D^{(s)}_{mn}(-\alpha) = mD^{(s)}_{mn}(-\alpha)$ and followed by a global change of sign it reduces to

$$-W_3(-\alpha)D^{(s)}_{mn}(-\alpha) = T_3(\alpha)D^{(s)}_{mn}(-\alpha) = -mD^{(s)}_{mn}(-\alpha),$$

so that the above spinors (4.70)-(4.71) are also eigenvectors of $T_3$.

With this notation, the four normalized spinors (4.70)-(4.71) become:

$$D^{(1/2)}_{1/2,-1/2} = \sqrt{2}\Psi^{1/2}_{1/2}, \qquad D^{(1/2)}_{-1/2,-1/2} = \sqrt{2}\Psi^{-1/2}_{1/2},$$

$$D^{(1/2)}_{1/2,1/2} = \sqrt{2}\widetilde{\Psi}^{1/2}_{1/2}, \qquad D^{(1/2)}_{-1/2,1/2} = \sqrt{2}\widetilde{\Psi}^{-1/2}_{1/2},$$

where spinors $\Psi$ are eigenvectors of $T_3$ with eigenvalue $-1/2$ while $\widetilde{\Psi}$ spinors are of eigenvalue 1/2. Because they span a four-dimensional vector space we shall choose as the four basis vectors the normalized spinors denoted by the corresponding eigenvalues $|s,m,n>$ as:

$$\Phi_1 = |1/2, 1/2, 1/2> = \sqrt{2}(\cos(\alpha/2) + i\cos\theta\sin(\alpha/2)), \qquad (4.76)$$

$$\Phi_2 = |1/2, -1/2, 1/2> = i\sqrt{2}\sin(\alpha/2)\sin\theta e^{-i\phi}, \qquad (4.77)$$

$$\Phi_3 = |1/2, 1/2, -1/2> = i\sqrt{2}\sin(\alpha/2)\sin\theta e^{i\phi}, \qquad (4.78)$$

$$\Phi_4 = |1/2, -1/2, -1/2> = \sqrt{2}(\cos(\alpha/2) - i\cos\theta\sin(\alpha/2)). \qquad (4.79)$$

They form an orthonormal set with respect to the normalized invariant measure (4.72) and with the scalar product defined in (4.73).

The important feature is that if the system has spin 1/2, although the $s = 1/2$ irreducible representations of the rotation group are two-dimensional, to describe the spin part of the wave function we need a function defined in the above four-dimensional complex Hilbert space, because to describe orientation we attach some local frame to the particle, and therefore in addition to the spin values in the laboratory frame we also have as additional observables the spin projections in the body axes, which can be included within the set of commuting operators.

## 3.2 MATRIX REPRESENTATION OF INTERNAL OBSERVABLES

The matrix representation of any observable $A$ that acts on the orientation variables or in this internal four-dimensional space spanned by these spin 1/2 wave functions $\Phi_i$, is obtained as $A_{ij} =< \Phi_i|A\Phi_j >$, $i,j = 1,2,3,4$. Once these four normalized basis vectors are fixed, when acting on the subspace they span, the differential operators $W_i$ and $T_i$ have the $4 \times 4$ block matrix representation

$$\boldsymbol{S} \equiv \boldsymbol{W} = \frac{\hbar}{2}\begin{pmatrix} \boldsymbol{\sigma} & 0 \\ 0 & \boldsymbol{\sigma} \end{pmatrix}, \tag{4.80}$$

$$T_1 = \frac{\hbar}{2}\begin{pmatrix} 0 & \mathbb{I} \\ \mathbb{I} & 0 \end{pmatrix}, \quad T_2 = \frac{\hbar}{2}\begin{pmatrix} 0 & i\mathbb{I} \\ -i\mathbb{I} & 0 \end{pmatrix}, \quad T_3 = \frac{\hbar}{2}\begin{pmatrix} \mathbb{I} & 0 \\ 0 & -\mathbb{I} \end{pmatrix}, \tag{4.81}$$

where $\boldsymbol{\sigma}$ are the three Pauli matrices and $\mathbb{I}$ represents the $2 \times 2$ unit matrix.

If we similarly compute the matrix elements of the nine components of the unit vectors $(\boldsymbol{e}_i)_j$, $i,j = 1,2,3$ we obtain the nine traceless hermitian matrices

$$\boldsymbol{e}_1 = \frac{1}{3}\begin{pmatrix} 0 & \boldsymbol{\sigma} \\ \boldsymbol{\sigma} & 0 \end{pmatrix}, \boldsymbol{e}_2 = \frac{1}{3}\begin{pmatrix} 0 & i\boldsymbol{\sigma} \\ -i\boldsymbol{\sigma} & 0 \end{pmatrix}, \boldsymbol{e}_3 = \frac{1}{3}\begin{pmatrix} \boldsymbol{\sigma} & 0 \\ 0 & -\boldsymbol{\sigma} \end{pmatrix}. \tag{4.82}$$

We see that the different components of the unit vectors $\boldsymbol{e}_i$, in general do not commute. The eigenvalues of every $e_{ij}$, in this matrix representation of definite spin, are $\pm 1/3$. However, the matrix representation of the square of any component is $(e_{ij})^2 = \mathbb{I}/3$, so that the magnitude of each vector $\boldsymbol{e}_i^2 = \sum_j (e_{ij})^2 = \mathbb{I}$ when acting on these wave functions. The eigenvalues of the squared operator $(e_{ij})^2$ are not the squared eigenvalues of $e_{ij}$. This is because the function $e_{ij}\Phi_k$ does not belong in general to the same space spanned by the $\Phi_k$, $k = 1, \ldots, 4$ although this space is invariant space for operators $W_i$ and $T_j$. In fact, each function $e_{ij}\Phi_k$ is a linear combination of a spin 1/2 and a spin 3/2 wave function.

## 3.3 PETER-WEYL THEOREM FOR COMPACT GROUPS

The above spinors can also be obtained by making use of an important theorem for representations of compact groups, known as the Peter-Weyl theorem, [12] which is stated without proof that can be read in any of the mentioned references.

**Theorem.-** Let $D^{(s)}(g)$ be a complete system of non-equivalent, unitary, irreducible representations of a compact group $G$, labeled by the parameter $s$. Let $d_s$ be the dimension of each representation and $D_{ij}^{(s)}(g)$, $1 \leq i,j \leq d_s$ the corresponding matrix elements. Then, the functions

$$\sqrt{d_s}\, D_{ij}^{(s)}(g), \quad 1 \leq i,j \leq d_s$$

form a complete orthonormal system on $G$, with respect to some normalized invariant measure $\mu_N(g)$ defined on this group, *i.e.*,

$$\int_G \sqrt{d_s}\, D_{ij}^{(s)*}(g)\, \sqrt{d_r}\, D_{kl}^{(r)}(g)\, d\mu_N(g) = \delta^{sr}\delta_{ik}\delta_{jl}. \qquad (4.83)$$

That the set is complete means that every square integrable function defined on $G$, $f(g)$, admits a series expansion, convergent in norm, in terms of the above orthogonal functions $D_{ij}^{(s)}(g)$, in the form

$$f(g) = \sum_{s,i,j} a_{ij}^{(s)}\, \sqrt{d_s}\, D_{ij}^{(s)}(g),$$

where the coefficients, in general complex numbers $a_{ij}^{(s)}$, are obtained by

$$a_{ij}^{(s)} = \int_G \sqrt{d_s}\, D_{ij}^{(s)*}(g)\, f(g) d\mu_N(g).$$

In our case $SU(2)$, as a group manifold, is the simply connected three-dimensional sphere of radius $2\pi$, with the normalized measure as seen before (4.72),

$$d\mu_N(\alpha,\theta,\phi) = \frac{1}{4\pi^2} \sin\theta \sin(\alpha/2)^2\, d\alpha d\theta d\phi.$$

In the normal parametrization, the two-dimensional representation of $SU(2)$ corresponds to the eigenvalue $s = 1/2$ of $S^2$ and the matrix representation is given by

$$D^{(1/2)}(\alpha) = \cos(\alpha/2)\mathbb{I} - i \sin(\alpha/2)(\boldsymbol{u}\cdot\boldsymbol{\sigma}),$$

*i.e.*,

$$D^{(1/2)}(\alpha) = \begin{pmatrix} \cos(\alpha/2) - i\cos\theta\sin(\alpha/2) & -i\sin\theta\sin(\alpha/2)\, e^{-i\phi} \\ -i\sin\theta\sin(\alpha/2)\, e^{i\phi} & \cos(\alpha/2) + i\cos\theta\sin(\alpha/2) \end{pmatrix}.$$

If we compare these four matrix components with the four orthogonal spinors given in (4.70)-(4.71) we see that

$$D^{(1/2)}(\alpha) = \begin{pmatrix} \Psi_{1/2}^{-1/2} & -\widetilde{\Psi}_{1/2}^{-1/2} \\ -\Psi_{1/2}^{1/2} & \widetilde{\Psi}_{1/2}^{1/2} \end{pmatrix} \qquad (4.84)$$

where $\Psi$ spinors stand on the first column and $\widetilde{\Psi}$ on the second. When multiplied by $\sqrt{2}$ they become the normalized spinors (4.76)-(4.79). We understand why, when considered by columns, the group action (4.43)

$$D^{(1/2)}(\rho') = D^{(1/2)}(\mu)\, D^{(1/2)}(\rho)$$

transforms the spinors $\Psi$ and $\widetilde{\Psi}$ among themselves, separately, according to the two-dimensional representation of the $SU(2)$ group, $D^{(1/2)}(\mu)$.

Operators $W_\pm$ transform among themselves these matrix components or spinors, within the same column, while operators $T_\pm$ within each row. The matrix (4.84) can be written in terms of the eigenvectors of the form $|w_3, t_3>$, as

$$D^{(1/2)}(\alpha) = \begin{pmatrix} |-1/2,-1/2> & |-1/2,1/2> \\ |1/2,-1/2> & |1/2,1/2> \end{pmatrix}.$$

In the three-dimensional representation of $SO(3)$, considered as a representation of SU(2)

$$D^{(1)}_{ij}(\alpha) = \delta_{ij}\cos\alpha + u_i u_j(1-\cos\alpha) + \epsilon_{ikj} u_k \sin\alpha \equiv e_{j_i}$$

we get another set of nine orthogonal functions. Multiplied by $\sqrt{3}$ they form another orthonormal set orthogonal to the previous four spinors. It is a good exercise to check this orthogonality among these functions. Let $|f_i> \equiv e_{1_i}$, $i = 1,2,3$ be the three functions corresponding to the first column of matrix $D^{(1)}(\alpha)$, $|g_i> \equiv e_{2_i}$ to the second and $|h_i> \equiv e_{3_i}$ to the third, then, explicitly

$$\begin{aligned}
|f_1> &= \sqrt{3}(\cos\alpha + \sin^2\theta\cos^2\phi(1-\cos\alpha)),\\
|f_2> &= \sqrt{3}(\cos\theta\sin\alpha + \sin^2\theta\sin\phi\cos\phi(1-\cos\alpha)),\\
|f_3> &= \sqrt{3}(-\sin\theta\sin\phi\sin\alpha + \sin\theta\cos\theta\cos\phi(1-\cos\alpha)),
\end{aligned}$$

$$\begin{aligned}
|g_1> &= \sqrt{3}(-\cos\theta\sin\alpha + \sin^2\theta\sin\phi\cos\phi(1-\cos\alpha)),\\
|g_2> &= \sqrt{3}(\cos\alpha + \sin^2\theta\sin^2\phi(1-\cos\alpha)),\\
|g_3> &= \sqrt{3}(\sin\theta\cos\phi\sin\alpha + \sin\theta\cos\theta\sin\phi(1-\cos\alpha)),
\end{aligned}$$

$$\begin{aligned}
|h_1> &= \sqrt{3}(\sin\theta\sin\phi\sin\alpha + \sin\theta\cos\theta\cos\phi(1-\cos\alpha)),\\
|h_2> &= \sqrt{3}(-\sin\theta\cos\phi\sin\alpha + \sin\theta\cos\theta\sin\phi(1-\cos\alpha)),\\
|h_3> &= \sqrt{3}(\cos\alpha + \cos^2\theta(1-\cos\alpha)).
\end{aligned}$$

These nine matrix components are eigenvectors of $W^2$ with eigenvalue $s=1$, but they are not eigenvectors of either $W_3$ or $T_3$ but rather they are linear combinations of them. Let $|y_i>$ be the three spinors obtained by tensor product of spinors $\Psi \otimes \Psi$, where the diferent components $\Psi^m_s$ are given in (4.70). They are explicitly given by

$$
\begin{aligned}
|y_1> &= \sin^2(\alpha/2)\sin^2\theta\, e^{2i\phi},\\
|y_2> &= \sin(\alpha/2)\sin\theta\,(\sin(\alpha/2)\cos\theta + i\cos(\alpha/2))e^{i\phi},\\
|y_3> &= (\sin(\alpha/2)\cos\theta + i\cos(\alpha/2))^2.
\end{aligned}
$$

They are eigenvectors of $T_3$ with eigenvalue $-1$. Let $|v_i>$ be those built from $\widetilde{\Psi} \otimes \widetilde{\Psi}$, where the $\widetilde{\Psi}^m_s$ components are given in (4.71), and thus

$$
\begin{aligned}
|v_1> &= (\sin(\alpha/2)\cos\theta - i\cos(\alpha/2))^2,\\
|v_2> &= \sin(\alpha/2)\sin\theta\,(\sin(\alpha/2)\cos\theta - i\cos(\alpha/2))e^{-i\phi},\\
|v_3> &= \sin^2(\alpha/2)\sin^2\theta\, e^{-2i\phi}.
\end{aligned}
$$

They are eigenvectors of $T_3$ with eigenvalue 1, and finally let $|w_i>$ be those coming from $\Psi \otimes \widetilde{\Psi}$,

$$
\begin{aligned}
|w_1> &= \sin(\alpha/2)\sin\theta(\sin(\alpha/2)\cos\theta - i\cos(\alpha/2))e^{i\phi},\\
|w_2> &= 2\sin^2(\alpha/2)\sin^2\theta - 1,\\
|w_3> &= \sin(\alpha/2)\sin\theta\,(\sin(\alpha/2)\cos\theta + i\cos(\alpha/2))e^{-i\phi},
\end{aligned}
$$

which are eigenvectors of $T_3$ with eigenvalue 0. Then we get the relationship

$$
\begin{aligned}
2\sqrt{3}|y_1> &= |f_1> + i|f_2> + i(|g_1> + i|g_2>),\\
2\sqrt{3}|y_2> &= |f_3> + i|g_3>,\\
-2\sqrt{3}|y_3> &= |f_1> - i|f_2> + i(|g_1> - i|g_2>),
\end{aligned}
$$

$$
\begin{aligned}
-2\sqrt{3}|v_1> &= |f_1> + i|f_2> - i(|g_1> + i|g_2>),\\
2\sqrt{3}|v_2> &= |f_3> - i|g_3>,\\
2\sqrt{3}|v_3> &= |f_1> - i|f_2> - i(|g_1> - i|g_2>),
\end{aligned}
$$

and

$$
\begin{aligned}
2\sqrt{3}|w_1> &= |h_1> + i|h_2>,\\
\sqrt{3}|w_2> &= -|h_3>,\\
2\sqrt{3}|w_3> &= |h_1> - i|h_2>.
\end{aligned}
$$

If we classify them according to the eigenvalues of the two operators $W_3$ and $T_3$, $|w_3, t_3>$, these nine spinors are respectively

$$
|y_1> \equiv |1,-1>, \quad |y_2> \equiv |0,-1>, \quad |y_3> \equiv |-1,-1>,
$$

$$|v_1> \equiv |1,1>, \quad |v_2> \equiv |0,1>, \quad |v_3> \equiv |-1,1>,$$
$$|w_1> \equiv |1,0>, \quad |w_2> \equiv |0,0>, \quad |w_3> \equiv |-1,0>.$$

## 3.4 GENERAL SPINORS

In the case that the zitterbewegung content of the spin is not vanishing we can also obtain spin 1/2 wave-functions as the irreducible representations contained in the tensor product of integer and half-integer spin states coming from the $U(\boldsymbol{u})$ and $V(\boldsymbol{\rho})$ part of the general wave function (4.66).

The total spin operator of the system is of the form

$$\boldsymbol{S} = \boldsymbol{u} \times \boldsymbol{U} + \boldsymbol{W} = \boldsymbol{Y} + \boldsymbol{W},$$

where $\boldsymbol{Y} = -i\hbar\nabla_u$ and $\boldsymbol{W}$ is given in (4.59). Spin projections on the body axes, *i.e.*, operators $T_i = \boldsymbol{e}_i \cdot \boldsymbol{W}$, are described in (4.64). They satisfy the commutation relations

$$[\boldsymbol{Y}, \boldsymbol{Y}] = i\boldsymbol{Y}, \quad [\boldsymbol{W}, \boldsymbol{W}] = i\boldsymbol{W}, \quad [\boldsymbol{T}, \boldsymbol{T}] = -i\boldsymbol{T},$$

$$[\boldsymbol{Y}, \boldsymbol{W}] = 0, \quad [\boldsymbol{Y}, \boldsymbol{T}] = 0, \quad [\boldsymbol{W}, \boldsymbol{T}] = 0.$$

These commutation relations are invariant under the change $\boldsymbol{\rho}$ by $-\boldsymbol{\rho}$ in the definition of the operators $\boldsymbol{W}$ and $\boldsymbol{T}$. The expression of the body frame unit vectors $\boldsymbol{e}_i$ is given in (4.61) and (4.62).

We can see that these unit vector components and spin operators $W_i$ and $T_j$ satisfy the following properties:

1) $e_{ij}(-\alpha, \theta, \phi) = -e_{j_i}(\alpha, \theta, \phi)$.

2) $\boldsymbol{e}_i \cdot \boldsymbol{W} \equiv \sum_j e_{ij} W_j = T_i$.

3) $\sum_j \boldsymbol{e}_j T_j = \boldsymbol{W}$.

4) For all $i, j$, the action $W_i e_{j_i} = 0$, with no addition on index $i$.

5) For all $i, j$, the action $T_i e_{ij} = 0$, with no addition on index $i$.

6) For all $i, j, k$, with $i \neq j$, we have that $W_i e_{kj} + W_j e_{ki} = 0$, and in the case that $i = j$, it leads to property 4.

7) For all $i, j, k$, with $i \neq j$, we have that $T_i e_{j_k} + T_j e_{ik} = 0$, and similarly as before in the case $i = j$ it leads to property 4.

This implies that $\boldsymbol{e}_i \cdot \boldsymbol{W} = \boldsymbol{W} \cdot \boldsymbol{e}_i = T_i$, because of property 4, since when acting on an arbitray function $f$,

$$(\boldsymbol{W} \cdot \boldsymbol{e}_i) f \equiv \sum_j W_j(e_{ij} f) = f \sum_j W_j(e_{ij}) + \sum_j e_{ij} W_j(f) = T_i(f),$$

because $\sum_j W_j e_{ij} = 0$.

In the same way $\sum_j \boldsymbol{e}_j T_j \equiv \sum_j T_j \boldsymbol{e}_j = \boldsymbol{W}$.

Now we fix the value of spin. Particles of different values of spin can be described. Let us consider then that elementary particles take the lowest admissible spin values. For spin 1/2 particles, if we take first for simplicity eigenfunctions $V(\rho)$ of $W^2$ with eigenvalue 1/2, and then since the total spin has to be 1/2, the orbital $\boldsymbol{Y}$ part can only contribute with spherical harmonics of value $y = 0$ and $y = 1$.

If there is no zitterbewegung spin, $y = 0$, and Wigner's functions as we have seen in Sec. 3.1 can be taken as simultaneous eigenfunctions of the three commuting $W^2$, $W_3$, and $T_3$ operators, and the normalized eigenvectors $|w, w_3, t_3>$ are explicitly given by the following functions in the usual ket notation:

$$\Phi_1 \equiv |1/2, 1/2, 1/2>= \sqrt{2}(\cos(\alpha/2) + i\cos\theta\sin(\alpha/2)), \quad (4.85)$$

$$\Phi_2 \equiv |1/2, -1/2, 1/2>= i\sqrt{2}\sin(\alpha/2)\,\sin\theta e^{-i\phi}, \quad (4.86)$$

$$\Phi_3 \equiv |1/2, 1/2, -1/2>= i\sqrt{2}\sin(\alpha/2)\,\sin\theta e^{i\phi}, \quad (4.87)$$

$$\Phi_4 \equiv |1/2, -1/2, -1/2>= \sqrt{2}(\cos(\alpha/2) - i\cos\theta\sin(\alpha/2)). \quad (4.88)$$

If we have a zitterbewegung spin of value $y = 1$, then the $U(\boldsymbol{u})$ part contributes with the spherical harmonics

$$Y_1^1(\tilde\theta, \tilde\phi) \equiv |1, 1>= -\sin(\tilde\theta)e^{i\tilde\phi}\sqrt{\frac{3}{8\pi}}, \quad (4.89)$$

$$Y_1^0(\tilde\theta, \tilde\phi) \equiv |1, 0>= \cos(\tilde\theta)\sqrt{\frac{3}{4\pi}}, \quad (4.90)$$

$$Y_1^{-1}(\tilde\theta, \tilde\phi) \equiv |1, -1>= \sin(\tilde\theta)e^{-i\tilde\phi}\sqrt{\frac{3}{8\pi}}, \quad (4.91)$$

normalized with respect to the measure

$$\int_0^\pi \int_0^{2\pi} \sin(\tilde\theta)d\tilde\theta d\tilde\phi,$$

which are the indicated eigenfunctions $|y, y_3>$ of $\boldsymbol{Y}^2$ and $Y_3$, and where the variables $\tilde\theta$ and $\tilde\phi$ determine the orientation of the velocity $\boldsymbol{u}$.

The tensor product representation of the rotation group constructed from the two irreducible representations **1** associated to the spherical harmonics (4.89)-(4.91) and **1/2** given in (4.85)-(4.88) is split into the direct sum $\mathbf{1} \otimes \mathbf{1/2} = \mathbf{3/2} \oplus \mathbf{1/2}$.

The following functions of five variables $\tilde\theta$, $\tilde\phi$, $\alpha$, $\theta$ and $\phi$, where variables $\tilde\theta$ and $\tilde\phi$ correspond to the ones of the spherical harmonics $Y_l^m$, and the remaining $\alpha$, $\theta$ and $\phi$, to the previous spinors $\Phi_i$, are normalized spin 1/2 functions $|s, s_3, t_3>$ that are eigenvectors of total spin $S^2$, and $S_3$

and $T_3$ operators

$$\Psi_1 \equiv |1/2, 1/2, 1/2> = \frac{1}{\sqrt{3}}\left(Y_1^0\Phi_1 - \sqrt{2}Y_1^1\Phi_2\right), \tag{4.92}$$

$$\Psi_2 \equiv |1/2, -1/2, 1/2> = \frac{1}{\sqrt{3}}\left(-Y_1^0\Phi_2 + \sqrt{2}Y_1^{-1}\Phi_1\right), \tag{4.93}$$

$$\Psi_3 \equiv |1/2, 1/2, -1/2> = \frac{1}{\sqrt{3}}\left(Y_1^0\Phi_3 - \sqrt{2}Y_1^1\Phi_4\right), \tag{4.94}$$

$$\Psi_4 \equiv |1/2, -1/2, -1/2> = \frac{1}{\sqrt{3}}\left(-Y_1^0\Phi_4 + \sqrt{2}Y_1^{-1}\Phi_3\right), \tag{4.95}$$

such that $\Psi_2 = S_-\Psi_1$ and similarly $\Psi_4 = S_-\Psi_3$, and also that $\Psi_3 = T_-\Psi_1$, and $\Psi_4 = T_-\Psi_2$. They are no longer eigenfunctions of the $W_3$ operator, although they span an invariant vector space for $S^2$, $S_3$ and $T_3$ operators. In the above basis (4.92)-(4.95) formed by orthonormal vectors $\Psi_i$, the matrix representation of the spin is

$$\boldsymbol{S} = \boldsymbol{Y} + \boldsymbol{W} = \frac{\hbar}{2}\begin{pmatrix} \boldsymbol{\sigma} & 0 \\ 0 & \boldsymbol{\sigma} \end{pmatrix}, \tag{4.96}$$

while the matrix representation of the $\boldsymbol{Y}$ and $\boldsymbol{W}$ part is

$$\boldsymbol{Y} = \frac{2\hbar}{3}\begin{pmatrix} \boldsymbol{\sigma} & 0 \\ 0 & \boldsymbol{\sigma} \end{pmatrix}, \quad \boldsymbol{W} = \frac{-\hbar}{6}\begin{pmatrix} \boldsymbol{\sigma} & 0 \\ 0 & \boldsymbol{\sigma} \end{pmatrix}, \tag{4.97}$$

which do not satisfy commutation relations of angular momentum operators because the vector space spanned by the above basis is not an invariant space for these operators $\boldsymbol{Y}$ and $\boldsymbol{W}$.

The spin projection of the $\boldsymbol{W}$ part on the body axis, *i.e.*, the $\boldsymbol{T}$ operator, takes the same form as before (4.81)

$$T_1 = \frac{\hbar}{2}\begin{pmatrix} 0 & \mathbb{I} \\ \mathbb{I} & 0 \end{pmatrix}, \quad T_2 = \frac{\hbar}{2}\begin{pmatrix} 0 & i\mathbb{I} \\ -i\mathbb{I} & 0 \end{pmatrix}, \quad T_3 = \frac{\hbar}{2}\begin{pmatrix} \mathbb{I} & 0 \\ 0 & -\mathbb{I} \end{pmatrix}, \tag{4.98}$$

because $\Psi_1$ and $\Psi_2$ functions are eigenfunctions of $T_3$ with eigenvalue 1/2, while $\Psi_3$ and $\Psi_4$ are of eigenvalue $-1/2$, and thus the spinors $\Psi_i$ span an invariant space for $S_i$ and $T_j$ operators. In fact the basis is formed by simultaneous eigenfunctions of total spin $S^2$, $S_3$ and $T_3$, and the ket representation is the same as in the case of the $\Phi_i$ given in (4.85)-(4.88).

The expression in this basis of the components of the unit vectors $\boldsymbol{e}_i$ are represented by

$$\boldsymbol{e}_1 = -\frac{1}{9}\begin{pmatrix} 0 & \boldsymbol{\sigma} \\ \boldsymbol{\sigma} & 0 \end{pmatrix}, \boldsymbol{e}_2 = -\frac{1}{9}\begin{pmatrix} 0 & i\boldsymbol{\sigma} \\ -i\boldsymbol{\sigma} & 0 \end{pmatrix}, \boldsymbol{e}_3 = -\frac{1}{9}\begin{pmatrix} \boldsymbol{\sigma} & 0 \\ 0 & -\boldsymbol{\sigma} \end{pmatrix}. \tag{4.99}$$

Then if we compute the projection of spin in some body axis $\boldsymbol{e}_i$,

$$\boldsymbol{S}\cdot\boldsymbol{e}_i = \boldsymbol{Y}\cdot\boldsymbol{e}_i + \boldsymbol{W}\cdot\boldsymbol{e}_i = \boldsymbol{e}_i\cdot\boldsymbol{Y} + \boldsymbol{e}_i\cdot\boldsymbol{W} = \boldsymbol{e}_i\cdot\boldsymbol{S} = R_i,$$

and similarly $\boldsymbol{Y}\cdot\boldsymbol{e}_i = M_i$, they are given by

$$R_1 = \frac{-\hbar}{6}\begin{pmatrix} 0 & \mathbb{I} \\ \mathbb{I} & 0 \end{pmatrix},\; R_2 = \frac{-\hbar}{6}\begin{pmatrix} 0 & i\mathbb{I} \\ -i\mathbb{I} & 0 \end{pmatrix},\; R_3 = \frac{-\hbar}{6}\begin{pmatrix} \mathbb{I} & 0 \\ 0 & -\mathbb{I} \end{pmatrix}, \tag{4.100}$$

and

$$M_1 = \frac{-2\hbar}{3}\begin{pmatrix} 0 & \mathbb{I} \\ \mathbb{I} & 0 \end{pmatrix},\; M_2 = \frac{-2\hbar}{3}\begin{pmatrix} 0 & i\mathbb{I} \\ -i\mathbb{I} & 0 \end{pmatrix},\; M_3 = \frac{-2\hbar}{3}\begin{pmatrix} \mathbb{I} & 0 \\ 0 & -\mathbb{I} \end{pmatrix}. \tag{4.101}$$

# 4. RELATIVISTIC PARTICLES

We can similarly quantize classical relativistic particles. We shall start again by considering the relativistic point particle to focus attention, in detail, on the kinematical space of particles traveling at the speed of light. We shall obtain the quantum mechanical description of the electron and its associated Dirac equation, considered as a one-particle wave equation, and showing the interpretation of its internal observables, that span Dirac's algebra. We shall also consider the quantization of the photon, obtaining the usual wave equation with two possible polarization states. Finally tachyons are quantized on the same footing showing the impossibility of having spin 1/2 tachyons.

## 4.1 RELATIVISTIC POINT PARTICLE

Kinematical variables of the point particle are position and time as in the nonrelativistic case. They transform under the Poincaré group as described in (3.33)-(3.34). Gauge functions for the Poincaré group are only trivial. Then wave functions are complex functions $\Phi(t,\boldsymbol{r})$ and therefore the representation of the ten generators when acting on these functions is

$$H = i\hbar\frac{\partial}{\partial t},\quad \boldsymbol{P} = \frac{\hbar}{i}\nabla,\quad \boldsymbol{K} = \frac{i\hbar\boldsymbol{r}}{c^2}\frac{\partial}{\partial t} - t\frac{\hbar}{i}\nabla, \tag{4.102}$$

$$\boldsymbol{J} = \boldsymbol{r}\times\frac{\hbar}{i}\nabla. \tag{4.103}$$

The two Casimir operators are $P_\mu P^\mu = m^2c^2$ and the square of the Pauli-Lubanski four-vector $w_\mu$ given in (3.14), which vanishes identically. The first invariant defines the mass of the system and every wave function must satisfy the Klein-Gordon equation

$$\left(\frac{1}{c^2}H^2 - \boldsymbol{P}^2\right)\Phi(t,\boldsymbol{r}) = m^2c^2\Phi(t,\boldsymbol{r}), \tag{4.104}$$

or

$$\left(\frac{1}{c^2}\frac{\partial^2}{\partial t^2} - \nabla^2 + \frac{m^2c^2}{\hbar^2}\right)\Phi(t,\boldsymbol{r}) = 0.$$

We can also find simultaneous eigenfunctions of the above Klein-Gordon operator, energy operator, $H\Phi(t,\boldsymbol{r}) = E\Phi(t,\boldsymbol{r})$ and linear momentum, $\boldsymbol{P}\Phi(t,\boldsymbol{r}) = \boldsymbol{p}\Phi(t,\boldsymbol{r})$ so that there exist solutions in the form of plane waves $e^{i(Et-\boldsymbol{p}\cdot\boldsymbol{r})/\hbar}$, with the condition $E^2/c^2 - \boldsymbol{p}^2 = m^2c^2$, and where the energy can be either positive or negative.

No spin operator can be defined for this system and as in the classical case the quantum mechanical point particle is a spinless particle.

In the case of interaction with an external electromagnetic field, the Poincaré invariant equation (4.104) is satisfied in the classical case (see (3.216)) by the mechanical parts $H_m = H - e\phi$ and $\boldsymbol{P}_m = \boldsymbol{P} - e\boldsymbol{A}$, and therefore the wave equation for a point particle is

$$\left(\frac{1}{c^2}(H - e\phi)^2 - (\boldsymbol{P} - e\boldsymbol{A})^2\right)\Phi(t,\boldsymbol{r}) = m^2c^2\Phi(t,\boldsymbol{r}), \tag{4.105}$$

where $H = i\hbar\partial/\partial t$ and $\boldsymbol{P} = -i\hbar\nabla$.

## 4.2 GENERAL RELATIVISTIC SPINNING PARTICLE

Quantization of the general relativistic particle will be done in the same way. In this case we have to distinguish the three maximal homogeneous spaces of the Poincaré group which give rise respectively to Bradyons, Luxons and Tachyons. We shall sketch now briefly the corresponding method for Bradyons and we shall devote the next section to the case of Luxons.

The ten kinematical variables transform under $\mathcal{P}$ according to

$$t'(\tau) = \gamma t(\tau) + \gamma(\boldsymbol{v}\cdot R(\boldsymbol{\mu})\boldsymbol{r}(\tau))/c^2 + b, \tag{4.106}$$

$$\boldsymbol{r}'(\tau) = R(\boldsymbol{\mu})\boldsymbol{r}(\tau) + \gamma\boldsymbol{v}t(\tau) + \frac{\gamma^2}{(1+\gamma)c^2}(\boldsymbol{v}\cdot R(\boldsymbol{\mu})\boldsymbol{r}(\tau))\boldsymbol{v} + \boldsymbol{a}, \tag{4.107}$$

$$\boldsymbol{u}'(\tau) = \frac{R(\boldsymbol{\mu})\boldsymbol{u}(\tau) + \gamma\boldsymbol{v} + (\boldsymbol{v}\cdot R(\boldsymbol{\mu})\boldsymbol{u}(\tau))\boldsymbol{v}\gamma^2/(1+\gamma)c^2}{\gamma(1 + \boldsymbol{v}\cdot R(\boldsymbol{\mu})\boldsymbol{u}(\tau)/c^2)}, \tag{4.108}$$

$$\boldsymbol{\rho}'(\tau) = \frac{\boldsymbol{\mu} + \boldsymbol{\rho}(\tau) + \boldsymbol{\mu}\times\boldsymbol{\rho}(\tau) + \boldsymbol{F}(\boldsymbol{v},\boldsymbol{\mu};\boldsymbol{u}(\tau),\boldsymbol{\rho}(\tau))}{1 - \boldsymbol{\mu}\cdot\boldsymbol{\rho}(\tau) + G(\boldsymbol{v},\boldsymbol{\mu};\boldsymbol{u}(\tau),\boldsymbol{\rho}(\tau))}, \tag{4.109}$$

where the functions $F$ and $G$ are given in (3.9) and (3.10), respectively. The wave function of the system is a function $\Phi(t, \boldsymbol{r}, \boldsymbol{u}, \boldsymbol{\rho})$ of these kinematical variables. The ten group generators are expressed as

$$H = i\hbar\frac{\partial}{\partial t}, \quad \boldsymbol{P} = \frac{\hbar}{i}\nabla, \quad \boldsymbol{K} = \frac{i\hbar\boldsymbol{r}}{c^2}\frac{\partial}{\partial t} - t\frac{\hbar}{i}\nabla - \boldsymbol{D}/c, \tag{4.110}$$

where $\boldsymbol{D}$ written in terms of $\boldsymbol{U}$ and $\boldsymbol{W}$ is given in (3.76), and $\boldsymbol{U} = -i\hbar\nabla_u$ and $\boldsymbol{W}$ takes the form (4.59). The total angular momentum is again

$$\boldsymbol{J} = \boldsymbol{r} \times \boldsymbol{P} + \boldsymbol{u} \times \boldsymbol{U} + \boldsymbol{W}. \tag{4.111}$$

We have as before

$$\left(\frac{1}{c^2}H^2 - \boldsymbol{P}^2\right)\Phi(t, \boldsymbol{r}, \boldsymbol{u}, \boldsymbol{\rho}) = m^2c^2\Phi(t, \boldsymbol{r}, \boldsymbol{u}, \boldsymbol{\rho}),$$

but now the wave function, that depends on more variables in this case, can be separated in the form

$$\Phi(t, \boldsymbol{r}, \boldsymbol{u}, \boldsymbol{\rho}) = \sum_i \psi_i(t, \boldsymbol{r})\chi_i(\boldsymbol{u}, \boldsymbol{\rho}). \tag{4.112}$$

By applying the Klein-Gordon operator to this expansion we find that each space-time function $\psi_i(t, \boldsymbol{r})$, irrespective of the accompanying function $\chi_i(\boldsymbol{u}, \boldsymbol{\rho})$, satisfies

$$\left(\frac{1}{c^2}H^2 - \boldsymbol{P}^2\right)\psi_i(t, \boldsymbol{r}) = m^2c^2\psi_i(t, \boldsymbol{r}). \tag{4.113}$$

The other functions $\chi_i(\boldsymbol{u}, \boldsymbol{\rho})$ are chosen as eigenfunctions of the angular momentum operators. These operators produce derivatives with respect to these additional variables $\boldsymbol{u}$ and $\boldsymbol{\rho}$ and therefore, commute with the Klein-Gordon operator. This is what justifies the above decomposition (4.112). We delay the analysis of these functions $\chi_i(\boldsymbol{u}, \boldsymbol{\rho})$, until the next section where we shall describe spin 1/2 particles.

Let us consider the classical kinematical momentum in explicit form

$$\boldsymbol{K} = \frac{H}{c^2}\boldsymbol{r} - t\boldsymbol{P} - \frac{1}{c}\boldsymbol{D}.$$

If we take the time derivative of the above expression we get a Poincaré invariant relation

$$\frac{H}{c^2}\boldsymbol{u} - \boldsymbol{P} - \frac{1}{c}\frac{d\boldsymbol{D}}{dt} = 0.$$

It shows a linear relationship between the energy $H$ and linear momentum $\boldsymbol{P}$. Taking the scalar product with the velocity, it yields

$$\frac{u^2}{c^2}H - \boldsymbol{P}\cdot\boldsymbol{u} - \frac{1}{c}\frac{d\boldsymbol{D}}{dt}\cdot\boldsymbol{u} = 0. \tag{4.114}$$

Now, in the quantum case, if we apply this operator to any wave function $\Phi(t, \boldsymbol{r}, \boldsymbol{u}, \boldsymbol{\rho})$, operators $H$ and $\boldsymbol{P}$ are expressed as the usual differential operators given in (4.110). What we need is to obtain the quantum mechanical representatives of the remaning observables like $\boldsymbol{u}$, $u^2$ and $d\boldsymbol{D}/dt$. These will act, in general, on the functions $\chi_i$ that depend on variables $\boldsymbol{u}$ and $\boldsymbol{\rho}$, so that each $\psi_i$ in addition to satisfying the Klein-Gordon equation (4.113), will satisfy a first order differential equation on time and position, in which the different $\psi_i$ will be linearly coupled because of the action of the mentioned operators on the $\chi_i$ part of the expansion.

If it happens that $u = c$ we see that the first two terms of expression (4.114) look quite similar to the corresponding ones of Dirac's Hamiltonian, but in the case of Luxons we have to replace operator $\boldsymbol{D}$ by the corresponding $\boldsymbol{Z} = \boldsymbol{u} \times \boldsymbol{U} + \boldsymbol{W}$ observable. We shall show that in the case of Luxons, quantization of the model gives rise to Dirac's equation. This is what is done in the next Section.

## 4.3 DIRAC'S EQUATION

For Luxons we have the nine-dimensional homogeneous space of the Poincaré group, spanned by the ten variables $(t, \boldsymbol{r}, \boldsymbol{u}, \boldsymbol{\alpha})$ similarly as before, but now $\boldsymbol{u}$ is restricted to $u = c$. For this system, since $\boldsymbol{u} \cdot \dot{\boldsymbol{u}} = 0$ and $\dot{\boldsymbol{u}} \neq 0$, we are describing particles with a circular internal orbital motion at the constant speed $c$, like the ones described in Section 4.2 of Chapter 3.

In the center of mass frame, (see Fig.4.1) the center of charge describes a circle of radius $R_0 = S/mc$ at the constant speed $c$, the spin being orthogonal to the charge trajectory plane and a constant of the motion in this frame.

The kinematical variables transform under $\mathcal{P}$ according to

$$t'(\tau) = \gamma t(\tau) + \gamma(\boldsymbol{v} \cdot R(\boldsymbol{\mu})\boldsymbol{r}(\tau))/c^2 + b, \tag{4.115}$$

$$\boldsymbol{r}'(\tau) = R(\boldsymbol{\mu})\boldsymbol{r}(\tau) + \gamma\boldsymbol{v}t(\tau) + \frac{\gamma^2}{(1+\gamma)c^2}(\boldsymbol{v} \cdot R(\boldsymbol{\mu})\boldsymbol{r}(\tau))\boldsymbol{v} + \boldsymbol{a}, \tag{4.116}$$

$$\boldsymbol{u}'(\tau) = \frac{R(\boldsymbol{\mu})\boldsymbol{u}(\tau) + \gamma\boldsymbol{v} + (\boldsymbol{v} \cdot R(\boldsymbol{\mu})\boldsymbol{u}(\tau))\boldsymbol{v}\gamma^2/(1+\gamma)c^2}{\gamma(1 + \boldsymbol{v} \cdot R(\boldsymbol{\mu})\boldsymbol{u}(\tau)/c^2)}, \tag{4.117}$$

$$\boldsymbol{\rho}'(\tau) = \frac{\boldsymbol{\mu} + \boldsymbol{\rho}(\tau) + \boldsymbol{\mu} \times \boldsymbol{\rho}(\tau) + \boldsymbol{F}_c(\boldsymbol{v}, \boldsymbol{\mu}; \boldsymbol{u}(\tau), \boldsymbol{\rho}(\tau))}{1 - \boldsymbol{\mu} \cdot \boldsymbol{\rho}(\tau) + G_c(\boldsymbol{v}, \boldsymbol{\mu}; \boldsymbol{u}(\tau), \boldsymbol{\rho}(\tau))}, \tag{4.118}$$

where the functions $\boldsymbol{F}_c$ and $G_c$ are given in (3.132) and (3.133), respectively. When quantized, the wave function of the system is a function $\Phi(t, \boldsymbol{r}, \boldsymbol{u}, \boldsymbol{\rho})$ of these kinematical variables. For the Poincaré group all exponents and thus all gauge functions on homogeneous spaces are equiv-

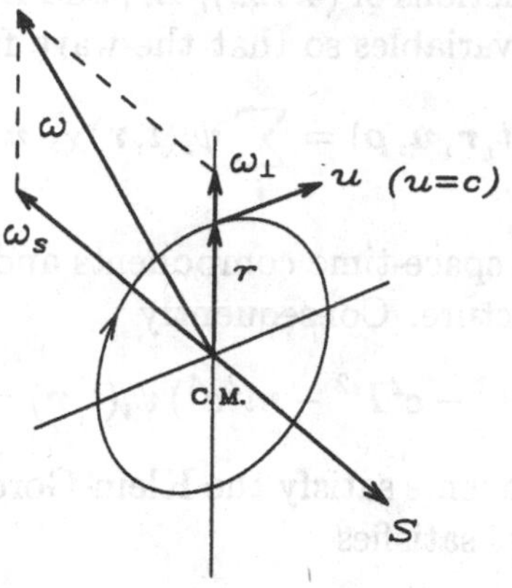


*Figure 4.1.* Motion of charge in the C.M. frame.

alent to zero, and the Lagrangians for free particles can thus be taken strictly invariant. Projective representations reduce to true representations so that the ten generators on the Hilbert space, taking into account (4.115)-(4.118) and (4.15) are given by:

$$H = i\hbar\frac{\partial}{\partial t}, \quad \boldsymbol{P} = \frac{\hbar}{i}\nabla, \quad \boldsymbol{K} = \boldsymbol{r}\frac{i\hbar}{c^2}\frac{\partial}{\partial t} - t\frac{\hbar}{i}\nabla - \frac{1}{c^2}\boldsymbol{Z}\times\boldsymbol{u}, \tag{4.119}$$

$$\boldsymbol{J} = \boldsymbol{r}\times\frac{\hbar}{i}\nabla + \boldsymbol{Z}, \tag{4.120}$$

where the angular momentum operator $\boldsymbol{Z}$ is the differential operator

$$\boldsymbol{Z} = \boldsymbol{u}\times\frac{\hbar}{i}\nabla_u + \frac{\hbar}{2i}\left\{\nabla_\rho + \boldsymbol{\rho}\times\nabla_\rho + \boldsymbol{\rho}(\boldsymbol{\rho}\cdot\nabla_\rho)\right\} = \boldsymbol{u}\times\boldsymbol{U} + \boldsymbol{W}, \tag{4.121}$$

and where the differential operators $\nabla_u$ and $\nabla_\rho$ are the corresponding gradient operators with respect to the $\boldsymbol{u}$ and $\boldsymbol{\rho}$ variables as in the Galilei case. As we shall see later, operator $\boldsymbol{Z}$ is not a constant of the motion even for the free particle, but it is the equivalent to Dirac's spin operator.

To obtain the complete commuting set of observables we start with the Casimir invariant operator, or Klein-Gordon operator

$$KG \equiv H^2 - c^2\boldsymbol{P}^2 - m^2c^4 = 0. \tag{4.122}$$

In the above representation, $H$ and $\boldsymbol{P}$ only differentiate the wave function with respect to time $t$ and position $\boldsymbol{r}$, respectively. Since the spin operator $\boldsymbol{Z}$ operates only on the velocity and orientation variables, it commutes with the Klein-Gordon operator (4.122). Thus, we can find simultaneous eigenfunctions of (4.122), $Z^2$, and $Z_3$. This allows us to try solutions in separate variables so that the wave function can be written

$$\Phi(t, \boldsymbol{r}, \boldsymbol{u}, \boldsymbol{\rho}) = \sum_i \psi_i(t, \boldsymbol{r})\chi_i(\boldsymbol{u}, \boldsymbol{\rho}), \tag{4.123}$$

where $\psi_i(t, \boldsymbol{r})$ are the space-time components and the $\chi_i(\boldsymbol{u}, \boldsymbol{\rho})$ represent the internal spin structure. Consequently

$$(H^2 - c^2\boldsymbol{P}^2 - m^2c^4)\,\psi_i(t, \boldsymbol{r}) = 0, \tag{4.124}$$

*i.e.*, space-time components satisfy the Klein-Gordon equation, while the internal structure part satisfies

$$Z^2\chi_i(\boldsymbol{u}, \boldsymbol{\rho}) = s(s+1)\hbar^2\chi_i(\boldsymbol{u}, \boldsymbol{\rho}), \tag{4.125}$$

$$Z_3\chi_i(\boldsymbol{u}, \boldsymbol{\rho}) = m_s\hbar\chi_i(\boldsymbol{u}, \boldsymbol{\rho}). \tag{4.126}$$

Eigenfunctions of the above type have been found in Section 3., in particular we are interested in solutions that give rise to spin 1/2 particles.

For spin 1/2 particles, if we take first for simplicity eigenfunctions $\chi(\boldsymbol{\rho})$ of $Z^2$ with eigenvalue 1/2, then since the total spin has to be 1/2, the orbital zitterbewegung part $\boldsymbol{Y} = \boldsymbol{u} \times \boldsymbol{U}$ can only contribute with spherical harmonics of value $y = 0$ and $y = 1$.

For $y = 0$, the spin 1/2 functions $\chi_i(\boldsymbol{\rho})$ are linear combinations of the four $\Phi$ functions (4.85)-(4.88) and in the case $y = 1$ they are linear combinations of the four $\Psi_i$ of (4.92)-(4.95), such that the factor function in front of the spherical harmonics is 1 because for this model $u = c$ is a constant. It turns out that the Hilbert space that describes the internal structure of this particle is isomorphic to the four-dimensional Hilbert space $\mathbb{C}^4$.

If we have two arbitrary directions in space characterized by the unit vectors $\boldsymbol{u}$ and $\boldsymbol{v}$ respectively, and $S_{\boldsymbol{u}}$ and $S_{\boldsymbol{v}}$ are the corresponding spin projections $S_{\boldsymbol{u}} = \boldsymbol{u} \cdot \boldsymbol{S}$ and $S_{\boldsymbol{v}} = \boldsymbol{v} \cdot \boldsymbol{S}$, then $S_{-\boldsymbol{u}} = -S_{\boldsymbol{u}}$, and $[S_{\boldsymbol{u}}, S_{\boldsymbol{v}}] = i\hbar S_{\boldsymbol{u}\times\boldsymbol{v}}$. In the case of the anomalous commutation relations of operators $T_i$, we have for instance for the spin projections $[T_1, T_2] = -i\hbar T_3$, suggesting that $\boldsymbol{e}_1 \times \boldsymbol{e}_2 = -\boldsymbol{e}_3$ and thus $\boldsymbol{e}_i$ vectors linked to the body behave in the quantum case as a left-handed system. In this case $\boldsymbol{e}_i$ vectors are not arbitrary vectors in space, but rather vectors linked to the rotating body and thus they are not compatible observables, in the sense that any measurement to determine, say the components of $\boldsymbol{e}_i$, will produce some interaction with the body that will mask the measurement of the others. We shall use this interpretation of a left-handed system later.

Operators $S_i$ and $T_i$ have the matrix representation obtained before which is just

$$\boldsymbol{S} \equiv \boldsymbol{W} = \frac{\hbar}{2}\begin{pmatrix} \boldsymbol{\sigma} & 0 \\ 0 & \boldsymbol{\sigma} \end{pmatrix}, \tag{4.127}$$

$$T_1 = \frac{\hbar}{2}\begin{pmatrix} 0 & \mathbb{I} \\ \mathbb{I} & 0 \end{pmatrix}, \quad T_2 = \frac{\hbar}{2}\begin{pmatrix} 0 & i\mathbb{I} \\ -i\mathbb{I} & 0 \end{pmatrix}, \quad T_3 = \frac{\hbar}{2}\begin{pmatrix} \mathbb{I} & 0 \\ 0 & -\mathbb{I} \end{pmatrix}, \tag{4.128}$$

where we represent by $\boldsymbol{\sigma}$ Pauli matrices and $\mathbb{I}$ is the $2 \times 2$ unit matrix.

Similarly, the matrix elements of the nine components of the unit vectors $(\boldsymbol{e}_i)_j$, $i, j = 1, 2, 3$ give rise to the two alternative sets of representations depending on whether the zitterbewegung contribution is $y = 0$ or $y = 1$. In the first case we get

$$\boldsymbol{e}_1 = \frac{1}{3}\begin{pmatrix} 0 & \boldsymbol{\sigma} \\ \boldsymbol{\sigma} & 0 \end{pmatrix}, \ \boldsymbol{e}_2 = \frac{1}{3}\begin{pmatrix} 0 & i\boldsymbol{\sigma} \\ -i\boldsymbol{\sigma} & 0 \end{pmatrix}, \ \boldsymbol{e}_3 = \frac{1}{3}\begin{pmatrix} \boldsymbol{\sigma} & 0 \\ 0 & -\boldsymbol{\sigma} \end{pmatrix}, \tag{4.129}$$

while in the $y = 1$ case the representation is

$$\boldsymbol{e}_1 = -\frac{1}{9}\begin{pmatrix} 0 & \boldsymbol{\sigma} \\ \boldsymbol{\sigma} & 0 \end{pmatrix}, \boldsymbol{e}_2 = -\frac{1}{9}\begin{pmatrix} 0 & i\boldsymbol{\sigma} \\ -i\boldsymbol{\sigma} & 0 \end{pmatrix}, \boldsymbol{e}_3 = -\frac{1}{9}\begin{pmatrix} \boldsymbol{\sigma} & 0 \\ 0 & -\boldsymbol{\sigma} \end{pmatrix}. \tag{4.130}$$

It must be remarked that the components of the observables $\boldsymbol{e}_i$ are not compatible in general, because they are represented by non-commuting operators.

We finally write the wave function for spin 1/2 particles in the following form:

$$\Phi(t, \boldsymbol{r}, \boldsymbol{u}, \boldsymbol{\alpha}) = \sum_{i=1}^{i=4} \psi_i(t, \boldsymbol{r})\Phi_i(\alpha, \theta, \phi), \tag{4.131}$$

for $y = 0$ or in the case $y = 1$ by

$$\Phi(t, \boldsymbol{r}, \boldsymbol{u}, \boldsymbol{\alpha}) = \sum_{i=1}^{i=4} \psi_i(t, \boldsymbol{r})\Psi_i(\widetilde{\theta}, \widetilde{\phi}; \alpha, \theta, \phi). \tag{4.132}$$

Then, once the $\Phi_i$ or $\Psi_j$ functions that describe the internal structure are identified with the four orthogonal unit vectors of the internal Hilbert space $\mathbb{C}^4$, the wave function becomes a four-component space-time wave function, and the six spin components $S_i$ and $T_j$ and the nine vector components $(\boldsymbol{e}_i)_j$, together the $4 \times 4$ unit matrix, completely exhaust the 16 linearly independent $4 \times 4$ hermitian matrices. They form a vector basis of Dirac's algebra, such that any other translation invariant internal observable that describes internal structure, for instance internal velocity and acceleration, angular velocity, etc., must necessarily be expressed as a real linear combination of the mentioned 16 hermitian matrices. We shall see in Sec. 4.4 that the internal orientation completely characterizes its internal structure.

From now on we shall represent by $\boldsymbol{S}$ the spin operator $\boldsymbol{S} = \boldsymbol{u}\times\boldsymbol{U}+\boldsymbol{W}$ which, as seen in (4.96) and (4.127), coincides with the usual matrix representation of Dirac's spin operator.

If we consider the expression of the kinematical momentum for $u = c$ particles

$$\boldsymbol{K} = \frac{H}{c^2}\boldsymbol{r} - t\boldsymbol{P} - \frac{1}{c^2}\boldsymbol{S}\times\boldsymbol{u}$$

and we take the time derivative of this expression followed by the scalar product with $\boldsymbol{u}$, it leads to the Poincaré invariant operator (Dirac's operator):

$$D \equiv H - \boldsymbol{P}\cdot\boldsymbol{u} - \frac{1}{c^2}(\frac{d\boldsymbol{u}}{dt}\times\boldsymbol{u})\cdot\boldsymbol{S} = 0, \qquad (4.133)$$

where we have used the fact that $\boldsymbol{S}$ is not in general a constant observable.

When Dirac's operator $D$ acts on a general wave function, we know that $H$ and $\boldsymbol{P}$ have the differential representation given by (4.44) and the spin the differential representation (4.59), or the equivalent matrix representation (4.127), but we do not know how to represent the action of the velocity $\boldsymbol{u}$ and the $(d\boldsymbol{u}/dt)\times\boldsymbol{u}$ observable. However, we know that for this particle $\boldsymbol{u}$ and $d\boldsymbol{u}/dt$ are orthogonal vectors and together with vector $\boldsymbol{u}\times d\boldsymbol{u}/dt$ they form an orthogonal right-handed system, and in the center of mass frame the particle describes a circle of radius $R_0 = \hbar/2mc$ for spin 1/2 particles in the plane spanned by $\boldsymbol{u}$ and $d\boldsymbol{u}/dt$ (see Fig.4.1).

Let us consider first the case $y = 0$, where the zitterbewegung part does not contribute to the total spin. Since $\boldsymbol{u}$ and $d\boldsymbol{u}/dt$ are translation invariant observables they will be elements of Dirac's algebra, and it turns out that we can relate these three vectors with the orthogonal left-handed system formed by vectors $\boldsymbol{e}_1$, $\boldsymbol{e}_2$ and $\boldsymbol{e}_3$ with representation (4.129). Then, as shown in part $(a)$ of Figure 4.2, we have $\boldsymbol{u} = a\boldsymbol{e}_1$ and $d\boldsymbol{u}/dt\times\boldsymbol{u} = b\boldsymbol{e}_3$, where $a$ and $b$ are constant positive real numbers. Then the third term in Dirac's operator is $-(b/c^2)\boldsymbol{e}_3\cdot\boldsymbol{S} = -(b/c^2)T_3$, and (4.133) operator becomes

$$D \equiv H - a\boldsymbol{P}\cdot\boldsymbol{e}_1 - \frac{b}{c^2}T_3 = 0. \qquad (4.134)$$

If we make the identification with the orthonormal system of part $(b)$ of Figure 4.2, the relation of the above observables is opposite to the previous one but now with the coefficients $-a$ and $-b$, respectively, *i.e.*, we get

$$D = H + a\boldsymbol{P}\cdot e_1 + \frac{b}{c^2}T_3 = 0. \tag{4.135}$$

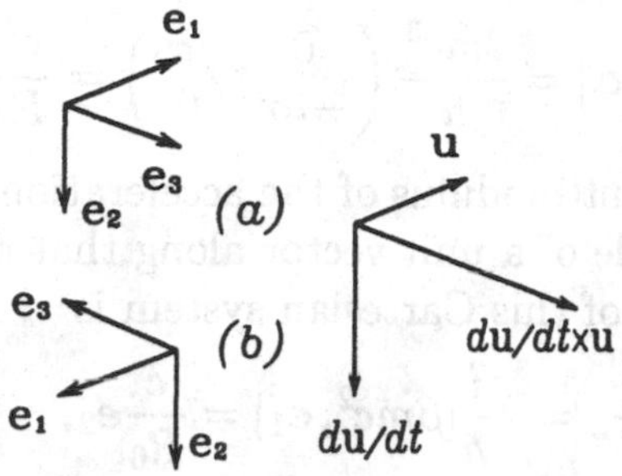


*Figure 4.2.* Orientation in the Pauli-Dirac representation.

Multiplying (4.135) by (4.134) we obtain

$$H^2 - \frac{a^2}{9}\boldsymbol{P}^2\mathbb{I} - \frac{b^2\hbar^2}{4c^4}\mathbb{I} = 0, \tag{4.136}$$

which is an algebraic relation between $H^2$ and $P^2$, and identification of this expression with the Klein-Gordon operator (4.122) leads to $a = 3c$ and $b = 2mc^4/\hbar = c^3/R_0$ and we obtain Dirac's operator:

$$H - c\boldsymbol{P}\cdot\boldsymbol{\alpha} - \beta mc^2 = 0, \tag{4.137}$$

where Dirac's matrices $\boldsymbol{\alpha}$ and $\beta$ are represented by

$$\boldsymbol{\alpha} = \begin{pmatrix} 0 & \boldsymbol{\sigma} \\ \boldsymbol{\sigma} & 0 \end{pmatrix}, \quad \beta = \begin{pmatrix} \mathbb{I} & 0 \\ 0 & -\mathbb{I} \end{pmatrix}, \tag{4.138}$$

and thus Dirac's gamma matrices are

$$\gamma^0 \equiv \beta = \begin{pmatrix} \mathbb{I} & 0 \\ 0 & -\mathbb{I} \end{pmatrix}, \quad \boldsymbol{\gamma} \equiv \gamma^0\boldsymbol{\alpha} = \begin{pmatrix} 0 & \boldsymbol{\sigma} \\ -\boldsymbol{\sigma} & 0 \end{pmatrix}, \tag{4.139}$$

*i.e.*, Pauli-Dirac representation, where $3e_1$ plays the role of a unit vector in the direction of the velocity.

This representation is compatible with vector $\boldsymbol{du}/dt$ lying along the third vector $\boldsymbol{e}_2$. In fact, in the center of mass frame and in the Heisenberg representation, Dirac's Hamiltonian reduces to $H = \beta mc^2$, and the time derivative of any observable $A$ is obtained as

$$\frac{dA}{dt} = \frac{i}{\hbar}[H, A] + \frac{\partial A}{\partial t}, \tag{4.140}$$

such that for the velocity operator $\boldsymbol{u} = c\boldsymbol{\alpha}$,

$$\frac{d\boldsymbol{u}}{dt} = \frac{i}{\hbar}[mc^2\beta, c\boldsymbol{\alpha}] = \frac{2mc^3}{\hbar}\begin{pmatrix} 0 & i\boldsymbol{\sigma} \\ -i\boldsymbol{\sigma} & 0 \end{pmatrix} = \frac{c^2}{R_0}3\boldsymbol{e}_2, \tag{4.141}$$

$c^2/R_0$ being the constant modulus of the acceleration in this frame, and where $3\boldsymbol{e}_2$ plays the role of a unit vector along that direction.

The time derivative of this Cartesian system is

$$\frac{d\boldsymbol{e}_1}{dt} = \frac{i}{\hbar}[\beta mc^2, \boldsymbol{e}_1] = \frac{c}{R_0}\boldsymbol{e}_2, \tag{4.142}$$

$$\frac{d\boldsymbol{e}_2}{dt} = \frac{i}{\hbar}[\beta mc^2, \boldsymbol{e}_2] = -\frac{c}{R_0}\boldsymbol{e}_1, \tag{4.143}$$

$$\frac{d\boldsymbol{e}_3}{dt} = \frac{i}{\hbar}[\beta mc^2, \boldsymbol{e}_3] = 0, \tag{4.144}$$

since $\boldsymbol{e}_3$ is orthogonal to the trajectory plane and does not change, and where $c/R_0 = \omega$ is the angular velocity of the internal orbital motion. This time evolution of the observables $\boldsymbol{e}_i$ is the correct one if assumed to be a rotating left-handed system of vectors as shown in Figure 4.2-($a$).

In general

$$\frac{d\boldsymbol{S}}{dt} = \frac{i}{\hbar}[H, \boldsymbol{S}] = \frac{i}{\hbar}[c\boldsymbol{P}\cdot\boldsymbol{\alpha} + \beta mc^2, \boldsymbol{S}] = c\boldsymbol{P}\times\boldsymbol{\alpha} \equiv \boldsymbol{P}\times\boldsymbol{u},$$

is not a constant of the motion, but for the center of mass observer, this spin operator $\boldsymbol{u}\times\boldsymbol{U} + \boldsymbol{W}$ reduces to the equivalent of the classical spin of the particle $\boldsymbol{S}$ and is constant in this frame:

$$\frac{d\boldsymbol{S}}{dt} = \frac{i}{\hbar}[\beta mc^2, \boldsymbol{S}] = 0. \tag{4.145}$$

Only the $T_3$ spin component on the body axis remains constant while the other two $T_1$ and $T_2$ change because of the rotation of the corresponding axis,

$$\frac{dT_1}{dt} = \frac{i}{\hbar}[\beta mc^2, T_1] = \frac{c}{R_0}T_2, \tag{4.146}$$

$$\frac{dT_2}{dt} = \frac{i}{\hbar}[\beta mc^2, T_2] = -\frac{c}{R_0}T_1, \tag{4.147}$$

$$\frac{dT_3}{dt} = \frac{i}{\hbar}[\beta mc^2, T_3] = 0. \tag{4.148}$$

When analyzed from the point of view of an arbitrary observer, the classical motion is a helix of elliptic cross section and the acceleration is not of constant modulus $c^2/R_0$, and the spin operator $\boldsymbol{S}$ is no longer a constant of the motion, because it is the total angular momentum $\boldsymbol{J} = \boldsymbol{r} \times \boldsymbol{P} + \boldsymbol{S}$ that is conserved.

Identification of the internal variables with different real linear combinations of the $\boldsymbol{e}_i$ matrices lead to different equivalent representations of Dirac's matrices, and thus to different expressions of Dirac's equation.

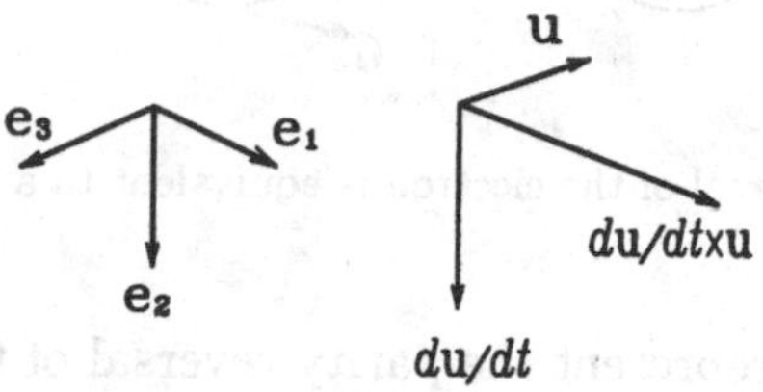


*Figure 4.3.* Orientation in the Weyl representation.

For instance if we make the identification suggested by Figure 4.3, $\boldsymbol{u} = -a\boldsymbol{e}_3$ and the observable $d\boldsymbol{u}/dt \times \boldsymbol{u} = b\boldsymbol{e}_1$ with positive constants $a$ and $b$, we obtain by the same method

$$\beta = \begin{pmatrix} 0 & \mathbb{I} \\ \mathbb{I} & 0 \end{pmatrix}, \qquad \boldsymbol{\alpha} = \begin{pmatrix} -\boldsymbol{\sigma} & 0 \\ 0 & \boldsymbol{\sigma} \end{pmatrix}, \tag{4.149}$$

and thus gamma matrices

$$\gamma^0 \equiv \beta = \begin{pmatrix} 0 & \mathbb{I} \\ \mathbb{I} & 0 \end{pmatrix}, \qquad \boldsymbol{\gamma} \equiv \gamma^0 \boldsymbol{\alpha} = \begin{pmatrix} 0 & \boldsymbol{\sigma} \\ -\boldsymbol{\sigma} & 0 \end{pmatrix}, \tag{4.150}$$

*i.e.*, Weyl's representation.

When we compare both representations, we see that Weyl's representation is obtained from Pauli-Dirac representation if we rotate the body frame $\pi/2$ around $\boldsymbol{e}_2$ axis. Then the corresponding rotation operator

$$R(\pi/2, \boldsymbol{e}_2) = \exp(\frac{i}{\hbar}\frac{\pi}{2}\boldsymbol{e}_2 \cdot \boldsymbol{S}) = \exp(\frac{i}{\hbar}\frac{\pi}{2}T_2) = \frac{1}{\sqrt{2}}\begin{pmatrix} \mathbb{I} & -\mathbb{I} \\ \mathbb{I} & \mathbb{I} \end{pmatrix}.$$

We can check that $R\gamma_{PD}^{\mu}R^{\dagger} = \gamma_W^{\mu}$, where $\gamma_{PD}^{\mu}$ and $\gamma_W^{\mu}$ are gamma matrices in the Pauli-Dirac and Weyl representation, respectively.

We can similarly obtain Dirac's equation in the case of zitterbewegung $y = 1$, by using the set of matrices (4.130) instead of (4.129), because they are multiples of each other and only some intermediate constant factor will change.

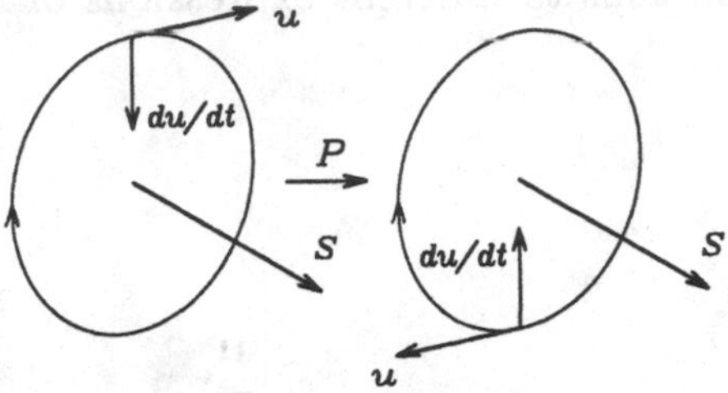


*Figure 4.4.* Space reversal of the electron is equivalent to a rotation of value $\pi$ along $\boldsymbol{S}$.

In Figure 4.4 we represent the parity reversal of the plausible description of the electron as given by this model of luxon which is circling around the center of mass at the velocity $c$ and that under $\mathbb{P}$ and in the center of mass frame it changes according to

$$\mathbb{P}: \{\boldsymbol{r} \to -\boldsymbol{r}, \boldsymbol{u} \to -\boldsymbol{u}, d\boldsymbol{u}/dt \to -d\boldsymbol{u}/dt, \boldsymbol{S} \to \boldsymbol{S}, H \to H\}.$$

In the Pauli-Dirac representation as we see in Figure 4.2, this amounts to a rotation of value $\pi$ around axis $\boldsymbol{e}_3$ and thus

$$\mathbb{P} \equiv R(\pi, \boldsymbol{e}_3) = \exp(i\pi \boldsymbol{e}_3 \cdot \boldsymbol{S}/\hbar) = \exp(i\pi T_3/\hbar) = i\gamma_0,$$

which is one of the possible representations of the parity operator $\pm\gamma_0$ or $\pm i\gamma_0$. In Weyl's representation this is a rotation of value $\pi$ around $\boldsymbol{e}_1$ which gives again $\mathbb{P} \equiv i\gamma_0$.

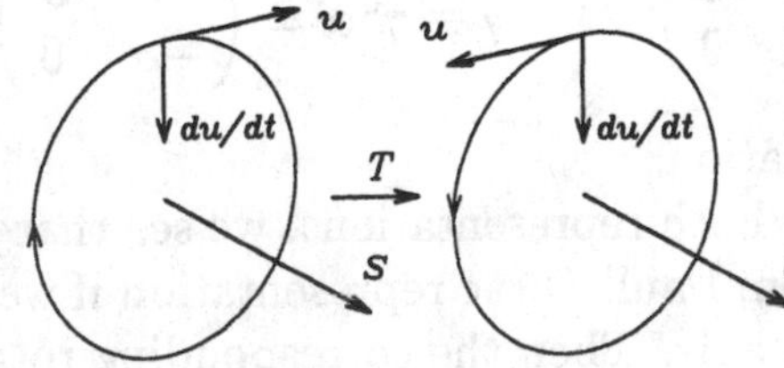


*Figure 4.5.* Time reversal of the electron produces a particle of negative energy.

In Figure 4.5 we represent its time reversal also in the center of mass frame

$$\mathbb{T} : \{\boldsymbol{r} \rightarrow \boldsymbol{r}, \boldsymbol{u} \rightarrow -\boldsymbol{u}, d\boldsymbol{u}/dt \rightarrow d\boldsymbol{u}/dt, \boldsymbol{S} \rightarrow \boldsymbol{S}, H \rightarrow -H\},$$

but this corresponds to a particle of negative energy, and thus $m < 0$, such that the relative orientation of spin, velocity and position, given by equation (3.168) agrees with the motion depicted in this figure.

## 4.4 DIRAC'S ALGEBRA

The three spatial spin components $S_i$, the three spin projections on the body frame $T_j$ and the nine components of the body frame $(\boldsymbol{e}_i)_j$, $i, j = 1, 2, 3$, whose matrix representations are given in the $y = 0$ case in (4.129) or in (4.130) in the $y = 1$ case, together with the $4 \times 4$ unit matrix $\mathbb{I}$, form a set of 16 linearly independent hermitian matrices. They are a linear basis of Dirac's algebra, and satisfy the following commutation relations:

$$[S_i, S_j] = i\hbar\epsilon_{ijk}S_k, \qquad [T_i, T_j] = -i\hbar\epsilon_{ijk}T_k, \qquad [S_i, T_j] = 0, \quad (4.151)$$

$$[S_i, (\boldsymbol{e}_j)_k] = i\hbar\epsilon_{ikr}(\boldsymbol{e}_j)_r, \qquad [T_i, (\boldsymbol{e}_j)_k] = -i\hbar\epsilon_{ijr}(\boldsymbol{e}_r)_k, \qquad (4.152)$$

and the scaled $3\boldsymbol{e}_i$ vectors in the $y = 0$ case

$$[(3\boldsymbol{e}_i)_k, (3\boldsymbol{e}_j)_l] = \frac{4i}{\hbar}\left(\delta_{ij}\epsilon_{klr}S_r - \delta_{kl}\epsilon_{ijr}T_r\right), \qquad (4.153)$$

showing that the $\boldsymbol{e}_i$ operators transform like vectors under rotations but they are not commuting observables. In the case $y = 1$, the scaled $-9\boldsymbol{e}_i$, satisfy the same relations.

If we fix the pair of indexes $i$, and $j$, then the set of four operators $S^2$, $S_i$, $T_j$ and $(\boldsymbol{e}_j)_i$ form a complete commuting set. In fact, the wave functions $\Phi_i$, $i = 1, \ldots, 4$, given before (4.76)-(4.79), are simultaneous eigenfunctions of $S^2$, $S_3$, $T_3$ and $(\boldsymbol{e}_3)_3$ with eigenvalues $s = 1/2$ and for $s_3$, $t_3$, and $e_{33}$ are the following ones:

$$\Phi_1 = |1/2, 1/2, 1/3>, \qquad \Phi_2 = |-1/2, 1/2, -1/3>, \qquad (4.154)$$

$$\Phi_3 = |1/2, -1/2, -1/3>, \qquad \Phi_4 = |-1/2, -1/2, 1/3>, \qquad (4.155)$$

and similarly for the $\Psi_j$ spinors of (4.92)-(4.95)

$$\Psi_1 = |1/2, 1/2, -1/9>, \qquad \Psi_2 = |-1/2, 1/2, 1/9>, \qquad (4.156)$$

$$\Psi_3 = |1/2, -1/2, 1/9>, \qquad \Psi_4 = |-1/2, -1/2, -1/9>. \qquad (4.157)$$

The basic observables satisfy the following anticommutation relations:

$$\{S_i, S_j\} = \{T_i, T_j\} = \frac{\hbar^2}{2}\,\delta_{ij}\mathbb{I}, \tag{4.158}$$

$$\{S_i, T_j\} = \frac{\hbar^2}{2}\,(3e_j)_i, \tag{4.159}$$

$$\{S_i, (3e_j)_k\} = 2\,\delta_{ik}T_j, \qquad \{T_i, (3e_j)_k\} = 2\,\delta_{ij}S_k, \tag{4.160}$$

$$\{(e_i)_j, (e_k)_l\} = \frac{2}{9}\,\delta_{ik}\delta_{jl}\mathbb{I} + \frac{2}{3}\epsilon_{ikr}\epsilon_{jls}(e_r)_s. \tag{4.161}$$

If we define the dimensionless normalized matrices:

$$a_{ij} = 3(e_i)_j, \; (\text{or } a_{ij} = -9(e_i)_j), \qquad s_i = \frac{2}{\hbar}S_i, \qquad t_i = \frac{2}{\hbar}T_i, \tag{4.162}$$

together with the $4 \times 4$ unit matrix $\mathbb{I}$, they form a set of 16 matrices $\Gamma_\lambda$, $\lambda = 1, \ldots, 16$ that are hermitian, unitary, linearly independent and of unit determinant. They are the orthonormal basis of the corresponding Dirac's Clifford algebra.

The set of 64 unitary matrices of determinant $+1$, $\pm\Gamma_\lambda$, $\pm i\Gamma_\lambda$, $\lambda = 1, \ldots, 16$ form a finite subgroup of $SU(4)$. Its composition law can be obtained from:

$$\begin{aligned}
a_{ij}\,a_{kl} &= \delta_{ik}\delta_{jl}\mathbb{I} + i\delta_{ik}\epsilon_{jlr}\,s_r - i\delta_{jl}\epsilon_{ikr}\,t_r + \epsilon_{ikr}\epsilon_{jls}\,a_{rs}, && (4.163)\\
a_{ij}\,s_k &= i\epsilon_{jkl}\,a_{il} + \delta_{jk}\,t_i, && (4.164)\\
a_{ij}\,t_k &= -i\epsilon_{ikl}\,a_{lj} + \delta_{ik}\,s_j, && (4.165)\\
s_i\,a_{jk} &= i\epsilon_{ikl}\,a_{jl} + \delta_{ik}t_j, && (4.166)\\
s_i\,s_j &= i\epsilon_{ijk}\,s_k + \delta_{ij}\mathbb{I}, && (4.167)\\
s_i\,t_j &= t_j\,s_i = a_{ji}, && (4.168)\\
t_i\,a_{jk} &= -i\epsilon_{ijl}\,a_{lk} + \delta_{ij}s_k, && (4.169)\\
t_i\,t_j &= -i\epsilon_{ijk}\,t_k + \delta_{ij}\mathbb{I}, && (4.170)
\end{aligned}$$

and similarly we can use these expressions to derive the commutation and anticommutation relations (4.151-4.161).

Dirac's algebra is generated by the four Dirac gamma matrices $\gamma^\mu$, $\mu = 0, 1, 2, 3$ that satisfy the anticommutation relations

$$\{\gamma^\mu, \gamma^\nu\} = 2\eta^{\mu\nu}\mathbb{I}, \tag{4.171}$$

$\eta^{\mu\nu}$ being Minkowski's metric tensor.

Similarly it can be generated by the following four observables, for instance: $S_1$, $S_2$, $T_1$ and $T_2$. In fact by (4.167) and (4.170) we obtain $S_3$ and $T_3$ respectively and by (4.168), the remaining elements.

Classically, the internal orientation of an electron is characterized by the knowledge of the components of the body frame $(\boldsymbol{e}_i)_j$, $i,j = 1,2,3$ that altogether constitute an orthogonal matrix. To completely characterize in a unique way this orthogonal matrix we need at least four of these components. In the quantum version, the knowledge of four $(\boldsymbol{e}_i)_j$ matrices and by making use of (4.163)-(4.170), allows us to recover the remaining elements of the complete Dirac algebra. It is in this sense that *internal orientation* of the electron completely characterizes its internal structure. Dirac's algebra of translation invariant observables of the electron can be generated by the orientation operators.

## 4.5 PHOTON QUANTIZATION

The kinematical space of the photon is the seven-dimensional manifold spanned by the variables $(t, \boldsymbol{r}, \boldsymbol{\rho})$, with the constraint $u = c$, so that the wave function of the photon will be a squared integrable function $\Phi(t, \boldsymbol{r}, \boldsymbol{\rho})$. On this manifold, the generators of the Poincaré group will be expressed as:

$$H = i\hbar\frac{\partial}{\partial t}, \quad \boldsymbol{P} = \frac{\hbar}{i}\nabla, \quad \boldsymbol{K} = \boldsymbol{r}\frac{i\hbar}{c^2}\frac{\partial}{\partial t} - t\frac{\hbar}{i}\nabla, \quad \boldsymbol{J} = \boldsymbol{r} \times \frac{\hbar}{i}\nabla + \boldsymbol{S},$$

where the spin operator $\boldsymbol{S}$ is given as in the case of the electron by (4.59). The two functionally invariant operators in this realization are:

$$(H/c)^2 - \boldsymbol{P}^2 \quad \text{and} \quad \boldsymbol{S}^2,$$

and because the photon is a massless particle of spin 1, when acting on the wave function they satisfy,

$$\Box\Phi(t, \boldsymbol{r}, \boldsymbol{\rho}) = 0, \tag{4.172}$$
$$S^2\Phi(t, \boldsymbol{r}, \boldsymbol{\rho}) = 1(1+1)\hbar^2\Phi(t, \boldsymbol{r}, \boldsymbol{\rho}), \tag{4.173}$$

where $\Box$ is the second-order d'Alambertian operator. We can try solutions in separate variables in the form $\Phi(t, \boldsymbol{r}, \boldsymbol{\rho}) = \sum \psi_j(t, \boldsymbol{r})\chi_j(\boldsymbol{\rho})$ in such a way that

$$\Box\psi_j(t, \boldsymbol{r}) = 0, \tag{4.174}$$
$$S^2\chi_j(\boldsymbol{\rho}) = 1(1+1)\hbar^2\chi_j(\boldsymbol{\rho}). \tag{4.175}$$

The space-time part satisfies the Klein-Gordon equation without the mass term, and the angular part is a function of spin 1, depending only on the orientation variables. According to the Peter-Weyl theorem mentioned before, [13] the most general spin 1 function will be a linear combination of the nine spin 1 spinors, corresponding to the nine matrix components of the $3 \times 3$ representation of $SU(2)$.

These nine functions are classified in the ket notation $|s_3, t_3>$, according to the eigenvalues of the spin operators $S_3$ and $T_3 = \boldsymbol{S} \cdot \boldsymbol{e}_3$, for all combinations of their eigenvalues $s_3$ and $t_3$ in $-1, 0, 1$. Vector $\boldsymbol{e}_3$ is one of the unit vectors linked to the particle to describe its orientation. However since the photon spin is not transversal, if we choose the laboratory axis $OZ$, along the photon trajectory, and also the unit vector $\boldsymbol{e}_3$ along that direction, then, only wave functions with the third spin component $S_3 = T_3 = \pm 1$ contribute, and therefore we are left with only two spinors, $|1,1> \equiv D^{(1)}_{1,1}(\rho)$ and $|-1,-1> \equiv D^{(1)}_{-1,-1}(\rho)$. They are given up to a normalization factor by $D^{(1)}_{1,1}(\rho) \simeq \Phi_1^2$ and $D^{(1)}_{-1,-1}(\rho) \simeq \Phi_4^2$, respectively, in terms of the $\Phi_i$ described in (4.76)-(4.79). If we choose $\boldsymbol{e}_3$ in the opposite direction, then $S_3$ and $T_3$ have opposite eigenvalues and consequently only $|1,-1> \simeq \Phi_3^2$ and $|-1,1> \simeq \Phi_2^2$ spinors contribute to the wave function.

If the state of the photon is of energy $\hbar\omega$ and linear momentum $\hbar\boldsymbol{k}$, then the most general function will be of the form

$$e^{-\hbar(\omega t - \boldsymbol{k}\cdot\boldsymbol{r})}\begin{pmatrix}\alpha\\ \beta\end{pmatrix},$$

where the two complex numbers $\alpha$ and $\beta$ represent the corresponding linear combination of the two basic states

$$|1,1> \equiv \begin{pmatrix}1\\ 0\end{pmatrix}, \quad |-1,-1> \equiv \begin{pmatrix}0\\ 1\end{pmatrix},$$

of helicity $+1$ and $-1$ respectively. Pure photon states of energy $\hbar\omega$ span a two-dimensional Hilbert space, such that the corresponding mixture of states gives rise to the so-called Poincaré sphere.

In this representation, the spin operators are represented by

$$S_3 = T_3 = \hbar\begin{pmatrix}1 & 0\\ 0 & -1\end{pmatrix}, \quad S_1 = S_2 = T_1 = T_2 = 0.$$

## 4.6 QUANTIZATION OF TACHYONS

The kinematical space of the general tachyon is the seven-dimensional manifold spanned by the variables $(t, \boldsymbol{r}, \boldsymbol{u})$, without orientation variables, and with the constraint $u > c$, so that the wave function of a tachyon will be a squared integrable function $\Phi(t, \boldsymbol{r}, \boldsymbol{u})$. The generators of the

Poincaré group will be expressed as:

$$H = i\hbar\frac{\partial}{\partial t}, \quad \boldsymbol{P} = \frac{\hbar}{i}\nabla, \quad \boldsymbol{K} = \boldsymbol{r}\,\frac{i\hbar}{c^2}\frac{\partial}{\partial t} - t\frac{\hbar}{i}\nabla - \boldsymbol{U} + (\boldsymbol{u}\cdot\boldsymbol{U})\boldsymbol{u}/c^2,$$

$$\boldsymbol{J} = \boldsymbol{r}\times\frac{\hbar}{i}\nabla + \boldsymbol{Z},$$

where the spin operator $\boldsymbol{Z}$ takes the form

$$\boldsymbol{Z} \equiv \boldsymbol{u}\times\boldsymbol{U} = \boldsymbol{u}\times\frac{\hbar}{i}\nabla_{\boldsymbol{u}},$$

and $\boldsymbol{U} = -i\hbar\nabla_{\boldsymbol{u}}$. Eigenvalues of $Z^2$ are spherical harmonics $Y^l_m(\theta,\phi)$ on the unit sphere, and therefore integer spin wave functions. Spin 1/2 particles can never have an internal tachyonic zitterbewegung. The remaining analysis is equivalent to the quantization of the previous models.

## Notes


1 R.P. Feynman and A.R. Hibbs, *Quantum Mechanics and Path Integrals*, MacGraw Hill, NY (1965), p. 36.

2 E.P. Wigner, *Group theory and its application to the quantum mechanics of atomic spectra*, Acad. Press, NY (1959).

3 V. Bargmann, *J. Math. Phys.* **5**, 862 (1964).

4 V. Bargmann, *Ann. Math.* **59**, 1 (1954).

5 J. M. Levy-Leblond, *Comm. Math. Phys.* **6**, 286 (1967).

6 E. Inönü and E.P. Wigner, *Nuovo Cimento* **9**, 705 (1952).

7 F. Bopp and R. Haag, *Z. Naturforschg.* **5a**, 644 (1950).

8 L.C. Biedenharn and J.D. Louck, *Angular Momentum in Quantum Physics. Theory and Application*, Cambridge U. P., Cambridge, England (1989).

9 A.R. Edmonds, *Angular Momentum in Quantum Mechanics*, Princeton U. P., Princeton NJ (1957).

10 N. Ja. Vilenkin, *Fonctions spéciales et Théorie de la représentation des groups*, Dunod, Paris (1969).

11 A.O. Barut and R. Raczka, *Theory of group representations and applications*, PWN, Warszawa (1980).

12 N. Ja. Vilenkin, *Fonctions spéciales et Théorie de la représentation des groupes*, Dunod, Paris (1969), p. 39.
A.O. Barut and R. Raczka, *Theory of group representations and applications*, PWN-Polish Scientific Publishers, Warszawa (1980), p. 174.
F. Peter and H. Weyl, *Math. Ann.* **7**, 735 (1927).

13 see ref. 10 and 11.

# Chapter 5

# OTHER SPINNING PARTICLE MODELS

In this chapter we shall review other models of classical elementary spinning particles, either relativistic or non-relativistic, that can be found in the literature. We shall discuss them in connection with the formalism we have developed in previous chapters to show the scope of the different proposals. Details of the different models are briefly sketched, paying attention mainly to the dynamical equations and the spin and dipole structure they provide. Whenever a Lagrangian is supplied we discuss the interaction terms, and in general, we have tried to write the different variables and observables with the same notation as in our treatment. For a more thorough analysis of these models the reader is referred to the original publications.

These models can be classified in many ways. We have collected them in two main sections according to the structure of their kinematical statements. One set is formed by those models whose kinematics has a certain group theoretical framework, while the other is restricted to phenomenological models or models which are defined without reference to any invariance principle, although their dynamics are subsequently restricted to fulfill the special relativity principle.

## 1. GROUP THEORETICAL MODELS

### 1.1 HANSON AND REGGE SPINNING TOP

The starting point is similar to the one we considered in Chapter 3 as a relativistic Bradyon. [1] At every instant of some arbitrary evolution parameter $\tau$, the particle is characterized by an element of the Poincaré group $g(\tau) \equiv (x^\mu(\tau), \Lambda^\mu{}_\nu(\tau))$. They are called *'Lagrangian coordinates'* of the system to the space-time position $x^\mu(\tau)$, and a tetrad or Lorentz matrix $\Lambda^\mu{}_\nu(\tau)$, which, at any instant $\tau$, satisfies the usual relation

$$\Lambda^\mu{}_\nu(\tau)\Lambda^{\nu\sigma}(\tau) = \eta^{\mu\sigma},$$

where $\eta^{\mu\sigma}$ is Minkowski's metric tensor. They state that these variables do not represent the independent degrees of freedom. Nevertheless, they consider that in $\Lambda^\mu{}_\nu$ there are spurious degrees of freedom so that some constraints will be established later in order that the only relevant independent variables will be rotation variables. We understand that these Lagrangian coordinates are the equivalent to the kinematical variables of our formalism, with some final constraints to eliminate the undesirable variables.

Since

$$\dot{\Lambda}^\mu{}_\nu(\tau)\Lambda^{\nu\sigma}(\tau) + \Lambda^\mu{}_\nu(\tau)\dot{\Lambda}^{\nu\sigma}(\tau) = 0,$$

the antisymmetric quantities

$$\Omega^{\mu\nu}(\tau) = \dot{\Lambda}^\mu{}_\alpha(\tau)\Lambda^{\alpha\nu}(\tau) = -\Omega^{\nu\mu}(\tau)$$

are the relativistic generalization of the three-dimensional angular velocity. However, the $\Omega^{\mu\nu}$ variables depend, by construction, on the velocity and orientation variables and also on the angular velocity and acceleration. It is claimed that this dependence on the acceleration will be suppressed later.

Then the basic statement is that the general free Lagrangian for describing a relativistic top will be an explicit function of the four-velocity $u^\mu \equiv \dot{x}^\mu$ and of this generalized angular velocity $\Omega^{\mu\nu}$, but independent of $x^\mu$ and $\Lambda^\mu{}_\nu$. There are some additional constraints to be defined properly, so that in addition to the four-velocity, the only relevant variables will be the angular variables and angular velocity. To establish a proper dynamical system of only six degrees of freedom, three for the position of a point, and the other three for some suitable angular variables to describe its orientation in space, we see that we need severe constraints to destroy the dependence on the acceleration variables. In the non-relativistic case this is simpler, as the example of the rigid body shows, but in the relativistic case the Lorentz boosts do not form a subgroup of the Poincaré group and the task is more difficult. Once acceleration variables are eliminated, the spin part of the system will not show zitterbewegung and therefore the dipole properties related to it have to be introduced in some *ad hoc* manner.

With these magnitudes the following four independent Poincaré invariant quantities can be formed:

$$a_1 = u^\mu u_\mu \equiv u^2, \quad a_2 = \Omega^{\mu\nu}\Omega_{\mu\nu} \equiv \Omega\cdot\Omega, \quad a_3 = u_\mu\Omega^{\mu\lambda}\Omega_{\lambda\nu}u^\nu \equiv u\Omega\Omega u,$$

and

$$a_4 = \frac{1}{16}\left(\Omega^{\mu\nu}\Omega^*_{\mu\nu}\right)^2 \equiv \frac{1}{16}\left(\Omega\cdot\Omega^*\right)^2,$$

where

$$\Omega^*_{\mu\nu} = \frac{1}{2}\epsilon_{\mu\nu\lambda\rho}\Omega^{\lambda\rho},$$

is the dual tensor with $\epsilon^{0123} = +1$, so that the free Lagrangian will be an explicit function of only these invariants $L_0(a_1, a_2, a_3, a_4)$.

Invariants $a_1$ and $a_2$ are homogeneous functions of second order in the derivatives of the Lagrangian coordinates while $a_3$ and $a_4$ are of fourth order. An additional requirement to ensure the arbitrary nature of the evolution parameter $\tau$ is that $L_0$ will be a homogeneous function of first order in these derivatives and thus Euler's theorem on homogeneous functions implies that

$$L_0 = 2(a_1L_1 + a_2L_2) + 4(a_3L_3 + a_4L_4) \tag{5.1}$$

where each $L_i = \partial L_0/\partial a_i$. Let us call $v_i$ the derivatives of the Lagrangian coordinates. Then $L_0 = v_i\partial L_0/\partial v_i$, while for $a_1$ and $a_2$, $v_i\partial a_1/\partial v_i = 2a_1$ holds and similarly $v_i\partial a_3/\partial v_i = 4a_3$ for $a_3$ and $a_4$ and we get the above result (5.1).

The canonical conjugate momenta of the Lagrangian coordinates (not of the independent degrees of freedom) are defined as

$$P^\mu = -\frac{\partial L_0}{\partial u_\mu}, \quad S^{\mu\nu} = -\frac{\partial L_0}{\partial \Omega_{\mu\nu}} = -S^{\nu\mu}.$$

Since it is assumed that $L_0$ does not depend explicitly on $x^\mu$ and $\Lambda^\mu{}_\nu$, then, before using any constraint, the following equations are considered as the dynamical equations of the system:

$$\dot{P}^\mu = 0, \tag{5.2}$$

$$\dot{S}^{\mu\nu} + \Omega^{\mu\lambda}S_\lambda{}^\nu - S^{\mu\lambda}\Omega_\lambda{}^\nu = 0. \tag{5.3}$$

Equation (5.2) is the energy-momentum conservation and equation (5.3) is the relativistic generalization of the three-dimensional dynamical equation $d\boldsymbol{W}/dt = \boldsymbol{\omega}\times\boldsymbol{W}$, with $\boldsymbol{W} = \partial L/\partial\boldsymbol{\omega}$, because also in this case all the dependence on the orientation variables is through the variables $\Omega^{\mu\nu}$. Because the tensor $S^{\mu\nu}$ is antisymmetric we have in (5.2-5.3) a set of ten independent dynamical equations.

It is not straightforward, but after using some tensor identities, this last equation can be rewritten as

$$\dot{S}^{\mu\nu} + u^\mu P^\nu - u^\nu P^\mu = 0. \tag{5.4}$$

Therefore the following magnitudes are constants of the motion

$$P^{\mu}, \tag{5.5}$$

$$J^{\mu\nu} = S^{\mu\nu} + x^{\mu}P^{\nu} - x^{\nu}P^{\mu}, \tag{5.6}$$

and will be identified with the generators of the Poincaré group.

In addition to the above equations, and in order to restrict the independent degrees of freedom, it is assumed that the orthogonality between the spin and linear momentum holds, *i.e.*,

$$S^{\mu\nu}P_{\nu} = 0. \tag{5.7}$$

These represent four conditions and it turns out that, of the above dynamical equations (5.2) and (5.4), only six are really independent, thus justifying in this way that the system has six degrees of freedom. There are three 'angular' variables which in addition to the spatial variables $\boldsymbol{r}$, characterize this spinning object. This condition (5.7) is usually known as the magnetic condition because in the center of mass frame it implies that components $S^{0i} = 0$ and therefore, for the particle at rest, no electric dipole term should appear.

However equation (5.4) is not correct because the invariant magnitudes $a_i$, $i = 2, 3, 4$, also depend on $\dot{u}^{\mu}$. The Lagrangian depends on second-order derivatives and thus the right definition of the conserved four-momentum is

$$P^{\mu} = -\frac{\partial L_0}{\partial u_{\mu}} + \frac{d}{d\tau}\left(\frac{\partial L_0}{\partial \dot{u}_{\mu}}\right). \tag{5.8}$$

It is not clear how the four equations (5.7) are equivalent to withdrawing the dependence on $\dot{u}^{\mu}$.

The last equation (5.6) can identified with equation (3.90) that defines the constant $J^{\mu\nu}$ as in Chapter 3. But remember that the $P^{\mu}$ in the case of Bradyons is defined according to (5.8).

Through a rather lengthy calculation, which is out of the scope of the analysis we want to present here, the authors show that the free Lagrangian can be cast into the reduced form

$$L_0 = (a_1)^{1/2}\mathcal{L}(\xi, \eta, \theta).$$

Here $\xi = a_2/a_1$, $\eta = a_3/a_1^2$ and $\theta = a_4/a_1^2$. To reach this conclusion, the previous constraints and the conjecture that the possible invariants of the system, mass $m$ and spin $S$, will satisfy what is known as a Regge trajectory, is used. Therefore, there must exist a functional relationship of the form $m^2 - f(S^2) = 0$.

Further analysis for $\mathcal{L}$, in the case $a_3 = a_4 = 0$, leads to the form

$$\mathcal{L}(\xi) = -m + S(\xi/2)^{1/2},$$

where the two invariant parameters $m$ and $S$ are separately fixed.

Because

$$(a_1\xi)^{1/2} = (a_2)^{1/2} = \sqrt{\omega^2 - \alpha^2},$$

where $\alpha$ and $\omega$ are the variables defined in (3.63) and (3.64) respectively, this leads to the Lagrangian

$$L_0 = -mc\sqrt{c^2\dot{t}^2 - \dot{r}^2} + S\sqrt{\omega^2 - \alpha^2}.$$

This looks like the bradyonic Lagrangian equivalent to the non-relativistic one (2.139) after some constraints are used to withdraw the dependence on the acceleration variables contained in the variables $\alpha$ and $\omega$. The zitterbewegung in our model corresponds to circular motions in the center of mass frame. When parameter $S = 0$ the usual point particle Lagrangian is obtained.

Interaction with an external electromagnetic field is introduced by assuming a minimal coupling of the usual form

$$L = L_0 - eu^\mu A_\mu(x),$$

where $e$ represents the charge of the particle and $A_\mu(x)$ the external potentials. The magnetic moment of the spinning top is introduced by assuming a new coupling term in the Lagrangian of the form

$$-\frac{g}{2}\Omega_{\mu\nu}F^{\mu\nu},$$

where parameter $g$ represents the intensity of the dipole of the system. It is clear that this coupling is the coupling of an object with an anomalous dipole structure of value $g\Omega^{\mu\nu}$.

## 1.2 KIRILLOV-KOSTANT-SOURIAU MODEL

To our knowledge, Bacry [2] was the first to consider the phase space of a dynamical system as a homogeneous space of the kinematical group. In this way he translated Wigner's irreducibility condition on Hilbert space into the homogeneity of the phase space manifold. The model we present here should also be called the Bacry model, but the considerations that follows were generalized by Kirillov, Kostant and Souriau giving rise to the so-called KKS theorem.

There is a theorem due to Lie [3] which states that in a canonical realization of a group on a manifold of dimension $2n$, with generators

$X_\alpha$, $\alpha = 1,\ldots,r$, it is always possible to choose the generating functions in the following way: $2h$ functions $Q_i(X_\alpha)$, $P_i(X_\alpha)$, $i = 1,\ldots,h$, $\alpha = 1,\ldots,r$ and another $k = r-2h$ functions, $I_j(X_\alpha)$, $j = 1,\ldots,k$ such that they satisfy the Poisson brackets

$$\{Q_i, Q_j\} = \{P_i, P_j\} = \{Q_i, I_j\} = \{P_i, I_j\} = \{I_i, I_j\} = 0, \tag{5.9}$$

$$\{Q_i, P_j\} = \delta_{ij}. \tag{5.10}$$

The theorem is based on the idea that when we have a symmetry group the generating functions in the group realization satisfy

$$\{X_\alpha, X_\beta\} = c^\gamma_{\alpha\beta}\, X_\gamma + d_{\alpha\beta},$$

where $c^\gamma_{\alpha\beta}$ are the structure constants of the group and the $d_{\alpha\beta}$ are constant real numbers that in general cannot be eliminated by redefining the $X_\alpha$ generators and are related to the existence of nontrivial central extensions of the group, or in an equivalent way to the existence of nontrivial exponents of the group. Then any two functions of the $X_\alpha$, $F(X_\alpha)$ and $G(X_\beta)$ satisfy

$$\{F(X_\alpha), G(X_\beta)\} = ({c_{\alpha\beta}}^\gamma X_\gamma + d_{\alpha\beta})\frac{\partial F}{\partial X_\alpha}\frac{\partial G}{\partial X_\beta}.$$

By considering that the right-hand side takes the form stated in (5.9-5.10) we get a system of partial differential equations for the unknown functions $F$ and $G$, of which we choose particular solutions that fulfill the above requirement. If the group has $k$ functionally independent invariants, then we can obtain up to $k$ independent functions of the $I_j$ type and the remaining solutions can be grouped in terms of pairs of canonical conjugate variables, because the difference between the dimension of the group and the number of independent invariants, $r-k$, is an even number.

In this way the set of generating functions is decomposed into $h$ pairs of canonical conjugate variables $Q_i$, $P_i$ and another set of $r-2h$ invariant functions $I_j$.

As an example, let us consider the rotation group with the three generators $J_i$ that in a canonical realization satisfy

$$\{J_i, J_k\} = \epsilon_{ikl}\, J_l.$$

Since this group has only one Casimir operator, it is possible to transform the above relations in terms of three new functions $Q = \arctan(J_2/J_1)$, $P = J_3$, $I = J^2 = J_1^2 + J_2^2 + J_3^2$, such that the new generating functions satisfy

$$\{Q, I\} = \{P, I\} = 0, \quad \{Q, P\} = 1.$$

**In this example $Q$ is precisely the angle that the projection of angular momentum on the plane $XOY$ forms with the $OX$ axis and its canonical conjugate momentum is $P$, *i.e.*, the component of the angular momentum perpendicular to this plane. The invariant $I$ is the Casimir operator of the group.**

When restricted to the Galilei group the generating functions satisfy the commutation relations of a central extension of $\mathcal{G}$, which is an 11-parameter group with three functionally independent invariants, and therefore we get four pairs of conjugate generating functions. In the Poincaré case it is a ten-parameter group with two independent Casimir invariants and we similarly get an underlying eight-dimensional phase space.

For the Galilei group all the $d_{\alpha\beta}$ coefficients can be absorbed, except the ones that arise in the Poisson brackets of the $K_i$ and $P_j$ which becomes

$$\{K_i, P_j\} = m\delta_{ij}.$$

In this realization we can go from the set of 11 generators $H$, $P_j$, $K_j$, $J_j$, $j = 1,2,3$ and $I$, that represents the unit function, to another set with three generating functions that commute with the others and the remaining can be grouped in a set of four pairs of conjugate variablcs. If we define the functions

$$S_i = J_i - \frac{1}{m}\,\epsilon_{ijk} K_j P_k$$

they satisfy

$$\{S_i, S_j\} = \epsilon_{ijk} S_k.$$

We thus have that the three invariant functions are $I_1 = I$, the constant unit function, $I_2 = S^2$, the absolute value of spin and $I_3 = H - P^2/2m$, the internal energy. The four pairs of conjugate variables are

$$q_i = \frac{1}{m} K_i, \quad p_i = P_i, \; i = 1,2,3, \quad q_4 = \arctan(S_2/S_1), \; p_4 = S_3.$$

In the Poincaré case all $d_{\alpha\beta}$ coefficients vanish and therefore the two invariant functions can be chosen as

$$I_1 = P^\mu P_\mu = m^2 c^2, \qquad I_2 = -W^\mu W_\mu,$$

with $P^\mu \equiv (H/c, \boldsymbol{P})$ and Pauli-Lubanski four-vector is

$$W^\mu \equiv (\boldsymbol{P}\cdot\boldsymbol{J}, H\boldsymbol{J}/c - \boldsymbol{K}\times\boldsymbol{P}) = (\boldsymbol{P}\cdot\boldsymbol{S}, H\boldsymbol{S}/c).$$

The spin observable is defined through

$$\boldsymbol{S} = \boldsymbol{J} - \frac{c}{H}\boldsymbol{K}\times\boldsymbol{P}$$

and the set of four conjugate variables are

$$q_i = \frac{c}{H}K_i, \quad p_i = P_i, \ i = 1,2,3, \quad q_4 = \arctan(S_2/S_1), \ p_4 = S_3.$$

Bacry thus arrives at the conclusion that the most general elementary particle either relativistic or non-relativistic, is a system of four degrees of freedom. Three represent the position of the particle, the linear momentum being their conjugate variables, and the fourth is an angle $\alpha$ whose conjugate momentum is a spin component $S_\alpha$ such that the Cartesian spin components are expressed in terms of these two variables and of an invariant value $S$, the absolute value of the spin, in the form:

$$S_x = (S^2 - S_\alpha^2)^{1/2}\cos\alpha, \quad S_y = (S^2 - S_\alpha^2)^{1/2}\sin\alpha, \quad S_z = S_\alpha.$$

This result was generalized independently by Kirillov, Kostant and Souriau and is known as the KKS theorem. They show that the coadjoint action of any Lie group defines on its orbits a symplectic structure. [4]

But the phase space, although interpreted as the state space of classical mechanics, does not play the same role as the Hilbert space in quantum mechanics, at least as far as the dynamics is concerned. In quantum mechanics the dynamics is stated in terms of initial $|\psi_i>$ and final $|\psi_f>$ states, such that the probability amplitude for the dynamical process $|\psi_i> \to |\psi_f>$ is given by the corresponding matrix element of the scattering operator $<\psi_f|S|\psi_i>$. But both $|\psi_i>$ and $|\psi_f>$ are elements of the same Hilbert space that, at the same time it plays the role of the space that describes all the particle states, it also represents the kinematical space where the dynamics is running. However, if we fix a point in phase space, the dynamics is completely determined. There is one and only one trajectory passing through that point and therefore phase space can no longer describe a two-end point dynamics, showing a clear difference with the quantum statements. This is one of the reasons why we believe that it is the kinematical space, rather than phase space, that has to be a homogeneous space of the kinematical group in order to have a dynamical formalism closer to the quantum one.

## 1.3 BILOCAL MODEL

We have described in previous chapters elementary spinning particles with magnetic moment. It has been shown that this property is directly related to the zitterbewegung and therefore to the existence of a separation between the center of mass and center of charge. We can try to describe this feature by describing the particle as a dynamical system characterized by two different points. One is the center of mass $q$ and the other the relative position $\boldsymbol{k}$ of the charge with respect to the center of mass. These kind of systems can be called bilocal systems. We are going to produce an equivalent Lagrangian description of a particular model of a nonrelativistic spinning particle, but considered as a bilocal system.

Let us consider, for instance, the simple nonrelativistic model of particle with zitterbewegung but no rotation. One free Lagrangian is, for instance,

$$L = \frac{m}{2}\frac{\dot{\boldsymbol{r}}^2}{\dot{t}} - \frac{m}{2}\frac{\dot{\boldsymbol{u}}^2}{\dot{t}}. \tag{5.11}$$

After the definition $\boldsymbol{q} = \boldsymbol{r} - \boldsymbol{k}$, with $m\boldsymbol{k} \equiv \partial L/\partial \dot{\boldsymbol{u}}$ we arrive at the second-order dynamical equations for these variables:

$$\frac{d^2\boldsymbol{q}}{dt^2} = 0, \qquad \frac{d^2\boldsymbol{k}}{dt^2} + \omega^2\boldsymbol{k} = 0.$$

In the case of a minimal coupling $L_I = -e\phi(t,\boldsymbol{r})\dot{t} + e\boldsymbol{A}(t,\boldsymbol{r})\cdot\dot{\boldsymbol{r}}$, with $\boldsymbol{r}$ replaced by $\boldsymbol{q}+\boldsymbol{k}$, the dynamical equations become

$$\frac{d^2\boldsymbol{q}}{dt^2} = \frac{1}{m}\boldsymbol{F}, \qquad \frac{d^2\boldsymbol{k}}{dt^2} + \omega^2\boldsymbol{k} = -\frac{1}{m}\boldsymbol{F}, \tag{5.12}$$

where the external Lorentz force $\boldsymbol{F} = e(\boldsymbol{E} + \boldsymbol{u}\times\boldsymbol{B})$ is defined at point $\boldsymbol{r}$. Now the question is: Is it possible to obtain a Lagrangian description in terms of two position vectors, that gives rise to the above dynamical equations (5.12), while preserving the definition of elementary particle?

We can see that the Lagrangian $L = L_0 + L_I$ with

$$L_0 = \frac{m}{2}\frac{\dot{\boldsymbol{q}}^2}{\dot{t}} - \frac{m}{2}\frac{\dot{\boldsymbol{k}}^2}{\dot{t}} + \frac{m\omega^2}{2}\boldsymbol{k}^2\dot{t} \tag{5.13}$$

and

$$L_I = -e\phi(t,\boldsymbol{q}+\boldsymbol{k})\dot{t} + e\boldsymbol{A}(t,\boldsymbol{q}+\boldsymbol{k})\cdot(\dot{\boldsymbol{q}}+\dot{\boldsymbol{k}}), \tag{5.14}$$

does. This is a regular first order Lagrangian with kinematical variables $(t,\boldsymbol{q},\boldsymbol{k})$ which transform under the Galilei group in the form

$$\begin{aligned} t'(\tau) &= t(\tau) + b, \\ \boldsymbol{q}'(\tau) &= R(\boldsymbol{\mu})\boldsymbol{q}(\tau) + \boldsymbol{v}t(\tau) + \boldsymbol{a}, \\ \boldsymbol{k}'(\tau) &= R(\boldsymbol{\mu})\boldsymbol{k}(\tau). \end{aligned} \tag{5.15}$$

Variable $\boldsymbol{k}$ is invariant under translation and Galilei boosts, as corresponds to a relative position vector. Their $\tau$-derivatives transform as

$$\begin{aligned} \dot{t}'(\tau) &= \dot{t}(\tau), \\ \dot{\boldsymbol{q}}(\tau)' &= R(\boldsymbol{\mu})\dot{\boldsymbol{q}}(\tau) + \boldsymbol{v}\dot{t}(\tau), \\ \dot{\boldsymbol{k}}(\tau)' &= R(\boldsymbol{\mu})\dot{\boldsymbol{k}}(\tau). \end{aligned}$$

Nevertheless, if these variables represent the kinematical variables of a classical elementary particle, according to our general definition they have to belong to a homogeneous space of $\mathcal{G}$. Since $\boldsymbol{k}$ transforms under $\mathcal{G}$ as in (5.15), we shall need to restrict variable $\boldsymbol{k}$ to have a constant magnitude $R$. This implies that, to be consistent with the consideration of our system as an elementary particle, we shall restrict ourselves to internal circular motions. This amounts to the internal motion being of constant radius and therefore at a velocity of constant magnitude in the center of mass frame.

The first term of Lagrangian (5.13) is not invariant under $\mathcal{G}$ but has a gauge function

$$\alpha(g,x) \equiv m\left(\frac{1}{2}\boldsymbol{v}^2 t + \boldsymbol{v}\cdot R(\boldsymbol{\mu})\boldsymbol{q}\right),$$

that depends on parameter $m$ which is considered as the mass of the system.

The free Lagrangian can be written as

$$L_0 = T\dot{t} + \boldsymbol{Q}\cdot\dot{\boldsymbol{q}} + \boldsymbol{N}\cdot\dot{\boldsymbol{k}},$$

where

$$\begin{aligned}
T = \frac{\partial L_0}{\partial \dot{t}} &= -\frac{m}{2}\left(\frac{d\boldsymbol{q}}{dt}\right)^2 + \frac{m}{2}\left(\frac{d\boldsymbol{k}}{dt}\right)^2 + \frac{m\omega^2}{2}\boldsymbol{k}^2,\\
\boldsymbol{Q} = \frac{\partial L_0}{\partial \dot{\boldsymbol{q}}} &= m\frac{d\boldsymbol{q}}{dt},\\
\boldsymbol{N} = \frac{\partial L_0}{\partial \dot{\boldsymbol{k}}} &= -m\frac{d\boldsymbol{k}}{dt}.
\end{aligned}$$

The Noether constants of the motion for the free particle are now

$$\begin{aligned}
H &= -T,\\
\boldsymbol{P} &= \boldsymbol{Q},\\
\boldsymbol{K} &= m\boldsymbol{q} - \boldsymbol{P}t,\\
\boldsymbol{J} &= \boldsymbol{q}\times\boldsymbol{P} + \boldsymbol{k}\times\boldsymbol{N} = \boldsymbol{L} + \boldsymbol{S}.
\end{aligned}$$

We therefore see that the following observables

$$H = \frac{\boldsymbol{P}^2}{2m} + H_0, \quad H_0 = -\frac{m}{2}\left(\frac{d\boldsymbol{k}}{dt}\right)^2 - \frac{m\omega^2}{2}\boldsymbol{k}^2,$$

and

$$\boldsymbol{S} = -m\boldsymbol{k}\times\frac{d\boldsymbol{k}}{dt},$$

are the same as in the previous exposition of Chapter 2. Nevertheless, the constraint $k = R$ has also to be suplemented by the constraint that the zitterbewegung velocity takes a constant value. We shall consider that this value is $|dk/dt| = c$. Here parameter $c$ plays no role of a limit velocity but only means that the velocity of internal zitterbewegung is of constant value, which agrees with the considered relativistic model of the electron. This constraint also fixes the value of the parameter $\omega = c/R$.

To solve the corresponding variational problem we have to consider those constraints which are written in the first-order homogeneous form:

$$L_C = \lambda_1(\tau)(\boldsymbol{k}^2 - R^2)\dot{t} + \lambda_2(\tau)\left(\frac{\dot{\boldsymbol{k}}^2}{\dot{t}} - c^2\dot{t}\right), \tag{5.16}$$

where $\lambda_1$ and $\lambda_2$ are the two Lagrange multipliers. With these two new terms the free dynamical equations for $\boldsymbol{q}$ are unchanged and for $\boldsymbol{k}$ variables they become

$$m\frac{d^2\boldsymbol{k}}{dt^2} + m\omega^2\boldsymbol{k} - 2\lambda_2\frac{d^2\boldsymbol{k}}{dt^2} + 2\lambda_1\boldsymbol{k} = 0. \tag{5.17}$$

Total energy is a constant of the motion and is given by

$$H = \frac{m}{2}\left(\frac{d\boldsymbol{q}}{dt}\right)^2 - \frac{m}{2}\left(\frac{d\boldsymbol{k}}{dt}\right)^2 - \frac{m\omega^2}{2}\boldsymbol{k}^2 - \lambda_1(\boldsymbol{k}^2 - R^2) + \lambda_2 c^2 + \lambda_2\left(\frac{d\boldsymbol{k}}{dt}\right)^2.$$

By using the two constraints $\partial L/\partial\lambda_i = 0$, $i = 1, 2$, it reduces to

$$H = \frac{m}{2}\left(\frac{d\boldsymbol{q}}{dt}\right)^2 - \frac{mc^2}{2} - \frac{mc^2}{2} + 2\lambda_2 c^2.$$

This implies that $\lambda_2$ has to be a constant and if we consider that the internal energy reduces to $mc^2$, then $\lambda_2 = m$. With this condition, if dynamical equations (5.17) remain the same as the original ones, because a circular motion of angular frequency $\omega$ is but a particular case of an isotropic harmonic motion, then it necessarily implies that $\lambda_1$ is also a constant of value $\lambda_1 = -m\omega^2$. Finally, the spin of the system is $S = mcR$. Therefore, the free particle is characterized by the three parameters $m$, $S$ and $c$, mass, spin and internal zitterbewegung velocity or internal energy $mc^2$, respectively.

If we compare Lagrangian (5.13) with the trivial expansion of Lagrangian (5.11), after the replacement of $\dot{\boldsymbol{r}} = \dot{\boldsymbol{q}} + \dot{\boldsymbol{k}}$ and the definition of $m\boldsymbol{k} \equiv \partial L/\partial\dot{\boldsymbol{u}}$, the final Lagrangian is

$$L_1 = \frac{m}{2}\frac{\dot{\boldsymbol{q}}^2}{\dot{t}} + \frac{m}{2}\frac{\dot{\boldsymbol{k}}^2}{\dot{t}} - \frac{m\omega^2}{2}\boldsymbol{k}^2\dot{t} + m\frac{\dot{\boldsymbol{q}}\cdot\dot{\boldsymbol{k}}}{\dot{t}}. \tag{5.18}$$

The difference with $L_0$ is the addition of the last term and that the second and the third term have opposite sign. But, if we add $L_0$ and the constraints $L_C$ with the accepted values for Lagrange's multipliers, we obtain in fact $L_1$ without the additional last term.

# 2. NON-GROUP BASED MODELS

There are basically two kind of models. One kind tries to justify a spin dynamical equation of the form

$$\frac{d\boldsymbol{S}}{dt} = \boldsymbol{P} \times \boldsymbol{u},$$

similar to the one Dirac's spin operator satisfies. Others establish a torque-like dynamical equation for spin, which is a constant of the motion for the free particle.

## 2.1 SPHERICALLY SYMMETRIC RIGID BODY

This is perhaps the first model created to incorporate the idea of rotation to the point particle. It was considered as the natural model in the Uhlenbeck and Goudsmit proposal of the spinning electron. [5] Because of its simplicity there is a huge literature that describes many applications using this model, such that a comprehensive description of the references is out of the scope of this book. Let us mention the widely known book by Rohrlich [6] where many references to this subject can be found. To mention the present interest in this model, it has been recently used by Kiessling [7] to show that, by assuming a Lorentz force interaction, the energy, linear and angular momentum are jointly conserved if and only if spin is associated to the point particle, in the form, at least, of rotation of a stiff matter distribution.

It consists of a rigid sphere of mass $m$, charge $e$ and radius $R$ rotating with certain angular velocity $\omega$ such that this rotation gives account of the spin. The spin is of the form $S = I\omega$, where $I$ is the moment of inertia that is defined provided a suitable mass density distribution is assumed. There are models where the mass density is uniform and others in which some form factor is introduced. We have a variety of models going from simple spherical shells of radius $R$ to hard core objects with a delta function or a continuous distribution of mass that may even have an infinite range with a Gaussian behaviour. This distribution is introduced in some *ad hoc* manner to reach some specific goals and in general the radius is constrained with an upper bound $\omega R < c$, to avoid the possibility of having matter moving with the speed of light.

Once these parameters $m$, $I$ and $S$ are fixed, the dynamics is reduced to the description of the evolution of the center of mass $\boldsymbol{q} \equiv \boldsymbol{r}$ and the rotation of a body frame with angular velocity along the spin direction. From the kinematical viewoint it is a system of seven kinematical variables, time $t$, position $\boldsymbol{r}$ and orientation $\boldsymbol{\alpha}$ that describes the orientation of the body frame. It turns out that this model agrees with the nonrelativistic particle with kinematical space $\mathcal{G}/\{\mathbb{R}^3,+\}$, where $\{\mathbb{R}^3,+\}$ is the three-parameter subgroup of Galilei boosts. But in our kinematical approach it is not necessary to assume either size and shape or any mass distribution. The knowledge of the extended Galilei group invariants, mass $m$, spin $S$ and internal energy $H_0$ and the seven kinematical variables, is sufficient to conjecture some Lagrangians.

From the relativistic point of view the main problem lies in the rigidity of this system. If it is an extended object, the velocity of an arbitrary point is changing depending on the relative situation to the observer and therefore taking into account the Lorentz contraction factor, its distance to the nearest neighbouring points is changing in time so that we have no rigid body at all. In our kinematical formulation this model is forbidden because the Lorentz boosts, and therefore the manifold equivalent to the nonrelativistic subgroup $\{\mathbb{R}^3,+\}$ does not form a subgroup of the Poincaré group and it follows that no homogeneous space with that kinematical structure can be defined.

The center of mass dynamics is subjected to Newton-like equations in terms of the external forces and fields, and for the spin dynamics a torque equation is assumed provided some relationship between the magnetic moment and spin is defined. To achieve this a charge distribution is introduced and its rotation gives rise to the magnetic dipole moment with a suitable gyromagnetic ratio depending on the companion mass density distribution. If the charge density is spherically symmetric there is no electric dipole but it has in general electric multipole momenta. This produces additional complications. If the charge is distributed we must glue it to matter to keep from repulsive deformation. Then matter is under some stress that defines its internal structure. Poincaré's stress tensor [8] solves this problem; its difficulties have been extensively discussed elsewhere and it will be unnecessary to reproduce them here.

The spinless version of this model is known as the Abraham-Lorentz-Poincaré model. [9] When applied to describe the electron it is assumed that it is a sphere only in the rest frame, the mass is of electromagnetic origin, and that there is a self-interaction which gives account of a very important feature: the radiation reaction term or the energy loss by radiation of an accelerated particle. This very important subject of analyzing the radiation reaction of a spinning particle has not been dealt with in our kinematical approach, and it deserves a much more thorough analysis.

## 2.2 WEYSSENHOFF-RAABE MODEL

This is a field theoretical model in which the system is considered as a continuous spin fluid, such that the integration of the different densities to a very small volume, in the limit when the size of the volume vanishes, define the corresponding properties of a spinning particle. [10]

The field is characterized by a four-momentum density $g^\mu(x)$ and also with an angular momentum density $s^{\mu\nu}(x) = -s^{\nu\mu}(x)$. Four-vector $u^\mu(x)$ represents the velocity of the fluid at point $x$. The fluid considered is of a magnetic type, *i.e.*, the components $s^{0i}(x)$ of tensor $s^{\mu\nu}(x)$ vanish for the rest frame observer at point $x$, so that the space-space components $s^{ij}$ define a three-vector density $\boldsymbol{s}$, the spin density of the system. For an arbitrary observer, the electric components $s^{0i}(x)$ define another vector density $\boldsymbol{k} = \boldsymbol{u} \times \boldsymbol{s}/c$. This magnetic condition is contained in the covariant expression

$$s^{\mu\nu}u_\nu = 0. \tag{5.19}$$

The energy-momentum tensor of a spinless fluid is usually defined as $T^{\mu\nu} = \rho u^\mu u^\nu = T^{\nu\mu}$, where $\rho(x)$ is the mass density. For a fluid with spin $T^{\mu\nu}$ is no longer symmetric and is defined as

$$T^{\mu\nu} = g^\mu u^\nu. \tag{5.20}$$

Dynamical equations are constructed by considering local conservation of momentum-energy and also of total angular momentum density, which is defined by

$$j^{\mu\nu} = x^\mu g^\nu - x^\nu g^\mu + s^{\mu\nu}.$$

If the dynamics is described in terms of some arbitrary evolution parameter $\tau$, then dynamical equations are

$$D_\tau g^\mu = 0, \quad D_\tau j^{\mu\nu} = 0,$$

where $D_\tau$ is a $\tau$-differential operator such that when applied to any field density $f(x)$ it gives

$$D_\tau f = \partial_\mu(f u^\mu).$$

Then $D_\tau g^\mu = 0$ amounts to $\partial_\mu T^{\mu\nu} = 0$, which is the usual form of energy-momentum conservation. The second equation, after using $D_\tau g^\mu = 0$, and that $D_\tau x^\mu = u^\mu$, gives

$$D_\tau s^{\mu\nu} = T^{\mu\nu} - T^{\nu\mu} = g^\mu u^\nu - g^\nu u^\mu. \tag{5.21}$$

Therefore, the existence of an internal angular momentum is related to the asymmetry of $T^{\mu\nu}$ and from its definition with the non-collinearity between the density of linear momentum $g^\mu$ and $u^\mu$. The fluid is moving at velocities below $c$, so that $u^\mu$ is time-like and satisfies $u^\mu u_\mu = c^2$. The invariant magnitude

$$\rho = \frac{1}{c^2} u^\mu g_\mu,$$

defines a mass density of the fluid, such that after contraction of (5.21) with $u_\nu$ we obtain

$$g^\mu = \rho u^\mu + \frac{1}{c^2} u_\nu D_\tau s^{\mu\nu} = \rho u^\mu - \frac{1}{c^2} s^{\mu\nu} \dot{u}_\nu,$$

where the magnetic condition (5.19) and $\dot{u}_\nu \equiv D_\tau u_\nu$, have been used. We see that for spinless systems the energy-momentum density reduces to the usual one $g^\mu = \rho u^\mu$ and $T^{\mu\nu}$ becomes symmetric.

Now a particle is defined as the integration of the above densities over a region of space at some instant $\tau$. This integration region is as small as possible, to consider that $u^\mu$ and $\dot{u}^\mu$ are constants during the integration. This allows us to write

$$P^\mu = \int g^\mu dV, \quad S^{\mu\nu} = \int s^{\mu\nu} dV, \quad m = \int \rho dV, \tag{5.22}$$

and also

$$S^{\mu\nu} u_\nu = 0, \tag{5.23}$$

and dynamical equations for a free particle become

$$\dot{P}^\mu = 0, \quad \dot{S}^{\mu\nu} = P^\mu u^\nu - P^\nu u^\mu. \tag{5.24}$$

Because $P^\mu$ can also be written as

$$P^\mu = m u^\mu - \frac{1}{c^2} S^{\mu\nu} \dot{u}_\nu = m u^\mu + \frac{1}{c^2} \dot{S}^{\mu\nu} u_\nu, \tag{5.25}$$

with $m = u_\nu P^\nu / c^2$, we can also write dynamical equations for the velocity and the spin components. Using (5.24) and differentiation of (5.25), we get

$$\dot{S}^{\mu\nu} = \frac{1}{c^2} S^{\mu\sigma} \dot{u}_\sigma u^\nu - \frac{1}{c^2} S^{\nu\sigma} \dot{u}_\sigma u^\mu, \tag{5.26}$$

$$m \dot{u}^\mu - \frac{1}{c^2} S^{\mu\nu} \ddot{u}_\nu = 0, \tag{5.27}$$

because $\dot{S}^{\mu\nu} \dot{u}_\nu = 0$. They obtain, therefore, for a spinning particle of a magnetic type, third-order differential equations for the position of the particle.

These dynamical equations were first obtained by Frenkel [11] and also by Mathisson [12] as particular cases and in the context of General Relativity.

Solutions of the above dynamical equations in the non-relativistic case were obtained by Mathisson, in the above mentioned references, and give rise for the particle to a circular trajectory at constant velocity in the center of mass frame.

In the center of mass frame, equations (5.27) are written as

$$m\dot{\boldsymbol{u}} - \frac{1}{c^2}\boldsymbol{S} \times \ddot{\boldsymbol{u}} = 0, \tag{5.28}$$

where the spin $\boldsymbol{S}$ is a constant vector in this frame. If we choose the spin along the $OZ$ axis and we call $\omega = mc^2/S$, the above equations for the velocity components are

$$\dot{u}_x = -\frac{1}{\omega}\ddot{u}_y, \quad \dot{u}_y = \frac{1}{\omega}\ddot{u}_x,$$

so that

$$\frac{d^3u_x}{dt^3} + \omega^2\frac{du_x}{dt} = 0, \quad \frac{d^3u_y}{dt^3} + \omega^2\frac{du_y}{dt} = 0,$$

and therefore we obtain the fourth-order dynamical equations for the position of the point, similar to the ones we found for the non-relativistic spinning particle with spin of (anti)orbital nature (2.81). The motion is clearly circular because velocity and acceleration are orthogonal vectors and thus the motion is with a velocity of constant absolute value. In a certain sense the condition (5.23) plays the role of a first integral that reduces the order of the above fourth-order system to system (5.28), thus selecting only the circular one, among all the possible elliptic trajectories. For a spin 1/2 particle the frequency of the motion is the zitterbewegung frequency of the spinning electron, so that an arbitrary velocity implies an arbitrary radius, and we have no way to fix it.

The relativistic solution found by Weyssenhoff and Raabe also gives a circular motion of the particle in the center of mass frame. Contraction of (5.26) with $S_{\mu\nu}$, using the magnetic condition (5.23), leads to $\dot{S}^{\mu\nu}S_{\mu\nu} = 0$ and thus $S^{\mu\nu}S_{\mu\nu}$ =constant, that in the center of mass frame defines a constant spin vector $\boldsymbol{S}$. Since $P^0 = mu^0 = m\gamma(u)c$ is also constant, this implies that the absolute value of the velocity $\boldsymbol{u}$ is another constant of the motion so that dynamical equations in the center of mass frame $\boldsymbol{P} = 0$, become

$$m\gamma(u)\boldsymbol{u} - \frac{1}{c^2}\boldsymbol{S} \times \gamma(u)\boldsymbol{a} = 0,$$

with $\boldsymbol{a} = d\boldsymbol{u}/dt$. Because now $u$ is constant, the motion is circular in a plane orthogonal to the spin and with angular frequency

$$\omega = \frac{mc^2}{S}.$$

To compare this model with the ones we have presented before we have to restrict ourselves to relativistic models with velocities below $c$, *i.e.*, to the spinning Bradyons. Whether or not spin variables are dependent on other kinematical variables, we shall now discuss in the example worked out in Section 3.1 of Chapter 3.

In this example the equivalent to the $S^{\mu\nu}$ tensor is the translation invariant part of the kinematical momentum $\boldsymbol{K}$ and angular momentum $\boldsymbol{J}$. They are given in the general case by

$$\boldsymbol{K} = \frac{H}{c^2}\boldsymbol{r} - \boldsymbol{P}t - \boldsymbol{D}/c, \quad \boldsymbol{J} = \boldsymbol{r} \times \boldsymbol{P} + \boldsymbol{u} \times \boldsymbol{U} + \boldsymbol{W},$$

where

$$\boldsymbol{D}/c = \boldsymbol{U} - \frac{\boldsymbol{u} \cdot \boldsymbol{U}}{c^2}\boldsymbol{u} - \frac{\gamma(u)}{(1+\gamma(u))c^2}\boldsymbol{u} \times \boldsymbol{W}.$$

Both $\boldsymbol{K}$ and $\boldsymbol{J}$ are, respectively, the space-time part and the space-space part of the tensor

$$J^{\mu\nu} = x^{\mu}P^{\nu} - x^{\nu}P^{\mu} + S^{\mu\nu},$$

so that the $S^{i0}$ define the three-vector $\boldsymbol{D}/c$ and the $S^{ij}$ the so-called spin part $\boldsymbol{S}_0 = \boldsymbol{u} \times \boldsymbol{U} + \boldsymbol{W}$.

Since $\boldsymbol{P}$ and $\boldsymbol{J}$ are conserved quantities for a free particle, the equations

$$\frac{d\boldsymbol{P}}{dt} = 0, \quad \frac{d\boldsymbol{S}_0}{dt} = \boldsymbol{P} \times \boldsymbol{u},$$

are consistent with dynamical equations (5.24).

For a reference frame in which the particle is instantaneously at rest, $\boldsymbol{D}$ and $\boldsymbol{S}_0$ observables reduce to $\boldsymbol{U}$ and $\boldsymbol{W}$ respectively, so that both parts, the electric and magnetic terms, are non-vanishing. Even more, in the particular case in which no orientation variables are involved it is the magnetic term that vanishes in this rest frame. Vanishing of $\boldsymbol{U}$ means, basically, the vanishing of the acceleration. This is not a coherent statement because it contradicts the inertia principle even in the relativistic approach, because in special relativity if the particle is accelerated in some reference frame it is not possible to find any other inertial frame in which acceleration vanishes. If the magnetic properties of the particle are related to its current, it is clear that in the rest frame we have a static charge and therefore no current and therefore no magnetic effects.

We thus arrive at the conclusion that the magnetic condition (5.23) is too strong to be sustainable as a basic statement for a spinning particle, because it is necessary to have the existence of both instantaneous electric and magnetic dipole momenta.

Nevertheless, it is not necessary to claim any spin fluid dynamics to obtain the above equations. It can be described by assuming that the elementary particle has an energy $H$ and linear momentum $\boldsymbol{P}$, that are conserved in a free motion, and other constants of the motion $J^{\mu\nu}$ defined by

$$J^{\mu\nu} = x^\mu P^\nu - x^\nu P^\mu + S^{\mu\nu}.$$

The antisymmetric tensor $S^{\mu\nu}$ is considered as an intrinsic property which contains the spin structure of the particle. In our notation, the essential components of $J^{\mu\nu}$ define the vectors

$$c\boldsymbol{K} = \frac{H}{c}\boldsymbol{r} - c\boldsymbol{P}t - \boldsymbol{D}, \quad \boldsymbol{J} = \boldsymbol{r} \times \boldsymbol{P} + \boldsymbol{S}_0,$$

so that $\boldsymbol{D}$ and $\boldsymbol{S}_0$ are the essential components of tensor $S^{\mu\nu}$. The magnetic condition (5.23) amounts to considering that in the frame in which the velocity $\boldsymbol{u} = 0$, these magnitudes take the values $\widetilde{\boldsymbol{S}}_0$ and $\widetilde{\boldsymbol{D}} = 0$, so that in any arbitrary frame, according to (3.81,3.82), they take the values

$$\boldsymbol{D} = -\frac{\gamma(u)}{c}\boldsymbol{u} \times \widetilde{\boldsymbol{S}}_0, \quad \boldsymbol{S}_0 = \gamma(u)\widetilde{\boldsymbol{S}}_0 - \frac{\gamma(u)^2}{(1+\gamma(u))c}(\boldsymbol{u} \cdot \widetilde{\boldsymbol{S}}_0)\boldsymbol{u}.$$

Thus,

$$\boldsymbol{D} = -\frac{\boldsymbol{u}}{c} \times \boldsymbol{S}_0$$

so that the above observables take the form for a pure magnetic particle as

$$c\boldsymbol{K} = \frac{H}{c}\boldsymbol{r} - c\boldsymbol{P}t + \frac{\boldsymbol{u}}{c} \times \boldsymbol{S}_0, \quad \boldsymbol{J} = \boldsymbol{r} \times \boldsymbol{P} + \boldsymbol{S}_0, \tag{5.29}$$

where $\boldsymbol{S}_0$ is considered the spin of the system. Because $\boldsymbol{K}$ and $\boldsymbol{J}$, in addition to $H$ and $\boldsymbol{P}$, are constants of the motion for the free particle we get from here, taking the time derivative, the spin dynamical equation

$$\frac{d\boldsymbol{S}_0}{dt} = \boldsymbol{P} \times \boldsymbol{u},$$

and

$$\frac{H}{c}\boldsymbol{u} - c\boldsymbol{P} + \frac{1}{c}\frac{d\boldsymbol{u}}{dt} \times \boldsymbol{S}_0 + \frac{\boldsymbol{u}}{c} \times \frac{d\boldsymbol{S}_0}{dt} = 0.$$

If we substitute here the derivative of $\boldsymbol{S}_0$, we get the expression of the linear momentum

$$\boldsymbol{P} = \gamma(u)^2 \left\{ \frac{1}{c^2} (H - \boldsymbol{u} \cdot \boldsymbol{P}) \boldsymbol{u} + \frac{1}{c^2} \frac{d\boldsymbol{u}}{dt} \times \boldsymbol{S} \right\}.$$

The term $\gamma(u)(H - \boldsymbol{u} \cdot \boldsymbol{P})/c^2 \equiv P^\mu u_\mu / c^2 = m$ is the mass term of Weyssenhoff and Raabe, assumed as a constant of the motion.

It is clear that the term $H - \boldsymbol{u} \cdot \boldsymbol{P}$, in the center of mass frame ($\boldsymbol{P} = 0$), takes the value $mc^2$ and the spin $\boldsymbol{S}_0$ is a constant vector $\boldsymbol{S}$ and thus we get in this frame the dynamical equation

$$mc^2 \boldsymbol{u} + \frac{d\boldsymbol{u}}{dt} \times \boldsymbol{S} = 0, \tag{5.30}$$

and because the velocity is orthogonal to the acceleration, it represents a circular motion with constant velocity $u$ in a plane orthogonal to the spin. It must be remarked here that the relative orientation of the different magnitudes correspond to an intrinsic spin oriented in the (anti)orbital direction corresponding to the particle motion as depicted in Figure 5.1.

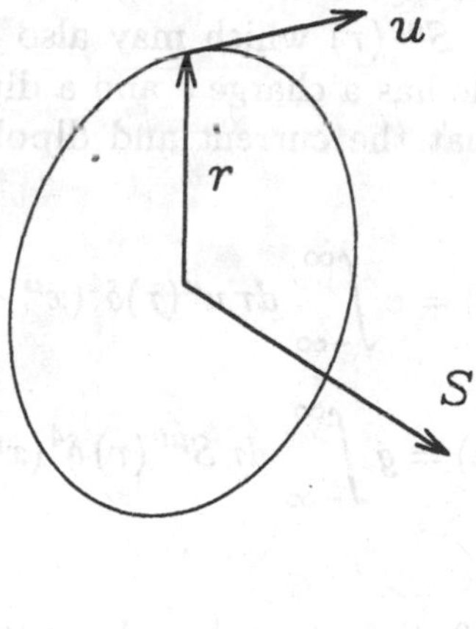


*Figure 5.1.* Motion of the particle in the C.M. frame.

To our knowledge no discussion of this relative orientation has been produced in this early literature about elementary spinning particles, but it agrees with the zitterbewegung part of all the models, either relativistic or non-relativistic, we have found in this monograph.

In the general case $\dot{\boldsymbol{P}} = 0$ for a free particle and this leads to a third-order dynamical equation for the position of the particle. But, if considered under our definition of the center of mass observer, *i.e.*, not only $\boldsymbol{P}=0$ but also $\boldsymbol{K}=0$, we get from (5.29)

$$mc^2\boldsymbol{r}+\boldsymbol{u}\times\boldsymbol{S}=0,$$

which also corresponds to a circular motion with constant velocity and radius $R_0=Su/mc^2$ and angular velocity $\omega=mc^2/S$.

We see that the magnetic condition leads to a particular case of our spinning Bradyon, but the idea is that this condition should be replaced by $S^{\mu\nu}P_\nu=0$, *i.e.*, a pure magnetic dipole in the center of mass frame, as is considered in the Hanson and Regge model (5.7). This amounts to $\boldsymbol{D}\cdot\boldsymbol{P}=0$ and $H\boldsymbol{D}+\boldsymbol{S}\times\boldsymbol{P}=0$, so that $\boldsymbol{D}=0$ in the center of mass frame. But if $\boldsymbol{D}=0$ this produces no separation between the center of mass and center of charge and thus no zitterbewegung.

## 2.3 BHABHA-CORBEN MODEL

Bhabha and Corben assume a model with a charge and dipole distribution with an anomalous coupling to the electromagnetic field. [13] The particle is a point with coordinates $z^\mu(\tau)$ which are functions of the proper time $\tau$ measured from an arbitrary point on the word line of the particle. The velocity is $u^\mu\equiv\dot{z}^\mu$ and the spin is described by an antisymmetric tensor $S^{\mu\nu}(\tau)$ which may also be considered as a function of $\tau$. The particle has a charge $e$ and a dipole characterized by the parameter $g$, such that the current and dipole distribution are given, respectively, by

$$j^\mu(x)=e\int_{-\infty}^{\infty}d\tau\, u^\mu(\tau)\delta^4(x^\mu-z^\mu(\tau)), \tag{5.31}$$

$$M^{\mu\nu}(x)=g\int_{-\infty}^{\infty}d\tau\, S^{\mu\nu}(\tau)\,\delta^4(x^\mu-z^\mu(\tau)), \tag{5.32}$$

where

$$\delta^4(x^\mu-z^\mu(\tau))\equiv\delta(x^0-z^0(\tau))\delta(x^1-z^1(\tau))\delta(x^2-z^2(\tau))\delta(x^3-z^3(\tau)).$$

The four-velocity $u^\mu\equiv(\gamma(u)c,\gamma(u)\boldsymbol{u})$ satisfies at any instant

$$u^2\equiv u^\mu(\tau)u_\mu(\tau)=c^2. \tag{5.33}$$

In addition to this, in order that the constant $g$ should have a meaning as an intrinsic property of the particle, they demand the constraint

$$S^2\equiv S^{\mu\nu}(\tau)S_{\mu\nu}(\tau)=\text{constant}, \tag{5.34}$$

at every instant $\tau$, and also all the $\tau$-derivatives of these expressions (5.33-5.34) up to a finite degree.

Final dynamical equations are written in the form

$$m\dot{u}_\mu + \frac{dR_\mu}{d\tau} = eF_{\mu\nu}u^\nu - \frac{g}{2}S^{\rho\sigma}\partial_\mu F_{\rho\sigma} + T_\mu^{(\text{reac.})}, \tag{5.35}$$

$$I_1 U_{\lambda\mu} + I_2 W_{\lambda\mu} = gZ_{\lambda\mu} + D_{\lambda\mu}^{(\text{reac.})}. \tag{5.36}$$

The different terms above are:

$$R_\mu = I_1\dot{S}_{\mu\nu}u^\nu + \frac{I_2}{4}(\dot{S})^2 u_\mu + I_2 S_{\mu\nu}\ddot{S}^{\nu\sigma}u_\sigma - \frac{g}{2}(S\cdot F)u_\mu - g\,S_\mu{}^\sigma F_{\sigma\rho}u_\rho,$$

and the antisymmetric tensors

$$U_{\lambda\mu} = c^2\dot{S}_{\lambda\mu} + u_\lambda\dot{S}_{\mu\nu}u^\nu - u_\mu\dot{S}_{\lambda\nu}u^\nu,$$

$$W_{\lambda\mu} = S_\lambda{}^\rho\left(c^2\ddot{S}_{\rho\mu} - \ddot{S}_{\rho\nu}u^\nu u_\mu\right) - S_\mu{}^\rho\left(c^2\ddot{S}_{\rho\lambda} - \ddot{S}_{\rho\nu}u^\nu u_\lambda\right),$$

$$Z_{\lambda\mu} = S_\lambda{}^\rho\left(c^2F_{\rho\mu} - F_{\rho\nu}u^\nu u_\mu\right) - S_\mu{}^\rho\left(c^2F_{\rho\lambda} - F_{\rho\nu}u^\nu u_\lambda\right),$$

are expressed in terms of the well-defined parameters $m$, $e$ and $g$ and two constant parameters $I_1$ and $I_2$. $T_\mu^{(\text{reac.})}$ and $D_{\lambda\mu}^{(\text{reac.})}$ contain the contributions of the self-fields and the radiation reaction modification to the linear and angular momentum, respectively.

In addition to (5.34) the magnetic condition $S_{\mu\nu}u^\nu = 0$ is assumed. In the case of electrons, compatibility with the scattering of light by electrons according to the Klein-Nishina formula, [14] require for the above parameters the values $g = 0$, $I_1 = 1/c^2$ and $I_2 = 0$. Therefore systems with $g \neq 0$ have a nonvanishing anomalous magnetic moment.

They admit the model is suitable for describing the electromagnetic interaction of protons, neutrons and charged mesons, by assuming a coupling with the electromagnetic field of the form

$$j^\mu(x)A_\mu(x) + \frac{1}{2}M^{\mu\nu}(x)F_{\mu\nu}(x),$$

so that the presence of the spin is only responsible for the anomalous coupling.

A particular analysis of the above dynamical equations was carried out by Corben. [15] With the above choice of constants and neglecting the radiation reaction terms, dynamical equations (5.35)-(5.36) for the electron ($g = 0$) become

$$\frac{d}{d\tau}\left(mu_\mu + \frac{1}{c^2}\dot{S}_{\mu\nu}u^\nu\right) = eF_{\mu\nu}u^\nu, \tag{5.37}$$

$$c^2\dot{S}_{\lambda\mu} + u_\lambda\dot{S}_{\mu\nu}u^\nu - u_\mu\dot{S}_{\lambda\nu}u^\nu = 0. \tag{5.38}$$

The right-hand side of (5.37) is just the Lorentz force and this allows us to define the four-momentum as

$$P_\mu = mu_\mu + \frac{1}{c^2}\dot{S}_{\mu\nu}u^\nu = mu_\mu - \frac{1}{c^2}S_{\mu\nu}\dot{u}^\nu,$$

by using the magnetic condition. Then we can write after some manipulations

$$\begin{aligned} \dot{P}_\mu &= eF_{\mu\nu}u^\nu, \\ \dot{S}_{\mu\nu} &= u_\nu P_\mu - u_\mu P_\nu. \end{aligned}$$

In terms of the components of four-vector $P^\mu \equiv (H/c, \boldsymbol{P})$, and the space-space part of $S^{\mu\nu}$, $S^{ij} \equiv \boldsymbol{S}$, the time-space part $S^{i0} \equiv \boldsymbol{D} = -\boldsymbol{u} \times \boldsymbol{S}/c$ the above equations look rather simple:

$$\boldsymbol{P} = m\gamma(u)\boldsymbol{u} - \frac{\gamma(u)^2}{c^2}\boldsymbol{S} \times \frac{d\boldsymbol{u}}{dt}, \tag{5.39}$$

$$H = m\gamma(u)c^2 - \frac{\gamma(u)^2}{c^2}\boldsymbol{u} \cdot \left(\boldsymbol{S} \times \frac{d\boldsymbol{u}}{dt}\right), \tag{5.40}$$

$$\frac{d\boldsymbol{P}}{dt} = e(\boldsymbol{E} + \boldsymbol{u} \times \boldsymbol{B}), \quad \frac{dH}{dt} = e\boldsymbol{u} \cdot \boldsymbol{E}, \tag{5.41}$$

$$\frac{d\boldsymbol{S}}{dt} = \boldsymbol{P} \times \boldsymbol{u}, \tag{5.42}$$

$$\frac{d\boldsymbol{D}}{dt} = \frac{H}{c}\boldsymbol{u} - c\boldsymbol{P}. \tag{5.43}$$

We can check that $P^\mu u_\mu = mc^2$ but $P^\mu P_\mu \neq m^2c^2$, and therefore for a free electron $P^\mu \dot{u}_\mu = 0$. It is doubtful whether parameter $m$ can be interpreted as the true mass of the particle, but nevertheless it is directly related to it.

In the case $\boldsymbol{P} = 0$, the energy takes the value $H = mc^2/\gamma(u)$. Equation (5.42) implies that the spin $\boldsymbol{S}$ is constant in time and thus (5.39) leads for the velocity to the equation

$$\boldsymbol{u} = \frac{\gamma(u)}{mc^2}\boldsymbol{S} \times \frac{d\boldsymbol{u}}{dt}. \tag{5.44}$$

Velocity and acceleration are orthogonal vectors and this represents a circular motion at constant velocity in a plane orthogonal to the spin. It is the magnitude $m/\gamma(u)$ which plays the role of the mass of the particle and with this interpretation (5.44) exactly becomes the Weyssenhoff-Raabe zitterbewegung (5.30).

In Corben's words *'the center of mass of the particle is not at the position of the particle, remaining instead fixed at the center of the circle around which the particle is moving. In this sense, the classical equations reflect the zitterbewegung of the Dirac equation and reveal the physical origin of the magnetic moment'.* [16]

We can deduce from the above sentences that for Corben's interpretation it is the charge of the particle which is moving around the center of mass. This motion is the origin of the magnetic moment. Nevertheless, he does not give any expression of the magnetic moment in terms of this motion and even its relationship with the spin (5.32), for the anomalous case, is an *ad hoc* one.

By inspection of equation (5.44), we find again that the zitterbewegung of Corben's electron is exactly the same as in Figure 5.1 or as in any of the (anti)orbital motions we have obtained in previous models.

## 2.4 BARGMANN-MICHEL-TELEGDI MODEL

This is a point particle system [17] with position four-vector $x^\mu$, and with some additional internal magnetic structure. Four-velocity $v^\mu = dx^\mu/d\tau \equiv (\gamma(v)c, \gamma(v)\boldsymbol{v})$, such that in a homogeneous external electromagnetic field $F^{\mu\nu}$ the position of the point satisfies the dynamical equations

$$\frac{dv^\mu}{d\tau} = \frac{e}{m} F^{\mu\nu} v_\nu. \tag{5.45}$$

We use here the symbol $v^\mu$ for the four-velocity because it represents the center of mass velocity of the particle, to distinguish from the $u^\mu$ four-velocity which we have reserved for the center of charge velocity in all the models.

The equivalent to the magnetic condition (5.23), or better to (5.7), is to assume the existence of an intrinsic spin vector $\boldsymbol{S}$ in the center of mass frame giving rise only to a magnetic dipole moment without any further electric dipole structure.

The difference with the previous model is, instead of assuming that the spin internal structure is characterized by an antisymmetric tensor $S^{\mu\nu}$, that the particle has associated a four-vector $S^\mu$ that in the rest frame of the particle takes the form $S^\mu \equiv (0, \boldsymbol{S})$. Therefore, for any arbitrary frame, it satisfies the orthogonality condition

$$S^\mu v_\mu = 0. \tag{5.46}$$

In this model they assume that spin is a constant of the motion for the free particle and its dynamics, in the rest frame, is governed by the torque equation

$$\frac{d\boldsymbol{S}}{dt} = \boldsymbol{\mu} \times \boldsymbol{B} \equiv \frac{ge}{2m}\boldsymbol{S} \times \boldsymbol{B}. \tag{5.47}$$

The relation between the magnetic moment and the spin is expressed in terms of the dimensionless parameter $g$, the gyromagnetic ratio of the particle. For an arbitrary observer (5.46) implies that if $S^\mu \equiv (S^0, \boldsymbol{S})$, then $S^0 = \boldsymbol{S} \cdot \boldsymbol{v}$, so that $\dot{S}^0 = \dot{\boldsymbol{S}} \cdot \boldsymbol{v} + \boldsymbol{S} \cdot \dot{\boldsymbol{v}}$, and for the rest frame observer it reduces to

$$\frac{dS^0}{dt} = \boldsymbol{S} \cdot \frac{d\boldsymbol{v}}{dt}. \tag{5.48}$$

Bargmann-Michel-Telegdi's dynamical equation (BMT for short), is the covariant generalization of (5.47) and (5.48) to arbitray frames. They consider as the spin four-vector the scaled Pauli-Lubansky four-vector $W^\mu/mc \equiv (\boldsymbol{P} \cdot \boldsymbol{S}, H\boldsymbol{S}/c)/mc$, given in (3.14), which in the center of mass frame reduces to $(0, \boldsymbol{S})$. Equation (5.46) is equivalent to $W^\mu P_\mu = 0$, a relation that holds between the two four-vector operators that define the Casimir operators of the Poincaré group, provided the momentum is related with the center of mass velocity by $P_\mu = m\gamma(v)v_\mu$.

A covariant expression that can be constructed out of tensors $S^\mu$, $v^\mu$, $dv^\mu/d\tau$ and $F^{\mu\nu}$, preserving the homogeneity of the different terms, linear in the spin components and consistent with the orthogonality condition (5.46) and its $\tau$-derivative, is

$$\frac{dS^\mu}{d\tau} = \frac{ge}{2m}(F^{\mu\nu}S_\nu + (S_\sigma F^{\sigma\rho}v_\rho)v^\mu) - \frac{dv^\nu}{d\tau}S_\nu v^\mu, \tag{5.49}$$

where they take $c = 1$. Using dynamical equations (5.45) in the last term they finally get

$$\frac{dS^\mu}{d\tau} = \frac{e}{m}\left(\frac{g}{2}F^{\mu\nu}S_\nu + \left(\frac{g}{2} - 1\right)(S_\sigma F^{\sigma\rho}v_\rho)v^\mu\right), \tag{5.50}$$

showing clearly the $g - 2$ anomaly. For strict $g = 2$ particles it reduces to the same linear dynamical equations for the spin components as in (5.45).

The last term of (5.49), which is not an explicit function of parameter $g$, is of pure kinematical origin and gives rise to the Thomas precession of spin even if there is no magnetic moment present, provided the acceleration $dv^\nu/d\tau$ is different from zero. [18]

Bacry [19] points out that equation (5.45) where the only external force is the Lorentz force, is only valid for uniform external fields. Otherwise the presence of the magnetic moment will produce on the particle an additional force related to the local variations of the fields of the form

$$\frac{ge}{2m}(S^\mu v^\nu - S^\nu v^\mu)\partial^\sigma F_{\sigma\nu},$$

which has to be included on the right-hand side of (5.49).

In previous models, spin dynamics has been guided by the idea that Dirac's electron spin is no longer a constant of the motion even for the free particle case, but it satisfies the equation $d\boldsymbol{S}/dt = \boldsymbol{P} \times c\boldsymbol{\alpha} \equiv \boldsymbol{P} \times \boldsymbol{u}$, which has as a covariant generalization the dynamical equation $\dot{S}^{\mu\nu} = P^\mu u^\nu - P^\nu u^\mu$. In Dirac's theory $P_\mu \neq m u_\mu$ but it is related to some average value of $u_\mu$. It is difficult to reconcile this dynamical equation with the idea that a pure magnetic moment has a torque-like dynamical equation, except if we consider that Dirac's spin operator does not represent the same angular momentum observable as the BMT spin $\boldsymbol{S}$ does.

In our formalism the spin observable is related more to BMT spin than to Dirac's but the velocity $\boldsymbol{v}$ represents the center of mass velocity. Even for non-relativistic particles we have obtained torque equations of the above form, but including an additional electric dipole torque

$$\frac{d\boldsymbol{S}}{dt} = \boldsymbol{d} \times \boldsymbol{E} + \boldsymbol{\mu} \times \boldsymbol{B}, \tag{5.51}$$

in which the presence of the instantaneous electric dipole $\boldsymbol{d}$ is unavoidable for a general spinning particle.

These authors also suggest to include some electric dipole related to the spin by

$$\boldsymbol{d} = \frac{g' e}{2m} \boldsymbol{S},$$

where $g'$ is a phenomenological parameter. But in the models presented in this book, the possible electric dipole structure is not lying along the spin direction but is orthogonal to it.

## 2.5 BARUT-ZANGHI MODEL

This is a semiclassical model because in addition to space-time variables the description of the internal degrees of freedom is expressed in terms of Dirac's spinors. In this way, when quantizing the system, the internal variables will give rise to a spin 1/2 object. [20]

The classical system is characterized by the kinematical variables $x_\mu$, $\mu = 0, 1, 2, 3$, where $x^\mu \equiv (ct, \boldsymbol{r})$ are the time and the position of the charge, and the internal structure is described by means of a dimensionless four-component complex Dirac spinor $z \in \mathbb{C}^4$, with conjugate momentum $-i\bar{z}$, where $\bar{z} = z^\dagger \gamma^0$, being $z^\dagger$ the hermitian conjugate Dirac spinor. The dynamics is expressed in terms of some invariant evolution parameter $\tau$, with dimensions of time.

The particle is a system characterized by the Poincaré invariant Lagrangian

$$L = -\frac{\hbar i}{2}(\dot{\bar{z}}z - \bar{z}\dot{z}) + p_\mu(\dot{x}^\mu - c\bar{z}\gamma^\mu z) + eA_\mu(x)\bar{z}\gamma^\mu z. \tag{5.52}$$

If we consider as independent dynamical variables $x^\mu$, $z$ and $\bar{z}$, the $p^\mu$ play the role of Lagrange multipliers that establish the constraint $\dot{x}^\mu = c\bar{z}\gamma^\mu z$, *i.e.*, the four-velocity of the charge is related to Dirac's current.

A predecessor of this use of spinors in classical dynamics leading to the same dynamics is the work by Proca, [21] who also obtains a classical Lorentz force by asuming a minimal coupling interaction of the current associated with the spinor variables with the external electromagnetic potentials. Proca also analyzed the case in which classical particles move at the speed of light, [22] but in that case the model leads to a spin tensor $S^{\mu\nu}$ of vanishing value $S^{\mu\nu}S_{\mu\nu} = 0$. This is inconsistent with a pure magnetic particle, but agrees with our spinning electron in which in addition to the magnetic part it also has an electric dipole part related to the nonvanishing $S^{0i}$ components, which define the three-vector $\boldsymbol{S}\times\boldsymbol{u}/c$, and because $u = c$ we get the vanishing of the invariant $S^{\mu\nu}S_{\mu\nu} = 0$.

Dynamical equations obtained from (5.52) are

$$\hbar\dot{z} = -i\pi z, \quad \hbar\dot{\bar{z}} = i\bar{z}\pi, \quad \dot{\pi}_\mu = eF_{\mu\sigma}u^\sigma,$$

where

$$u_\sigma \equiv \dot{x}_\sigma = c\bar{z}\gamma_\sigma z, \quad \pi = \gamma^\mu\pi_\mu, \quad \pi_\mu = p_\mu - eA_\mu, \quad F_{\mu\nu} = \partial_\mu A_\nu - \partial_\nu A_\mu.$$

In the case of a free particle ($A_\mu = 0$), $p_\mu$ is constant and $p_\mu p^\mu = m^2c^2$. The Hamiltonian of the system is

$$H = p_\mu\dot{x}^\mu + \dot{z}\,\frac{\partial L}{\partial\dot{z}} + \dot{\bar{z}}\,\frac{\partial L}{\partial\dot{\bar{z}}} - L = p_\mu\dot{x}^\mu,$$

and the solution for the Dirac spinor, in units $\hbar = 1$ and $c = 1$, is

$$z(\tau) = \left(\cos(\omega\tau/2) + \frac{2i}{\omega}\gamma^\mu p_\mu \sin(\omega\tau/2)\right) z(0),$$

and

$$\bar{z}(\tau) = \bar{z}(0)\left(\cos(\omega\tau/2) - \frac{2i}{\omega}\gamma^\mu p_\mu \sin(\omega\tau/2)\right),$$

where $\omega = 2m$ in these units. In our units $\omega = 2mc/\hbar$, and is the same as our zitterbewegung frequency.

The spin is defined as

$$S_{\mu\nu} = \frac{i}{4}\,\bar{z}\,[\gamma_\mu, \gamma_\nu]z,$$

and is neither a constant of the motion nor the orbital angular momentum. But, the total angular momentum

$$J^{\mu\nu} = x^{\mu}p^{\nu} - x^{\nu}p^{\mu} + S^{\mu\nu}$$

is a conserved quantity, $\dot{J}^{\mu\nu} = 0$.

The velocity of the particle becomes

$$u_{\mu} = \frac{H}{m^2}p_{\mu} + \left(\dot{x}_{\mu}(0) - \frac{H}{m^2}p_{\mu}\right)\cos\omega\tau + \frac{2}{m}S^{\mu\nu}(0)p_{\nu}\sin\omega\tau \quad (5.53)$$

which is not lying along $p^{\mu}$ and shows a clear oscillatory zitterbewegung motion of frequency $\omega$, in addition to the translational motion along $p_{\mu}$.

Instead of using the spinor variables $z$ and $\bar{z}$, it is also possible to describe the dynamics in terms of the spin variables. Then the equivalent dynamical equations are:

$$\dot{x}_{\sigma} = u_{\sigma}, \quad \dot{u}_{\mu} = 4\,S_{\mu\nu}\pi^{\nu},$$

$$\dot{\pi}_{\mu} = eF_{\mu\sigma}u^{\sigma}, \quad \dot{S}_{\mu\nu} = \pi_{\mu}u_{\nu} - \pi_{\nu}u_{\mu},$$

where the dynamical variables are $x^{\mu}, u^{\mu}, \pi^{\mu}$ and $S^{\mu\nu}$.

Recently, Salesi and Recami [23] have applied this model to the interaction with a uniform external magnetic field. Taking the magnetic field $\boldsymbol{B}$ along the $OZ$ axis, the dynamical equations become

$$\begin{aligned}
\ddot{u}_x - 4m\pi_x + 4(m^2 - 3eS_zB)u_x &= 0,\\
\ddot{u}_y - 4m\pi_y + 4(m^2 - 3eS_zB)u_y &= 0,
\end{aligned}$$

where $S_z = \pm 1/2$ is the component of spin along the $OZ$ axis. Taking the next order derivative and using the expression for $\dot{\pi}_i = eF_{ij}u^j$, we get

$$\begin{aligned}
\dddot{u}_x + 4(m^2 - 3eS_zB)\dot{u}_x - 4emBu_y &= 0,\\
\dddot{u}_y + 4(m^2 - 3eS_zB)\dot{u}_y + 4emBu_x &= 0.
\end{aligned}$$

This system of equations has circular motions of angular frequency $\lambda$ around the $OZ$ axis as particular solutions. Then, $\dot{u}_x = -\lambda u_y$ and $\dot{u}_y = \lambda u_x$ and similarly for their derivatives. Therefore, the normal frequencies satisfy the third order algebraic equation

$$\lambda^3 - 4(m^2 - 3eS_zB)\lambda + 4meB = 0. \quad (5.54)$$

If we set $a = \omega_c/\omega$, where $\omega_c = eB/m$ is the cyclotron frequency and $\omega = 2m$ the internal zitterbewegung frequency, when expressing the above equation (5.54) in terms of the variables $z = \lambda/\omega$ and $b = S_z/2$, it leads to

$$z^3 - (1 - 3ab)z + a = 0.$$

This equation with $b = 0$ is exactly the equation for the normal frequencies we obtained in the case of the non-relativistic model of (anti)orbital spin in Sec. 4.3 of Chapter 2. However our model cannot give rise to spin 1/2 particles when quantized. It needs another contribution to spin coming from the orientation variables.

To lowest order in the parameter $a$, the normal frequencies in terms of the dimensionless time $\theta \equiv \omega\tau$, as in the mentioned example, are

$$\omega_1 = a + 3a^2b + a^3(1 + 9b^2), \quad \omega_2 = -1 - \frac{a}{2}(1 - 3b) + \frac{3a^2}{8}\left(1 - 4b + 3b^2\right),$$

$$\omega_3 = 1 - \frac{a}{2}(1 + 3b) - \frac{3a^2}{8}\left(1 + 4b + 3b^2\right),$$

that can be compared with the ones in (2.114). According to this model the dominant macroscopic mode corresponds to the frequency $\omega_1$, which is basically a modification of the cyclotron frequency, while the other two modes $\omega_2$ and $\omega_3$ correspond to slight corrections of the internal zitterbewegung frequency. But frequency $\omega_1$ depends on the orientation of spin $S_z$ and therefore spin up electrons move a little bit faster than the spin down ones. To lowest order

$$\omega_\uparrow - \omega_\downarrow = \frac{3a^2}{2},$$

and therefore an electron beam injected into a cyclotron will produce spatial separations of polarized electrons in the beam, after waiting for a sufficient number of turns.

The only objection to this interpretation comes from the meaning of the velocity $u_\mu$ in (5.53). If, as in our models, it represents the velocity of the charge, as can be deduced from its general expression (5.53), this will produce a difference in the precession frequency of spin but not in the difference between the center of mass velocity of spin up and down electrons and thus the predicted effect will not be observed.

## 2.6 ENTRALGO-KURYSHKIN MODEL

There are many phenomenological models in the literature. We include only Entralgo-Kuryshkin's model in our review, because of the importance given to the intrinsic electric dipole in spite of the magnetic moment of the particle.

The basic idea is that the zitterbewegung of elementary particles reflects the fact that an elementary object, although very small, is no longer point-like but rather it is an extended object. Its dynamics can be described by the evolution of two different points: the center of mass $\boldsymbol{q}(t)$ and the center of charge $\boldsymbol{r}(t)$.[24] It looks like the bilocal model, analyzed in Section 1.3, but it needs some more degrees of freedom to describe the electromagnetic structure of the particle.

This model is non-relativistic, and an elementary particle is considered as a 'structural point particle' in the sense that it is a cluster of point particles of finite, but arbitrary number. These points have masses $m_i$, charges $e_i$ and are placed at positions $\boldsymbol{r}_i$, $i = 1, \ldots, n$, with linear momenta $\boldsymbol{p}_i = m_i \dot{\boldsymbol{r}}_i$, where the dot means time derivative, and in general all these internal features will be unobservable.

The physical observable attributes of the particle are its total mass $m = \sum m_i$ and total charge $e = \sum e_i$. The position of the center of mass of the particle is, as usual, defined by

$$\boldsymbol{q} = \frac{1}{m} \sum_i m_i \boldsymbol{r}_i,$$

and the total linear momentum $\boldsymbol{p} = m\dot{\boldsymbol{q}} = \sum \boldsymbol{p}_i$.

Due to the internal motion of the cluster of point particles, this system also has internal angular momentum $\boldsymbol{s}$, in terms of which the spin $\boldsymbol{S}$ will be defined, and an electric $\boldsymbol{d}$ and magnetic $\boldsymbol{\mu}$ dipole momenta. If the relative position of the internal points with respect to the center of mass is defined by $\boldsymbol{k}_i = \boldsymbol{r}_i - \boldsymbol{q}$, then these momenta are given, respectively, by

$$\boldsymbol{s} = \sum m_i \boldsymbol{k}_i \times \dot{\boldsymbol{k}}_i, \quad \boldsymbol{d} = \sum e_i \boldsymbol{k}_i, \quad \boldsymbol{\mu} = \frac{1}{2} \sum e_i \boldsymbol{k}_i \times \dot{\boldsymbol{k}}_i. \qquad (5.55)$$

For all these particles to be held together, an unobservable confinement potential $W(k_1, \ldots, k_n)$ must exist, in such a way that the relative position variables are restricted to $|\boldsymbol{k}_i(t)| \leq l_0$, where the parameter $l_0$ is very small. In addition to the above intrinsic properties of the particle as a whole, there are other internal mechanical and electromagnetic properties, like internal kinetic energy $\mathcal{T}$, internal potential energy $W_0$ given by

$$\mathcal{T} = \sum \frac{m_i}{2} \dot{\boldsymbol{k}}_i^2, \quad W_0 = W + \sum M^{ij}_{\alpha\beta} k^\alpha_i k^\beta_j = W + E_p,$$

which contain the confinement potential and the elastic potential energy $E_p$, and the different electric and magnetic multipoles produced by the cluster of unobservable points.

Thus, from a global viewpoint, in addition to the point characteristics $m$, $e$, $\boldsymbol{q}$ and $\boldsymbol{p}$, there is a chain of structural characteristics $\boldsymbol{s}$, $\boldsymbol{d}$, $\boldsymbol{\mu}$, $\mathcal{T}, \ldots$. Therefore dynamical equations contain an undetermined number of degrees of freedom so that to produce a finite dynamical system it is necessary to introduce some plausible constraints. Their claim is to reduce the phase space of the system to a manifold of dimension 15 with variables $a_i$, $i = 1, \ldots, 15$, and chosen as the basic ones the variables $(\boldsymbol{q}, \boldsymbol{p}, \boldsymbol{d}, \dot{\boldsymbol{d}}, \boldsymbol{s})$, such that the remaining characteristics will be expressed as functions of them, $\boldsymbol{\mu}(a)$, $\mathcal{T}(a)$, $W_0(a)$, etc.

The spin is defined as

$$\boldsymbol{S} = \boldsymbol{s} - \frac{1}{\chi}\boldsymbol{d} \times \dot{\boldsymbol{d}},$$

with $\chi = lc^2$, $l$ being a constant parameter with dimensions of length, and therefore is different from the internal angular momentum $\boldsymbol{s}$ and is not proportional to the magnetic moment $\boldsymbol{\mu}$ with the usual relation.

This phenomenological model contains at least five intrinsic parameters: the mass $m$, charge $e$, a gyromagnetic ratio $g$, a characteristic length $l$ and finally an internal frequency $\omega$ that is interpreted as the zitterbewegung frequency, and represents the frequency of oscillation of the electric dipole $\boldsymbol{d}$ in the center of mass frame. Then they suply dynamical equations for the above 15 independent dynamical variables, where $\boldsymbol{s}$ is replaced by the total spin $\boldsymbol{S}$. We are not going to discuss the dynamics of this system for which, although it is non-relativistic, one of the difficulties is to properly justify the constraints that lead to the proposed dynamical equations. On the contrary, the model should be justified as a particular model to meet the needs of some experiment with the subsequent determination of the free parameters, but not as a general framework for describing elementary spinning particles.

## Notes


1 A.J. Hanson and T. Regge, *Ann. Phys.* **87**, 498 (1974).

2 H. Bacry, *Commun. Math. Phys.* **5**, 97 (1967).

3 M. Pauri, G.M. Prosperi, *J. Math. Phys.* **7**, 366 (1966); **8**, 2256 (1967); **9**, 1146 (1968).

4 A.A. Kirillov, *Cours sur la thèorie des reprèsentations des groupes*, Ed. Univ. Moscow, Moscow (1971); B. Kostant, *Lecture Notes in Mathematics* **170**, Springer-Verlag, Berlin (1970); J. Souriau, *Structure des systèmes dynamiques*, Dunod, Paris (1970).

5 G.E. Uhlenbeck and S.A. Goudsmit, *Nature* (London) **117**, 264 (1926).

6 F. Rohrlich, *Classical charged particles*, Addison-Wesley, Reading, Mass. (1965).

7 M.K.-H. Kiessling, *Phys. Lett*, **A 258**, 197 (1999).

8 H. Poincaré, *Comptes Rendues* **5**, 1504 (1905); *Rend. Circ. Mat. Palermo*, **21**, 129 (1906).

9 M. Abraham, *Ann. Physik*, **10**, 105 (1903); *Phys. Zeits.* **5**, 576 (1904); *Theorie der Elektrizität*, vol. II, Springer-Verlag, Leipzig, 1905. H.A. Lorentz, *The theory of electrons and its applications to the phenomena of light and radiant heat*, 1909; second edition Dover, New York 1952. H. Poincaré, *Comptes Rendues*, **140**, 1504 (1905).

10 M.H.L. Weyssenhoff, *Acta Phys. Pol.* **9**, 46 (1947); M.H.L. Weyssenhoff and A. Raabe, *Acta Phys. Pol.* **9**, 7 (1947); **9**, 19 (1947).

11 J. Frenkel, *Zeits. f. Phys.* **37**, 243 (1926).

12 M. Mathisson, *Acta Phys. Pol.* **6**, 163 (1937); **6**, 218 (1937).

13 H.J. Bhabha and H.C. Corben, *Proc. Roy. Soc. London*, **A 278**, 273 (1941).

14 J.J. Sakurai, *Advanced Quantum Mechanics*, Addison-Wesley, Reading, Mass. (1967), ch. 3.

15 H.C. Corben, *Nuovo Cimento*, **20**, 529 (1961); *Phys. Rev.* **121**, 1833 (1961).

16 H.C. Corben, *Nuovo Cimento*, **20**, 529 (1961).

17 V. Bargmann, L. Michel and V.L. Telegdi, *Phys. Rev. Lett.* **2**, 435 (1959).

18 J.D. Jackson, *Classical Electrodynamics*, John Wiley and Sons, NY 3rd. ed. (1998), ch. 11.

19 H. Bacry, *Nuovo Cimento*, **26**, 1164 (1962).

20 A.O. Barut and N. Zanghi, *Phys. Rev. Lett.* **52**, 2009 (1984).

21 A. Proca, *J. Phys. Rad.* **15**, 5 (1954).

22 A. Proca, *Nuovo Cimento*, **2**, 962 (1955).

23 G. Salesi and E. Recami, *Phys. Lett.* **A 267**, 219 (2000).

24 E.E. Entralgo and V.V. Kuryshkin, *Nuovo Cimento*, **A 103**, 561 (1990); E.E. Entralgo, B. Cabrera and J. Portieles, *Nuovo Cimento*, **A 111**, 1185 (1998).

Chapter 6

# SPIN FEATURES AND RELATED EFFECTS

We review in this chapter some features related to the spin structure of the particle models. We analyze first the electromagnetic structure of the electron, in particular the electromagnetic field associated to a charge with circular zitterbewegung and the magnetic and electric dipole structure. The fields are not static but a time average over a complete turn of the charge shows a nice behavior with the fields of a point charge $e$ and an intrinsic magnetic dipole $\boldsymbol{\mu}$ at rest in the center of mass.

One salient feature of the classical spinning models with zitterbewegung is the possibility of obtaining a nonvanishing crossing through a potential barrier, for kinetic energies below the potential wall, which in quantum mechanics is termed the Tunnel Effect. This feature, like the Darwin term of Dirac's equation, is related to the instantaneous electric dipole of a charged spinning particle. A section is devoted to previous works on the concept of center of mass or position operator in quantum mechanics and its relationship with the center of mass and spin analyzed in this monograph.

From the geometrical point of view, we also show the relation between the kinematical approach and the realization of the kinematical space as a Finsler space, where particle trajectories are geodesics on this manifold. Some hints are given to consider spin in the context of General Relativity. We analyze the extension of the kinematical groups by including local rotations of the body frame, local Lorentz transformations and space and time dilations, and to the analysis of the Conformal group and the different observables obtained from it. We consider finally the so-called classical limit of quantum mechanics that, basically, leads to the classical mechanics of spinless systems.

# 1. ELECTROMAGNETIC STRUCTURE OF THE ELECTRON

Let us consider that the classical electron is described by the model whose charge is moving in circles at the speed of light in the center of mass frame.

One of the immediate questions concerning the classical structure of the electron is, what is the associated electromagnetic field of the particle? We see that the charge is accelerated and according to the classical electromagnetic theory, the particle must necessarily radiate continuously. However, from the mechanical point of view we have produced a classical free system, such that properties like mechanical energy and mechanical linear and angular momentum are conserved in time. The Lagrangian that describes the system is Poincaré invariant, and if we think about a free system, the corresponding field structure cannot produce loss of energy and linear momentum. The free particle has to have associated an electromagnetic field without radiation. Radiation has to be produced whenever the center of mass of the particle is accelerated, *i.e.*, when the particle is no longer free.

There must exist radiationless solutions of Maxwell's equations, associated to point charges moving in circles at the speed of light. One possibility is to consider solutions derived from the Liénard-Wiechert potentials $(A^\mu_{ret} + A^\mu_{adv})/2$, where $A^\mu_{ret}$ and $A^\mu_{adv}$ are the corresponding retarded and advanced potentials. But, even if we take as the probable electric field $(\boldsymbol{E}_{ret} + \boldsymbol{E}_{adv})/2$, it is neither static nor Coulomb-like, and therefore it does not look like the estimated electric field of a point electron. We shall consider next a particular static solution: the time average field during a complete turn of the charge.

## 1.1 THE TIME AVERAGE ELECTRIC AND MAGNETIC FIELD

Let us assume that we have a test charge in the neighborhood of the electron. The frequency of the zitterbewegung is very high, of order $\sim 10^{21}\ \mathrm{s}^{-1}$. If our test particle is moving slowly, then presumably the detected electric field will be some time average field during a complete turn of the charge.

The retarded (or advanced) electric field of a point charge at the observation point $\boldsymbol{x}$ at time $t$ is given by [1]

$$\boldsymbol{E} = \boldsymbol{E}_\beta + \boldsymbol{E}_a,$$

where

$$\boldsymbol{E}_\beta(t, \boldsymbol{x}) = \frac{e(1-\beta^2)}{R^2(1-\boldsymbol{n}\cdot\boldsymbol{\beta})^3}\,(\boldsymbol{n}-\boldsymbol{\beta})\,, \tag{6.1}$$

$$\boldsymbol{E}_a(t,\boldsymbol{x}) = \frac{e}{Rc^2(1-\boldsymbol{n}\cdot\boldsymbol{\beta})^3}\,\boldsymbol{n}\times((\boldsymbol{n}-\boldsymbol{\beta})\times\boldsymbol{a}), \tag{6.2}$$

are the velocity and acceleration fields, respectively. Observables $\boldsymbol{r}$, $\boldsymbol{u} = d\boldsymbol{r}/dt$ and $\boldsymbol{a} = d\boldsymbol{u}/dt$, are the position, velocity and acceleration of the charge, evaluated at the retarded (or advanced) time $\tilde{t} = t - R/c$, (or $\tilde{t} = t + R/c$). Vector $\boldsymbol{\beta} = \boldsymbol{u}/c$, and

$$\boldsymbol{n} = \frac{\boldsymbol{x}-\boldsymbol{r}}{|\boldsymbol{x}-\boldsymbol{r}|}, \quad R = |\boldsymbol{x}-\boldsymbol{r}|.$$

The corresponding magnetic field is $\boldsymbol{B} = \boldsymbol{n}\times\boldsymbol{E}/c$. Because for the electron the charge is moving at the speed of light $\beta = 1$, the velocity field $\boldsymbol{E}_\beta$ vanishes, and it seems that the only field contribution behaves as $1/R$.

The complete analytical expression of a time average field at any arbitrary point has not yet been obtained. However, to obtain an estimate, let us compute the average field on some particular point. Let us consider that the electron is at rest, with the center of mass at the origin of a reference frame. The constant spin is pointing along the $OZ$ axis. We shall try to calculate this average field at a point $P$ of coordinate $z$ in this $OZ$ axis. In Figure 6.1, we represent the different magnitudes at the retarded time $\tilde{t}$, needed to apply equation (6.2).

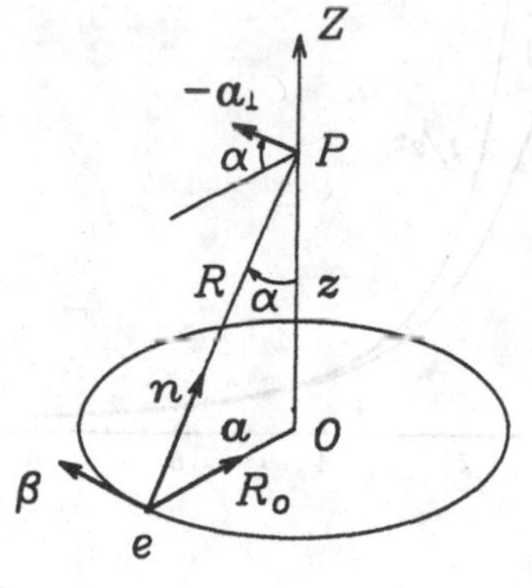


***Figure 6.1.*** **Instantaneous electric field of the electron at point $P$ has a component along $-\boldsymbol{a}_\perp$ and $-\boldsymbol{\beta}$.**

In that particular point shown in the figure, $\boldsymbol{n}\cdot\boldsymbol{\beta} = 0$, and thus

$$\boldsymbol{E} = \frac{e}{c^2R}(\boldsymbol{n}\times(\boldsymbol{n}\times\boldsymbol{a}) - \boldsymbol{n}\times(\boldsymbol{\beta}\times\boldsymbol{a})) = \frac{e}{c^2R}(-\boldsymbol{a}_\perp - \boldsymbol{\beta}(\boldsymbol{n}\cdot\boldsymbol{a})),$$

where vector $\boldsymbol{a}_\perp = \boldsymbol{a} - \boldsymbol{n}(\boldsymbol{a}\cdot\boldsymbol{n})$, is the component of the acceleration orthogonal to the unit vector $\boldsymbol{n}$. For the observation point $P$, the expression $\boldsymbol{n}\cdot\boldsymbol{a}$ is constant at any retarded point, and the time average of $\boldsymbol{\beta}$ during a complete turn is zero, and for the vector $\boldsymbol{a}_\perp$ it reduces to its $z$-component $a_\perp \sin\alpha$. Since the acceleration in this frame is $a = c^2/R_0$, $a_\perp = a\cos\alpha$ and $\sin\alpha = R_0/R$ and $\cos\alpha = z/R$, the time average electric field at point $P$ is

$$\boldsymbol{E}(z) = \frac{ez}{(R_0^2+z^2)^{3/2}}\hat{\boldsymbol{z}}, \tag{6.3}$$

where $\hat{\boldsymbol{z}}$ is a unit vector along the $OZ$ axis. The advanced field has exactly the same expression. This is a radial field from the origin of the reference frame with a Coulomb-like behaviour $1/z^2$, but it does not diverge at the origin. We depict this field in Figure 6.2, for comparison with the Coulomb field of a point charge at the origin, where we take as a unit of length the radius $R_0$ of the internal motion.

We can clearly see the fitting of the average field and the Coulomb field for large $z$. The maximum of the average field takes place at $z = R_0/\sqrt{2}$. If we consider that the static field of a pointlike electron is this time average field, then the electrostatic energy does not diverge. But to obtain the expression for the energy is necessary to find the field at any point, a goal that we have not attained.

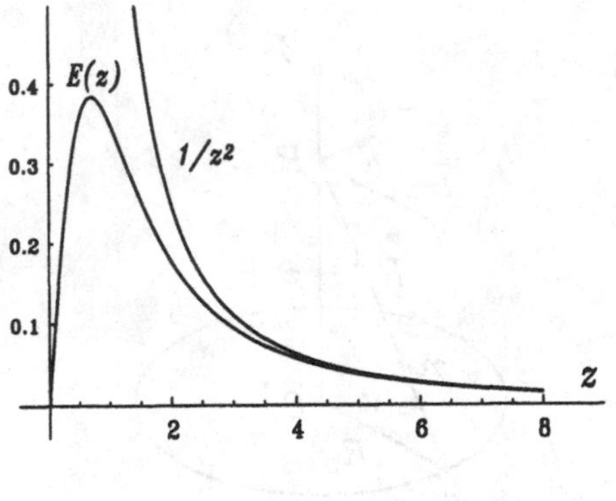


*Figure 6.2.* Average retarded (or advanced) electric field (6.3) and Coulomb field along the $OZ$ axis.

However, if we are involved in high energy processes, our test particle is moving sufficiently fast relative to the electron, then the field it senses is the instantaneous $1/R$ field, which is greater than the average field, and becomes important for points closer to the electron. This means that the average energy density of the local instantaneous field is greater than the average Coulomb-like energy density, and we can naively interpret this difference, from the classical point of view, as the energy associated to the cloud of virtual photons in the surroundings of the particle. Is this the corresponding infinite energy which is usually cancelled out in the renormalization of quantum electrodynamics?

To compute numerically the average field at an arbitrary position, let us consider the different magnitudes depicted in Figure 6.3.

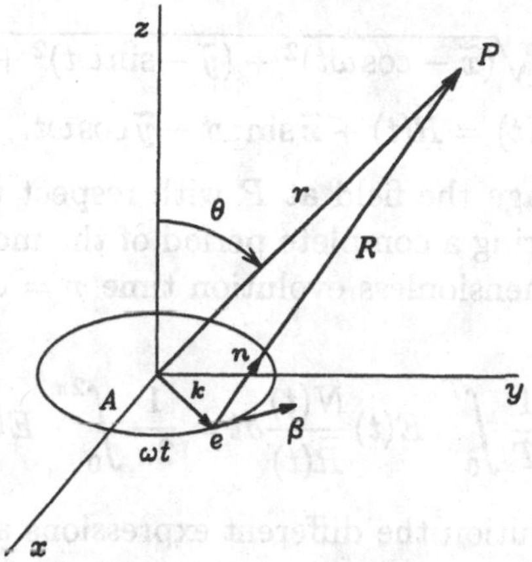


*Figure 6.3.* Charge motion and observation point $P$.

If at time $t = 0$ the charge is located at point $A$ on the $OX$ axis, then at time $t$ the different observables shown in the figure are described in Cartesian coordinates and in the laboratory frame by

$$\boldsymbol{k} = R_0[\cos\omega t, \sin\omega t, 0] \equiv R_0\widetilde{\boldsymbol{k}}, \quad \boldsymbol{\beta} = \frac{\boldsymbol{u}}{c} = [-\sin\omega t, \cos\omega t, 0],$$

$$\boldsymbol{r} = [x, y, z], \quad \boldsymbol{a} = \frac{d\boldsymbol{u}}{dt} = \frac{c^2}{R_0}[-\cos\omega t, -\sin\omega t, 0] = \frac{c^2}{R_0}\widehat{\boldsymbol{a}},$$

$$\boldsymbol{R} = \boldsymbol{r} - \boldsymbol{k} = R_0(\widetilde{\boldsymbol{r}} - \widetilde{\boldsymbol{k}}), \quad \boldsymbol{n} = \frac{\boldsymbol{R}}{R}, \quad R = |\boldsymbol{R}| = R_0\widetilde{R}.$$

With these definitions, field (6.2) can be written as

$$\boldsymbol{E}(t, \boldsymbol{r}) = \left(\frac{e}{R_0^2}\right) \frac{\boldsymbol{n} \times ((\boldsymbol{n} - \boldsymbol{\beta}) \times \widehat{\boldsymbol{a}})}{(1 - \boldsymbol{n}\cdot\boldsymbol{\beta})^3\widetilde{R}}.$$

We want to compare the time average value of this field with the static Coulomb field of a point charge $e$ at the center of mass

$$\boldsymbol{E}_0(\boldsymbol{r}) = \left(\frac{e}{R_0^2}\right) \frac{\widehat{\boldsymbol{r}}}{r^2},$$

where $\widehat{\boldsymbol{r}}$ is a unit vector in the radial direction. The constant factor in brackets in front of these formulae will be dropped out from now on. In this way the unit of length is the zitterbewegung radius $R_0$.

When the charge is at the point indicated in Figure 6.3, the retarded field it produces at point $P$ is evaluated at the observation time $t_o = t + R/c$. Thus $dt_o = dt + d\widetilde{R}/\omega$, because $R_0/c = 1/\omega$. If we express $d\widetilde{R}$ in terms of $dt$, we get $dt_o = (N(t)/\widetilde{R}(t))dt$, where $N$ and $\widetilde{R}$ are explicitly given by

$$\widetilde{R}(t) = \sqrt{(\widetilde{x} - \cos\omega t)^2 + (\widetilde{y} - \sin\omega t)^2 + \widetilde{z}^2},$$

$$N(t) = \widetilde{R}(t) + \widetilde{x}\sin\omega t - \widetilde{y}\cos\omega t.$$

We are going to average the field at $P$ with respect to the observation time at that point during a complete period of the motion of the charge $T$. If we define a dimensionless evolution time $\tau = \omega t$, then $\omega T = 2\pi$ and thus

$$\frac{1}{T}\int_0^T E(t_o)\,dt_o = \frac{1}{T}\int_0^T E(t)\frac{N(t)}{\widetilde{R}(t)}dt = \frac{1}{2\pi}\int_0^{2\pi} E(\tau)\frac{N(\tau)}{\widetilde{R}(\tau)}d\tau. \quad (6.4)$$

In terms of the $\tau$ evolution the different expressions are

$$\boldsymbol{n}\times(\boldsymbol{n}\times\widehat{\boldsymbol{a}}) = \boldsymbol{n}(\boldsymbol{n}\cdot\widehat{\boldsymbol{a}}) - \widehat{\boldsymbol{a}},$$

and

$$\boldsymbol{n}(\boldsymbol{n}\cdot\widehat{\boldsymbol{a}}) = \frac{1 - \widetilde{x}\cos\tau - \widetilde{y}\sin\tau}{\widetilde{R}^2}[\widetilde{x} - \cos\tau, \widetilde{y} - \sin\tau, \widetilde{z}],$$

$$\widehat{\boldsymbol{a}} = [-\cos\tau, -\sin\tau, 0],$$

while

$$\boldsymbol{n}\times(\boldsymbol{\beta}\times\widehat{\boldsymbol{a}}) = \frac{1}{\widetilde{R}}[\widetilde{y} - \sin\tau, -\widetilde{x} + \cos\tau, 0],$$

and

$$1 - \boldsymbol{n}\cdot\boldsymbol{\beta} = \frac{1}{\widetilde{R}}\left(\widetilde{R} + \widetilde{x}\sin\tau - \widetilde{y}\cos\tau\right).$$

We are interested in the radial and transversal part of the field $E_r = \boldsymbol{E}\cdot\widehat{\boldsymbol{r}}$, $E_\theta = \boldsymbol{E}\cdot\widehat{\boldsymbol{\theta}}$, and $E_\phi = \boldsymbol{E}\cdot\widehat{\boldsymbol{\phi}}$, respectively. Here $\widehat{\boldsymbol{r}}$, $\widehat{\boldsymbol{\theta}}$ and $\widehat{\boldsymbol{\phi}}$ are respectively the usual unit vectors in polar spherical coordinates. If we consider that the observation point $P$ is on the plane $XOZ$, then we have to take $\widetilde{x} = r\sin\theta$, $\widetilde{y} = 0$ and $\widetilde{z} = r\cos\theta$, where $r$ is the radial separation from the origin in units of $R_0$.

The final expressions for the field components are

$$E_r(r,\theta,\tau) = \frac{(\widetilde{R}^2 - r^2 - 1)\sin\theta\cos\tau + \widetilde{R}\sin\theta\sin\tau + r(1 + \sin^2\theta\cos^2\tau)}{\left(\widetilde{R} + r\sin\theta\sin\tau\right)^3},$$

$$E_\theta(r,\theta,\tau) = \frac{\left[(\widetilde{R}^2-1)\cos\tau + \widetilde{R}\sin\tau + r\sin\theta\cos^2\tau\right]\cos\theta}{\left(\widetilde{R} + r\sin\theta\sin\tau\right)^3},$$

$$E_\phi(r,\theta,\tau) = \frac{(\widetilde{R}^2-1)\sin\tau + \widetilde{R}(r\sin\theta - \cos\tau) + r\sin\theta\sin\tau\cos\tau}{\left(\widetilde{R} + r\sin\theta\sin\tau\right)^3},$$

with

$$\widetilde{R} = \sqrt{r^2 - 2r\sin\theta\cos\tau + 1}.$$

To take the time average value of the above fields we have to perform the integration (6.4) so that the above expressions of $E_r$, $E_\theta$ and $E_\phi$ have to be multiplied by $N(\tau)/\widetilde{R}(\tau)$, where now

$$N(\tau) = \widetilde{R} + r\sin\theta\sin\tau.$$

The average retarded radial electric field for $\theta = 0$ is already depicted in Figure 6.2 but we also include it in the next Figure 6.4. We see the Coulomb behavior of the radial component for the directions $\theta = 0, \pi/3, \pi/4, \pi/6$. Similarly, in Figure 6.5 is displayed the transversal component of the average retarded electric field $< E_\theta(r,\theta) >$ for the same directions, that goes to zero very quickly. For $\theta = \pi/2$, we see that $< E_\theta(r,\pi/2) >= 0$. The average $< E_\phi(r,\theta) >$ vanishes everywhere for any $\theta \neq \pi/2$. On the plane $\theta = \pi/2$ the numerical routine fails.

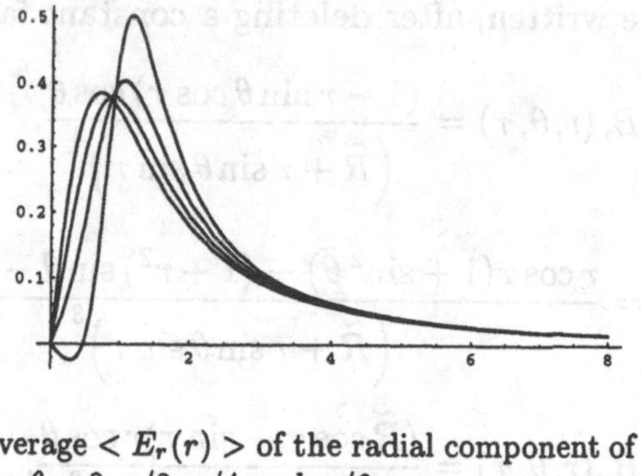


*Figure 6.4.* Time average $< E_r(r) >$ of the radial component of the retarded electric field in the directions $\theta = 0, \pi/3, \pi/4$ and $\pi/6$.

The average magnetic field can be computed in the same way. Here we shall consider only the retarded solution and we will compare it with the magnetic field produced by an intrinsic magnetic moment $\boldsymbol{\mu}$ placed at the center of mass. This magnetic field is [2]

$$\boldsymbol{B}_0(\boldsymbol{r}) = \frac{3\widehat{\boldsymbol{r}}(\widehat{\boldsymbol{r}}\cdot\boldsymbol{\mu}) - \boldsymbol{\mu}}{c^2 r^3}.$$

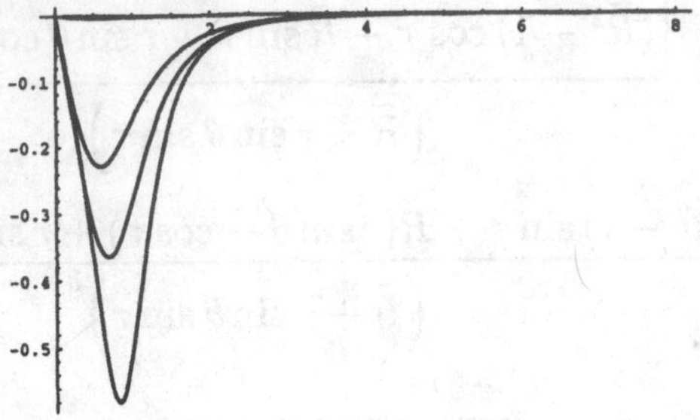


*Figure 6.5.* Time average of the component $< E_\theta(r) >$ of the retarded electric field in the directions $\theta = 0, \pi/3, \pi/4$ and $\pi/6$. It goes to zero very quickly. For $\theta = \pi/2$ it vanishes everywhere.

For our system the magnetic moment produced by the moving charge is of value $ecR_0/2$ in the direction of $OZ$, so that in units of $R_0$ it can be written as

$$\boldsymbol{B}_0(\boldsymbol{r}) = \left(\frac{e}{2cR_0^2}\right)\frac{3\widehat{\boldsymbol{r}}(\widehat{\boldsymbol{r}}\cdot\widehat{\boldsymbol{z}}) - \widehat{\boldsymbol{z}}}{\widetilde{r}^3}.$$

The nonvanishing components are

$$B_{0r}(r,\theta) = \left(\frac{e}{cR_0^2}\right)\frac{\cos\theta}{\widetilde{r}^3}, \quad B_{0\theta}(r,\theta) = \left(\frac{e}{cR_0^2}\right)\frac{\sin\theta}{2\widetilde{r}^3}. \qquad (6.5)$$

In our model, the instantaneous magnetic field is $\boldsymbol{B} = \boldsymbol{n}\times\boldsymbol{E}/c$. Their components can be written, after deleting a constant factor $e/cR_0^2$, as:

$$B_r(r,\theta,\tau) = \frac{(1 - r\sin\theta\cos\tau)\cos\theta}{\left(\widetilde{R} + r\sin\theta\sin\tau\right)^3},$$

$$B_\theta(r,\theta,\tau) = \frac{r\cos\tau(1+\sin^2\theta) - (1+r^2)\sin\theta - \widetilde{R}r\sin\tau}{\left(\widetilde{R} + r\sin\theta\sin\tau\right)^3},$$

$$B_\phi(r,\theta,\tau) = \frac{(\widetilde{R}\cos\tau + \sin\tau)r\cos\theta}{\left(\widetilde{R} + r\sin\theta\sin\tau\right)^3}.$$

To proceed with the retarded time average integral we have to multiply the above fields by $N(t)/\widetilde{R}(t)$, as before. The numerical integration is compared with the analytical expression of the magnetic field of a dipole (6.5) for different directions.

The magnetic dipole field (6.5) goes to infinity when $r \to 0$. In Figures 6.6-6.8 we show the matching of the $B_{0r}(r)$ components of the dipole and the computed time average value $< B_r(r,\theta) >$, for $r > R_0$ and in the directions given by $\theta = \pi/6, \pi/4$ and $\pi/3$. Similarly, in Figures 6.9-6.11, for the corresponding $B_{0\theta}(r,\theta)$ and $< B_\theta(r,\theta) >$ components.

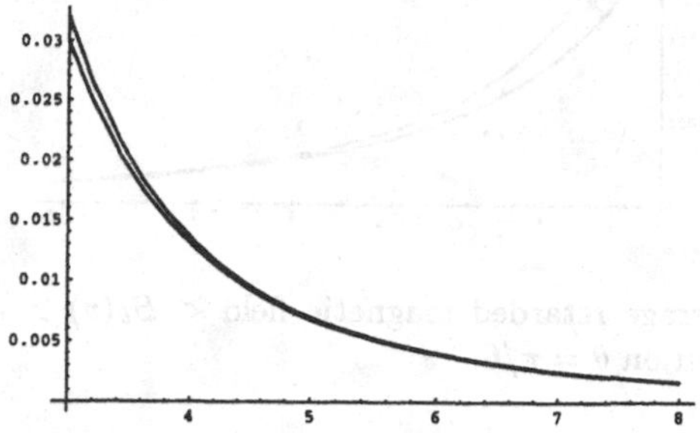


*Figure 6.6.* Radial components of the dipole field $B_{0r}(r)$ and the time average retarded magnetic field $< B_r(r) >$, along the direction $\theta = \pi/6$.

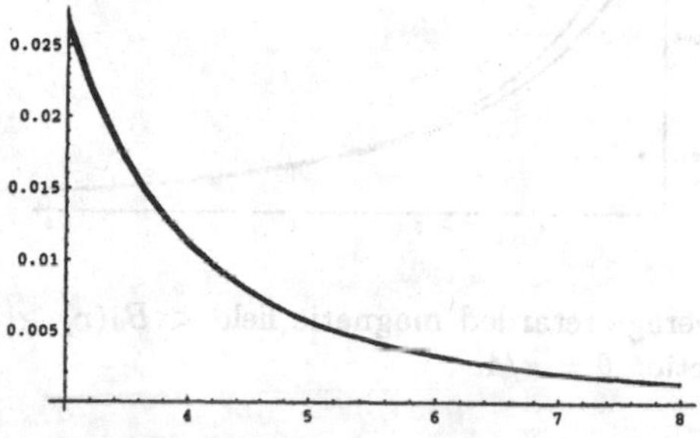


*Figure 6.7.* Radial components of the dipole field $B_{0r}(r)$ and the time average retarded magnetic field $< B_r(r) >$, along the direction $\theta = \pi/4$.

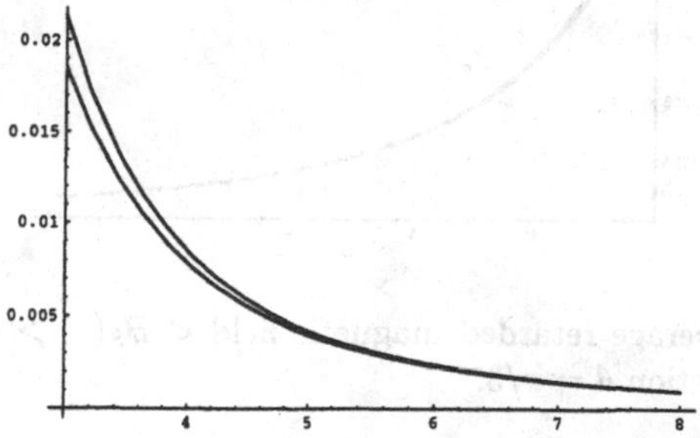


*Figure 6.8.* Radial components of the dipole field $B_{0r}(r)$ and the time average retarded magnetic field $< B_r(r) >$, along the direction $\theta = \pi/3$.

The computed time averages $< B_r(r) >$ and $< B_\theta(r) >$ do not diverge at the origin but have the behavior depicted in 6.12 and 6.13, respec-

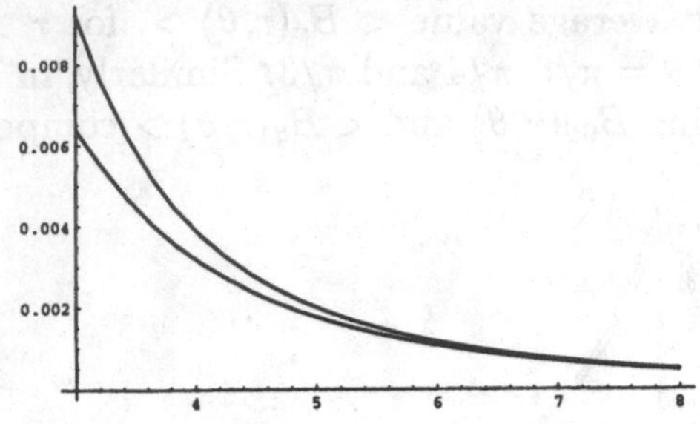


*Figure 6.9.* Time average retarded magnetic field $< B_\theta(r) >$ and the dipole field $B_{0\theta}(r)$, along the direction $\theta = \pi/6$.

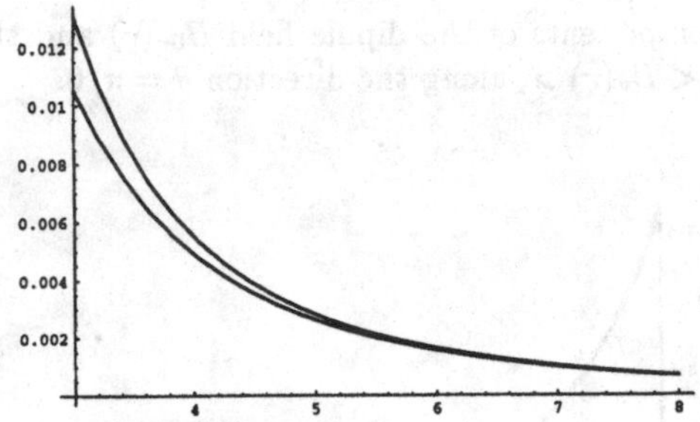


*Figure 6.10.* Time average retarded magnetic field $< B_\theta(r) >$ and the dipole field $B_{0\theta}(r)$, along the direction $\theta = \pi/4$.

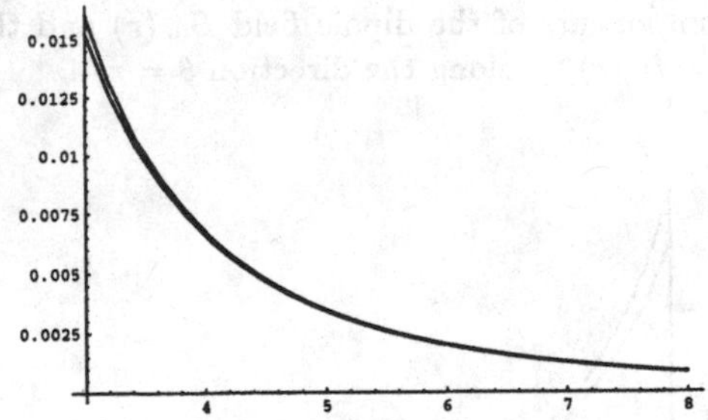


*Figure 6.11.* Time average retarded magnetic field $< B_\theta(r) >$ and the dipole field $B_{0\theta}(r)$, along the direction $\theta = \pi/3$.

tively, when represented along the directions $\theta = 0, \pi/3, \pi/4$ and $\pi/6$, and they take the values $\cos\theta$ and $-\sin\theta$ respectively, at point $r = 0$.

The time average value of the transversal component $< B_\phi(r, \theta) >$ vanishes everywhere for all directions.

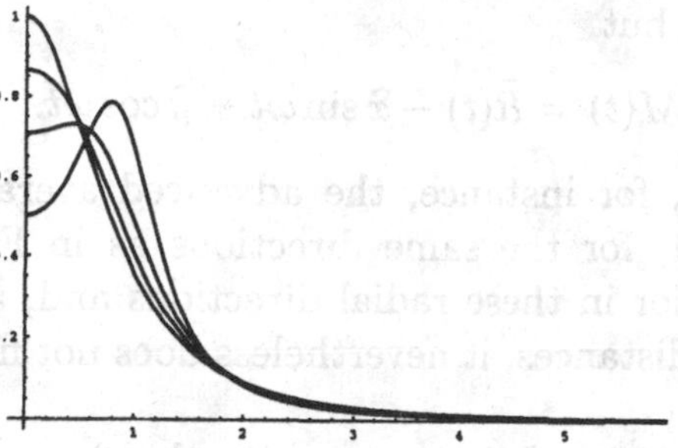


*Figure 6.12.* **Time average retarded magnetic field $< B_r(r) >$ along the directions $\theta = 0, \pi/3, \pi/4$ and $\pi/6$ and its behavior at $r = 0$. For $\theta = \pi/2$ it vanishes everywhere.**

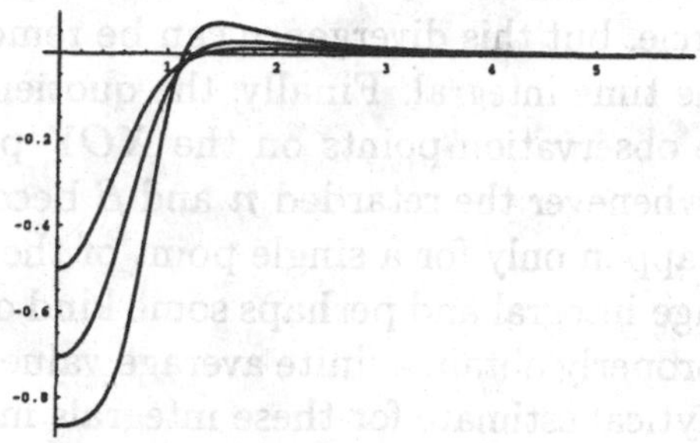


*Figure 6.13.* **Time average retarded magnetic field $< B_\theta(r) >$ along the directions $\theta = 0, \pi/3, \pi/4$ and $\pi/6$ and its behavior at $r = 0$.**

To end this section we can think about the possibility of computing the average fields using the advanced solutions in spite of the retarded ones.

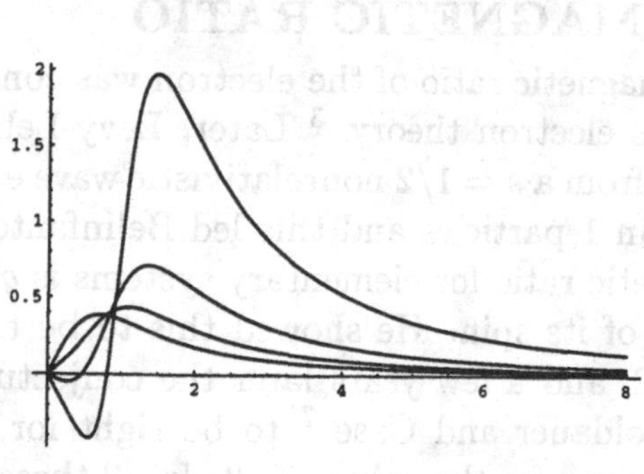


*Figure 6.14.* **Time average radial component $< E_r(r) >$ of the advanced electric field in the directions $\theta = 0, \pi/3, \pi/4$ and $\pi/6$.**

In that case the observation time will be related with the laboratory time by $t_o = t - R/c$, and therefore $dt_o = (M(t)/\widetilde{R}(t))dt$, where $\widetilde{R}(t)$ is the same as before, but

$$M(t) = \widetilde{R}(t) - \widetilde{x}\sin\omega t + \widetilde{y}\cos\omega t.$$

Then, if we depict, for instance, the advanced average radial electric field in Figure 6.14, for the same directions as in Figure 6.4, we see the different behavior in these radial directions and, although the field decreases for large distances, it nevertheless does not fit with a Coulomb field.

The numerical routine fails to compute the corresponding integrals for $\theta = \pi/2$ where we have some indefiniteness of the integrands for observation points lying on the $XOY$ plane. There are no singularities for points inside the circle of radius $R_0$. We have a divergence of order $1/r$ for points on this circle, but this divergence can be removed by taking a principal value of the time integral. Finally, the quotient term $1 - \boldsymbol{n}\cdot\boldsymbol{\beta}$ can vanish for some observation points on the $XOY$ plane outside the circle of radius $R_0$, whenever the retarded $\boldsymbol{n}$ and $\boldsymbol{\beta}$ become parallel vectors. But this can happen only for a single point of the retarded charge position in the average integral and perhaps some kind of principal value should be taken to properly obtain a finite average value. The difficulties of obtaining an analytical estimate for these integrals make this analysis incomplete. Nevertheless, the nice fitting of the average electric field with a Coulomb field and the average magnetic field with the field of a magnetic dipole, for distances of a few Compton wave lengths away, except on the $\theta = \pi/2$ plane where we have not been able to obtain an estimate, suggests that we devote some effort to renormalize and improve the model at a classical level.

## 1.2 GYROMAGNETIC RATIO

The $g = 2$ gyromagnetic ratio of the electron was considered for years a success of Dirac's electron theory. [3] Later, Levy-Leblond [4] obtained similarly $g = 2$ but from a $s = 1/2$ nonrelativistic wave equation. Proca [5] found $g = 1$ for spin 1 particles and this led Belinfante [6] to conjecture that the gyromagnetic ratio for elementary systems is $g = 1/s$, irrespective of the value $s$ of its spin. He showed this to be true for quantum systems of spin 3/2, and a few years later the conjecture was analyzed and checked by Moldauer and Case [7] to be right for any half-integer spin, and by Tumanov [8] for the value $s = 2$. In all these cases a minimal electromagnetic coupling was assumed.

Weinberg [9] made the prediction $g = 2$ for the intermediate bosons of the weak interactions when analyzing the interaction of $W$ bosons with the electromagnetic field by requiring a good high-energy behavior of the scattering amplitude. The discovery of the charged $W^{\pm}$ spin 1 bosons with $g = 2$, contradictory to Belinfante's conjecture, corroborated Weinberg's prediction and raised the question as to whether $g = 2$ for any elementary particle of arbitrary spin.

Jackiw [10] has given another dynamical argument confirming that the gyromagnetic ratio of spin-1 fields is $g = 2$, provided a nonelectromagnetic gauge invariance is accepted. He also gives some *ad hoc* argument for $s = 2$ fields, consistent with the $g = 2$ prescription.

Ferrara *et al.* [11] in a Lagrangian approach for massive bosonic and fermionic strings, by the requirement of a smooth fixed-charge $M \to 0$ limit, get $g = 2$ as the most natural value for particles of arbitrary spin. However the only known particles which fulfill this condition are leptons and charged $W^{\pm}$ bosons, *i.e.*, charged fermions and bosons of the lowest admissible values of spin. No other higher spin charged elementary particles have been found.

The aim of this section, instead of using dynamical arguments as in the previous attempts, is to give a kinematical description of the gyromagnetic ratio of elementary particles [12] which is based upon the double content of their spin operator structure as mentioned in the classical analysis of Section 6.2 of Chapter 2.

The general structure of the quantum mechanical angular momentum operator in either relativistic or nonrelativistic approach is

$$\boldsymbol{J} = \boldsymbol{r} \times \frac{\hbar}{i}\nabla + \boldsymbol{S} = \boldsymbol{r} \times \boldsymbol{P} + \boldsymbol{S}, \tag{6.6}$$

where the spin operator is

$$\boldsymbol{S} = \boldsymbol{u} \times \frac{\hbar}{i}\nabla_{u} + \boldsymbol{W}, \tag{6.7}$$

and $\nabla_u$ is the gradient operator with respect to the velocity variables and $\boldsymbol{W}$ is a linear differential operator that operates only on the orientation variables $\boldsymbol{\alpha}$ and therefore commutes with the other. For instance, in the $\boldsymbol{\rho} = \boldsymbol{n}\tan(\alpha/2)$ parametrization $\boldsymbol{W}$ is written as

$$\boldsymbol{W} = \frac{\hbar}{2i}\left[\nabla_{\rho} + \boldsymbol{\rho} \times \nabla_{\rho} + \boldsymbol{\rho}(\boldsymbol{\rho} \cdot \nabla_{\rho})\right]. \tag{6.8}$$

The first part in (6.7), related to the zitterbewegung spin, has integer eigenvalues because it has the form of an orbital angular momentum in terms of the $\boldsymbol{u}$ variables. Half-integer eigenvalues come only from the operator (6.8). This operator $\boldsymbol{W}$ takes into account the change of orientation, *i.e.*, the rotation of the particle.

We have seen in either relativistic or non-relativistic examples that if the only spin content of the particle $\boldsymbol{S}$ is related to the zitterbewegung part $\boldsymbol{Z} = \boldsymbol{u} \times \boldsymbol{U}$, then the relationship between the magnetic moment and zitterbewegung spin is given by

$$\boldsymbol{\mu} = \frac{e}{2}\,\boldsymbol{k} \times \frac{d\boldsymbol{k}}{dt} = -\frac{e}{2m}\boldsymbol{Z}, \qquad (6.9)$$

*i.e.*, with a normal up to a sign gyromagnetic ratio $g = 1$. If the electron has a gyromagnetic ratio $g = 2$, this implies necessarily that another part of the spin is coming from the angular velocity of the body, but producing no contribution to the magnetic moment.

Therefore for the electron, both parts $\boldsymbol{W}$ and $\boldsymbol{Z}$ contribute to the total spin. But the $\boldsymbol{W}$ part is related to the angular variables that describe orientation and does not contribute to the separation $\boldsymbol{k}$ between the center of charge and the center of mass. It turns out that the magnetic moment of a general particle is still related to the motion of the charge by the expression (6.9), *i.e.*, in terms of the $\boldsymbol{Z}$ part but not to the total spin $\boldsymbol{S}$. It is precisely when we try to express the magnetic moment in terms of the total spin that the concept of gyromagnetic ratio arises.

Now, let us assume that both $\boldsymbol{Z}$ and $\boldsymbol{W}$ terms contribute to the total spin $\boldsymbol{S}$ with their lowest admissible values.

For Dirac's particles, the classical zitterbewegung is a circular motion at the speed of light of radius $R = S/mc$ and angular frequency $\omega = mc^2/S$, in a plane orthogonal to the total spin. The total spin $\boldsymbol{S}$ and the $\boldsymbol{Z}$ part, are both orthogonal to this plane and can be either parallel or antiparallel. Let us define the gyromagnetic ratio as in Section 6.2 of Chapter 2 by $Z = gS$. For the lowest admissible values of the quantized spins $z = 1$ and $w = 1/2$ in the opposite direction this gives rise to a total $s = 1/2$ perpendicular to the zitterbewegung plane and then $g = 2$.

For $s = 1$ particles the lowest possible values compatible with the above relative orientations are $z = 2$ and $w = 1$ in the opposite direction, thus obtaining again $g = 2$. The possibility $z = 1$ and $w = 0$ is forbidden in the relativistic case because necessarily $w \neq 0$ to describe vector bosons with a multicomponent wave-function.

No higher spin charged elementary particles are known. The predictions of this formalism for hypothetical particles of $s = 3/2$ are $z = 1$ and $w = 1/2$ in the same direction, and thus $g = 2/3$, or $z = 2$ and $w = 1/2$ in the opposite direction, and therefore $g = 4/3$. Similarly, for $s = 2$ particles the lowest values are $z = 1$ and $w = 1$ in the same direction, and thus $g = 1/2$, compatible with Belinfante's conjecture.

## 1.3 INSTANTANEOUS ELECTRIC DIPOLE

The internal motion of the charge of the electron in the center of mass frame is a circle at the speed of light. We call this particle a Dirac particle because when quantizing this system with the additional condition that the spin $s = 1/2$, its wave function satisfies Dirac's equation. Nevertheless in the classical version the value of the spin $s$ and the angular velocity of the system is unrestricted and therefore there are infinitely many classical systems that when quantized satisfy Dirac's equation.

The position of the charge in this frame is related to the total spin by eq. (3.167), *i.e.*,

$$\boldsymbol{k} = \frac{1}{mc^2}\boldsymbol{S} \times \boldsymbol{u}, \tag{6.10}$$

where $\boldsymbol{S}$ is the total constant spin and $\boldsymbol{u} = d\boldsymbol{k}/dt$, with $u = c$ is the velocity of the charge. In addition to this motion there is a rotation of a local frame linked to the particle that gives rise to some angular velocity, but this rotation has no effect on the electric dipole structure. (See Fig. 6.15 where the angular velocity and the local frame are not depicted).

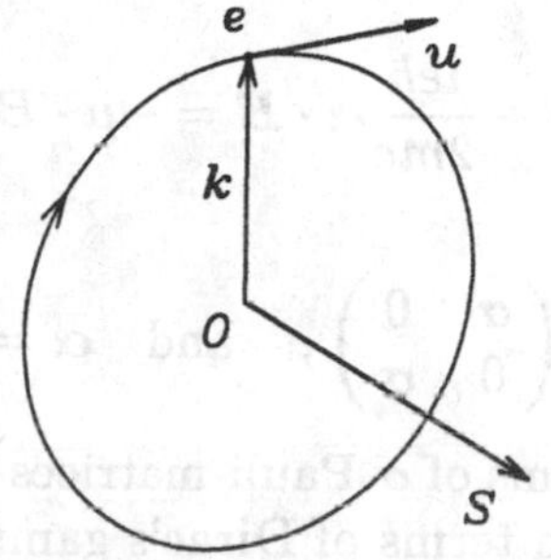


*Figure 6.15.* Electron charge motion in the C.M. frame.

Now, from the point of view of the center of mass observer, the particle behaves as though it has a magnetic moment related to the particle current by the usual classical expression

$$\boldsymbol{\mu} = \frac{1}{2}\int \boldsymbol{k} \times \boldsymbol{j}\, d^3\boldsymbol{r} = \frac{e}{2}\boldsymbol{k} \times \frac{d\boldsymbol{k}}{dt},$$

where $e$ is the charge and $\boldsymbol{j}(\boldsymbol{r}-\boldsymbol{k}) = e\, d\boldsymbol{k}/dt\ \delta^3(\boldsymbol{r}-\boldsymbol{k})$ is the particle current density. The orbital term $\boldsymbol{k} \times d\boldsymbol{k}/dt$ is related to the zitterbewegung part of spin that quantizes with integer values and which for spin 1/2 and spin 1 charged particles is twice the total spin $\boldsymbol{S}$, giving rise to a pure kinematical interpretation of the gyromagnetic ratio $g = 2$ for this model as seen in the previous section.

**But also in the center of mass frame the particle has an oscillating instantaneous electric dipole moment $\boldsymbol{d} = e\boldsymbol{k}$, that is thus related to the total spin by**

$$\boldsymbol{d} = \frac{e}{mc^2}\,\boldsymbol{S} \times \boldsymbol{u}. \tag{6.11}$$

**This instantaneous electric dipole, which fulfills the usual definition of the momentum of the point charge $e$ with respect to the origin of the reference frame, is translation invariant because it is expressed in terms of a relative position vector $\boldsymbol{k}$. It can never be interpreted as some kind of fluctuation of a spherical symmetry of a charge distribution. Even in this kind of model, it is not necessary to talk about charge distributions, because all particle attributes are defined at single points.**

**In his original 1928 article, [13] Dirac obtains that the Hamiltonian for the electron has, in addition to the Hamiltonian of a free point particle of mass $m$, two new terms that in the presence of an external electromagnetic field are**

$$\frac{e\hbar}{2m}\boldsymbol{\Sigma}\cdot\boldsymbol{B} + \frac{ie\hbar}{2mc}\boldsymbol{\alpha}\cdot\boldsymbol{E} = -\boldsymbol{\mu}\cdot\boldsymbol{B} - \boldsymbol{d}\cdot\boldsymbol{E}, \tag{6.12}$$

**where**

$$\boldsymbol{\Sigma} = \begin{pmatrix} \boldsymbol{\sigma} & 0 \\ 0 & \boldsymbol{\sigma} \end{pmatrix}, \quad \text{and} \quad \boldsymbol{\alpha} = \gamma_0\boldsymbol{\gamma},$$

***i.e.*, $\boldsymbol{\Sigma}$ is expressed in terms of $\boldsymbol{\sigma}$ Pauli-matrices and $\boldsymbol{\alpha}$ is Dirac's velocity operator when written in terms of Dirac's gamma matrices.**

**We shall show that the quantum counterpart of expression (6.11) is in fact the electric dipole term of Dirac's Hamiltonian (6.12). The remaining part of this section is to consider the representation of the 'cross' product in (6.11) in terms of the matrix (or geometric) product of the elements of Dirac's algebra that represent the quantum version of the above observables, so that a short explanation to properly interpret these observables as elements of a Clifford algebra is given in what follows.**

**Both, velocity operator $\boldsymbol{u} = c\boldsymbol{\alpha}$ and spin operator $\boldsymbol{S}$ are bivectors in Dirac's algebra, considered as elements of the Geometric or Clifford algebra of space-time in the sense of Hestenes. [14]**

**In fact, Dirac's alpha matrices are written as a product of two gamma matrices $\alpha_i = \gamma_0\gamma_i$ and also the spin components $S_j = (i\hbar/2)\,\gamma_k\gamma_l$, $j,k,l$ cyclic $1,2,3$, and where the four gamma matrices, $\gamma_\mu$, $\mu = 0,1,2,3$ are interpreted as the four basic vectors of Minkowski's space-time that generate Dirac's Clifford algebra. They satisfy $\gamma_\mu\cdot\gamma_\nu = \eta_{\mu\nu}$, *i.e.*, $\gamma_0^2 = 1$ and $\gamma_i^2 = -1$, where the dot means the inner product in Dirac's Clifford algebra. We thus see that velocity and spin belong to the even**

subalgebra of Dirac's algebra and therefore they also belong to Pauli algebra or geometric algebra of three-dimensional space. Under spatial inversions $\gamma_0 \to \gamma_0$ and $\gamma_i \to -\gamma_i$, the velocity operator changes its sign and it is thus a spatial vector, while the spin is invariant under this transformation as it corresponds to a spatial bivector or pseudovector.

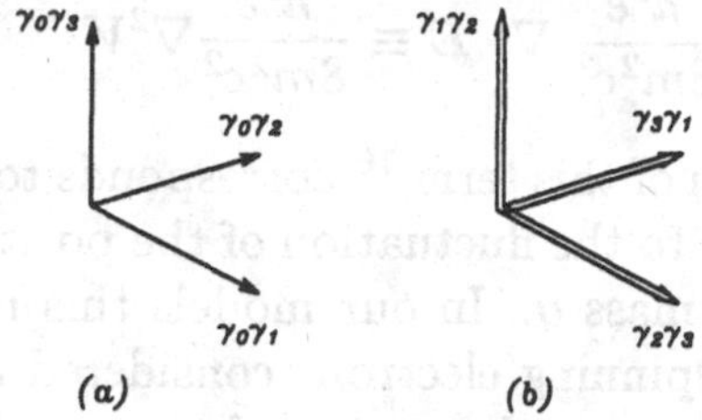


*Figure 6.16.* A basis for vectors (*a*) and bivectors (pseudovectors) (*b*) of Pauli algebra.

The relationship between the cross product and the outer and inner product of two vectors **a** and **b** in Pauli algebra is,

$$\boldsymbol{a} \times \boldsymbol{b} = -i\boldsymbol{a} \wedge \boldsymbol{b} = \boldsymbol{b} \cdot (i\boldsymbol{a}), \tag{6.13}$$

where $\wedge$ represents the symbol for the outer product in geometric algebra, the imaginary unit $i$ represents the unit three-vector or pseudoscalar and $i\boldsymbol{a}$ is the dual bivector of vector $\boldsymbol{a}$.

The inner product of a vector $\boldsymbol{b}$ and a bivector $A$ is expressed in terms of the geometric product in the form

$$\boldsymbol{b} \cdot A = \frac{1}{2}(\boldsymbol{b}A - A\boldsymbol{b}) \tag{6.14}$$

where in Dirac's or Pauli algebra the geometric product $\boldsymbol{b}A$ is just the ordinary multiplication of matrices.

If we choose a basis of vectors and pseudovectors as in Fig. 6.16, where the double-lined objects of part (*b*) represent the dual vectors of the corresponding spatial bivectors, and express in these bases the observables of Fig. 6.15, then the spatial velocity vector $\boldsymbol{u} = c\gamma_0\gamma_2$ and the pseudovector $\boldsymbol{S} = (\hbar/2)\gamma_2\gamma_3$ and therefore, using (6.13) and (6.14) we get

$$\boldsymbol{S} \times \boldsymbol{u} = \boldsymbol{u} \cdot (i\boldsymbol{S}) = \frac{ic\hbar}{2}\left(\frac{1}{2}(\gamma_0\gamma_2\gamma_2\gamma_3 - \gamma_2\gamma_3\gamma_0\gamma_2)\right) = \frac{-ic\hbar}{2}\gamma_0\gamma_3.$$

Now vector $\boldsymbol{k} = R\gamma_0\gamma_3$ with $R = \hbar/2mc$, and substituting in (6.11) we get the desired result.

## 1.4 DARWIN TERM OF DIRAC'S HAMILTONIAN

When analyzing Dirac's equation in the presence of an external electric field $\boldsymbol{E}$, Darwin [15] found in the expansion of the Hamiltonian an energy term, that bears his name, of the form

$$-\frac{\hbar^2 e}{8m^2c^2}\,\nabla\cdot\boldsymbol{E} \equiv \frac{\hbar^2 e}{8m^2c^2}\nabla^2 V. \tag{6.15}$$

The usual interpretation of this term [16] corresponds to the idea of zitterbewegung and therefore to the fluctuation of the position of the electron $\boldsymbol{r}$ around the center of mass $\boldsymbol{q}$. In our models this is very well understood because for the spinning electron, considered as a Luxon, there is a separation $S/mc$ between the center of mass and center of charge. Thus, by expanding the interaction potential around the center of mass $\boldsymbol{q}$ we get

$$V(\boldsymbol{q}+\delta\boldsymbol{r}) = V(\boldsymbol{q}) + \delta\boldsymbol{r}\cdot\nabla V + \frac{1}{2}\delta r_i \delta r_j \frac{\partial^2 V}{\partial r_i \partial r_j} + \cdots.$$

The fluctuation of the relative coordinates of the center of charge position vanish and thus $< \delta r_i >= 0$. Similarly $< \delta r_i \delta r_j >= 0$ for $i \neq j$ and the lowest order non-vanishing terms come from the fluctuations of

$$< (\delta r_1)^2 >=< (\delta r_2)^2 >=< (\delta r_3)^2 >= \frac{1}{3} < |\delta \boldsymbol{r}|^2 >= \frac{1}{3}\frac{S^2}{m^2c^2}.$$

We thus get

$$V(\boldsymbol{q}+\delta\boldsymbol{r}) = V(\boldsymbol{q}) + \frac{S^2}{6m^2c^2}\nabla^2 V,$$

but for a spin 1/2 particle $S^2 = 3\hbar^2/4$ and by multiplying the above expression by the electric charge $e$ we get the electrostatic potential energy of a charge at point $\boldsymbol{q}$, $eV(\boldsymbol{q})$, and the additional Darwin term (6.15). One important feature is that the Darwin term, related to the separation between the center of mass and center of charge, can also be used within a non-relativistic context, as shown by Fushchich et al. [17]

# 2. CLASSICAL SPIN CONTRIBUTION TO THE TUNNEL EFFECT

As a consequence of the zitterbewegung and therefore of the above analysis of the separation between the center of mass and center of charge, we shall see that spinning particles can have a non-vanishing crossing of potential barriers.

Let us consider a spinning particle with spin of (anti)orbital type, as described in Section 4. of Chapter 2, under the influence of a potential barrier. The Langrangian of this system is given by:

$$L = \frac{m}{2}\frac{\dot{r}^2}{\dot{t}} - \frac{m}{2\omega^2}\frac{\dot{u}^2}{\dot{t}} - eV(r)\dot{t}. \tag{6.16}$$

Sharp walls correspond classically to infinite forces so that we shall consider potentials that give rise to finite forces like those of the shape depicted in Fig. 6.17, where $V_0$ represents the top of the potential.

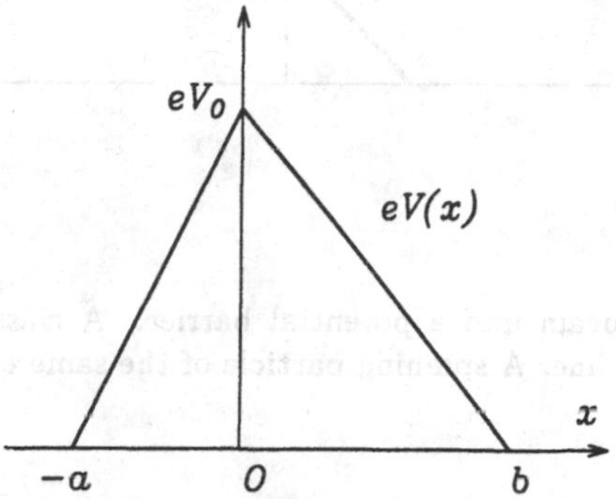


*Figure 6.17.* Triangular potential barrier.

Then the external force $F(x)$, is constant and directed leftwards in the region $x \in (-a, 0)$ and rightwards for $x \in (0, b)$, vanishing outside these regions.

Potentials of this kind can be found for instance in the simple experiment depicted in Figure 6.18 in which an electron beam, accelerated with some acceleration potential $V_a$, is sent into the uniform field region of potential $V_0$ contained between the grids or plates $A$, $C$ and $B$. Similarly in the $\alpha$-decay process, the estimated potential of the nucleus depicted in Figure 6.19 has a Coulomb-like behaviour for large distances and an unknown dotted part, and where $R_0$ is the radius of the nucleus. Distance $R_1 \simeq 10^{-14}$m is the estimated position of the top of the potential.

In Figure 6.18 from a strict classical viewpoint a spinless electron stops at the dotted line and is rejected backwards. But a classical spinning electron can cross the barrier provided its kinetic energy is above some minimum value, although below the top of the potential. This minimum value depends on the separation between plates.

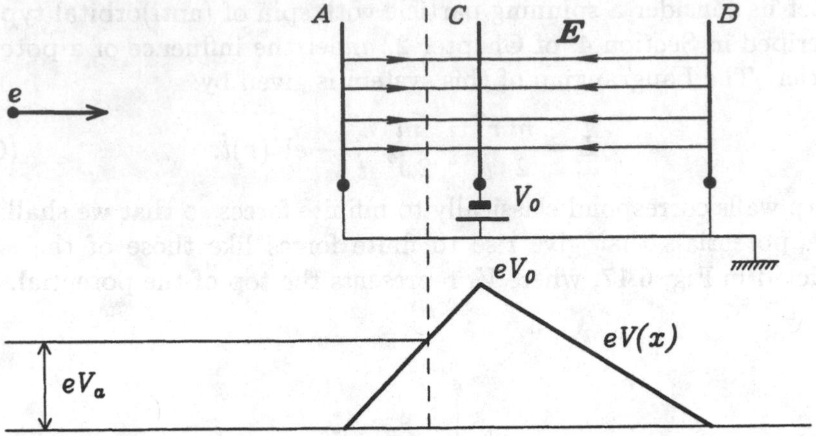


*Figure 6.18.* **Electron beam into a potential barrier. A classical spinless electron never crosses the dotted line. A spinning particle of the same energy might cross the barrier.**

Let us assume for simplicity that the spin is pointing up or down in the $z$ direction such that the point charge motion takes place in the $XOY$ plane. Let $q_x$, $q_y$ and $q_z = 0$, be the coordinates of the center of mass and $x$, $y$ and $z = 0$, the position of the charge.

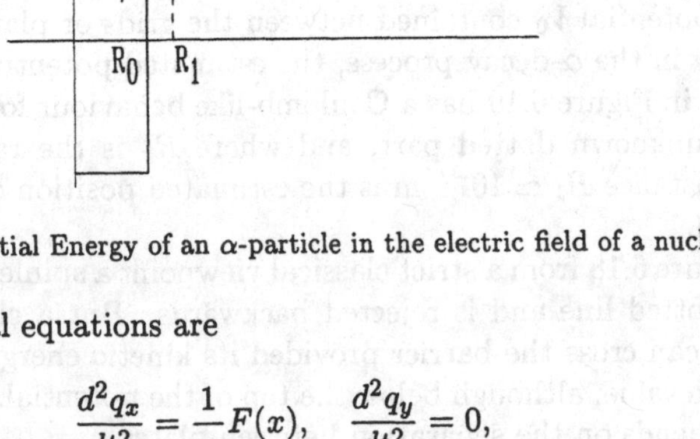

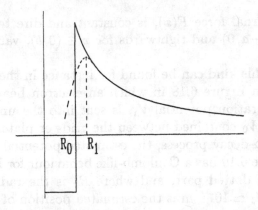

*Figure 6.19.* **Potential Energy of an $\alpha$-particle in the electric field of a nucleus.**

The dynamical equations are

$$\frac{d^2 q_x}{dt^2} = \frac{1}{m} F(x), \quad \frac{d^2 q_y}{dt^2} = 0, \tag{6.17}$$

$$\frac{d^2x}{dt^2} + \omega^2(x - q_x) = 0, \quad \frac{d^2y}{dt^2} + \omega^2(y - q_y) = 0, \tag{6.18}$$

where

$$F(x) = \begin{cases} -eV_0/a, & \text{for } x \in (-a, 0), \\ eV_0/b, & \text{for } x \in (0, b), \\ 0, & \text{otherwise.} \end{cases}$$

Equations (6.17) are nonlinear and we have not been able to obtain an analytical solution in closed form. We shall try to find a numerical solution. To make the corresponding numerical analysis we shall define different dimensionless variables. Let $R$ be the average separation between the center of charge and center of mass. In the case of circular internal motion, it is just the radius $R_0$ of the zitterbewegung. Then we define the new dimensionless position variables:

$$\hat{q}_x = q_x/R, \quad \hat{q}_y = q_y/R, \quad \hat{x} = x/R, \quad \hat{y} = y/R, \quad \hat{a} = a/R, \quad \hat{b} = b/R.$$

The new dimensionless time variable $\alpha = \omega t$ is just the phase of the internal motion, such that the dynamical equations become

$$\frac{d^2\hat{q}_x}{d\alpha^2} = A(\hat{x}), \quad \frac{d^2\hat{q}_y}{d\alpha^2} = 0,$$

$$\frac{d^2\hat{x}}{d\alpha^2} + \hat{x} - \hat{q}_x = 0, \quad \frac{d^2\hat{y}}{d\alpha^2} + \hat{y} - \hat{q}_y = 0,$$

where $A(\hat{x})$ is given by

$$A(\hat{x}) = \begin{cases} -eV_0/\hat{a}m\omega^2R^2, & \text{for } \hat{x} \in (-\hat{a}, 0), \\ eV_0/\hat{b}m\omega^2R^2, & \text{for } \hat{x} \in (0, \hat{b}), \\ 0, & \text{otherwise.} \end{cases}$$

In the case of the relativistic electron, the internal velocity of the charge is $\omega R = c$, so that the parameter $e/mc^2 = 1.9569 \times 10^{-6}\mathrm{V}^{-1}$, and for potentials of order of 1 volt we can take the dimensionless parameter $eV_0/m\omega^2R^2 = 1.9569 \times 10^{-6}$.

If we choose as initial conditions for the center of mass motion

$$\hat{q}_y(0) = 0, \quad d\hat{q}_y(0)/d\alpha = 0,$$

then the center of mass is moving along the $OX$ axis. The above system reduces to the analysis of the one-dimensional motion where the only variables are $\hat{q}_x$ and $\hat{x}$. Let us call from now on these variables $q$ and $x$ respectively and remove all hats from the dimensionless variables. Then the dynamical equations to be solved numerically are just

$$\frac{d^2q}{d\alpha^2} = A(x), \quad \frac{d^2x}{d\alpha^2} + x - q = 0, \tag{6.19}$$

where $A(x)$ is given by

$$A(x) = \begin{cases} -1.9569 \times 10^{-6}\, a^{-1} V_0, & \text{for } x \in (-a, 0), \\ 1.9569 \times 10^{-6}\, b^{-1} V_0, & \text{for } x \in (0, b), \\ 0, & \text{otherwise.} \end{cases} \tag{6.20}$$

Numerical integration has been performed by means of the computer package *Dynamics Solver.* [18] The quality of the numerical results is tested by using the different integration schemes this program allows, ranging from the very stable embedded Runge-Kutta code of eight order due to Dormand and Prince to very fast extrapolation routines. All codes have adaptive step size control and we check that smaller tolerances do not change the results.

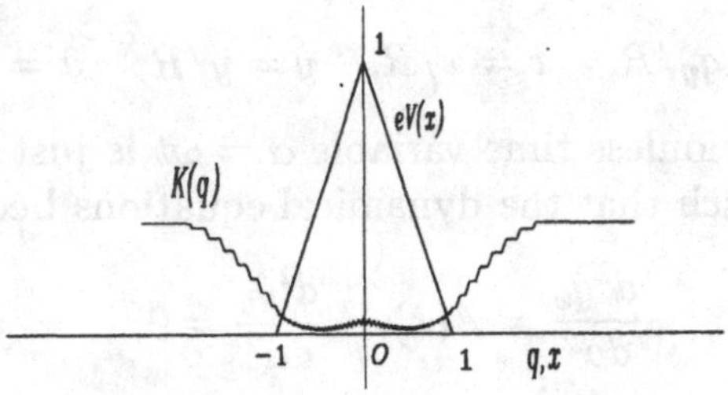


*Figure 6.20.* Kinetic Energy during the crossing for the values $a = b = 1$.

With $a = b = 1$, and in energy units such that the top of the barrier is 1, if we take an initial kinetic energy $K$ below this threshold, $K = m\dot{q}(0)^2/2eV_0 = 0.41$ we obtain for the center of mass motion the graphic depicted in Fig. 6.20, where is shown the variation of the kinetic energy of the particle $K(q)$, with the center of mass position during the crossing of the barrier. There is always crossing with a kinetic energy above this value. In Fig. 6.21, the same graphical evolution with $a = 1$ and $b = 10$ and $K = 0.9055$ for a potential of $10^3$ Volts in which the different stages in the evolution are evident. Below the initial values for the kinetic energy of 0.4 and 0.9 respectively, the particle does not cross these potential barriers and it is rejected backwards.

If in both examples the parameter $a$ is ranged from 1 to 0.05, thus making the left slope sharper, there is no appreciable change in the crossing energy, so that with $a = 1$ held fixed we can compute the minimum crossing kinetic energies for different $b$ values, $K_c(b)$.

To compare this model with the quantum tunnel effect, let us quantize the system. In the quantization of generalized Lagrangians developed in Chapter 4, the wave function for this system is a squared-integrable function $\psi(t, \boldsymbol{r}, \boldsymbol{u})$, of the seven kinematical variables and the generators

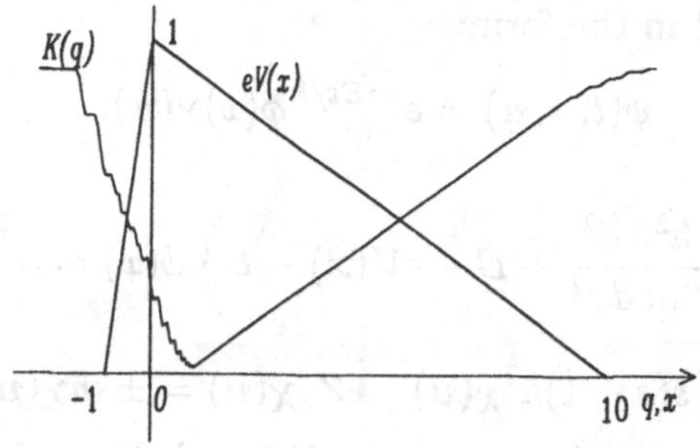


*Figure 6.21.* Kinetic Energy during the crossing for the values $a = 1$, $b = 10$.

of the Galilei group have the form (see Sec. 2.):

$$H = i\hbar\frac{\partial}{\partial t}, \; \boldsymbol{P} = -i\hbar\nabla, \; \boldsymbol{K} = m\boldsymbol{r} - t\boldsymbol{P} + i\hbar\nabla_u, \; \boldsymbol{J} = \boldsymbol{r} \times \boldsymbol{P} + \boldsymbol{Z}, \tag{6.21}$$

where $\nabla_u$ is the gradient operator with respect to the $\boldsymbol{u}$ variables. These generators satisfy the commutation relations of the extended Galilei group,[19] and the spin operator is given by $\boldsymbol{Z} = -i\hbar\boldsymbol{u} \times \nabla_u$.

One Casimir operator of this extended Galilei group is the Galilei invariant internal energy of the system $\mathcal{E}$, which in the presence of an external electromagnetic field and with the minimal coupling prescription is written as,

$$\mathcal{E} = H - eV - \frac{1}{2m}(\boldsymbol{P} - e\boldsymbol{A})^2, \tag{6.22}$$

where $V$ and $\boldsymbol{A}$ are the external scalar and vector potentials, respectively.

In our system $\boldsymbol{A} = 0$, and $V$ is only a function of the $x$ variable. It turns out that because of the structure of the above operators we can find simultaneous eigenfunctions of the following observables: the Casimir operator (6.22), $H$, $P_y$, $P_z$, $Z^2$ and $Z_z$. The particle moves along the $OX$ axis, with the spin pointing in the $OZ$ direction, and we look for solutions which are eigenfunctions of the above operators in the form:

$$\left(H - eV(x) - \frac{1}{2m}\boldsymbol{P}^2\right)\psi = \mathcal{E}\psi, \; H\psi = E\psi, \; P_y\psi = 0, \; P_z\psi = 0, \tag{6.23}$$

$$Z^2\psi = s(s+1)\hbar^2\psi, \quad Z_z\psi = \pm s\hbar\psi, \tag{6.24}$$

so that $\psi$ is independent of $y$ and $z$, and its time dependence is of the form $\exp(-iEt/\hbar)$. Since the spin operators produce derivatives only with respect to the velocity variables, we can look for solutions with the variables separated in the form:

$$\psi(t, x, u) = e^{-iEt/\hbar}\phi(x)\chi(u),$$

and thus

$$\left(\frac{\hbar^2}{2m}\frac{d^2}{dx^2} + E - eV(x) - \mathcal{E}\right)\phi(x) = 0, \tag{6.25}$$

$$Z^2\chi(u) = s(s+1)\hbar^2\chi(u), \quad Z_z\chi(u) = \pm s\hbar\chi(u), \tag{6.26}$$

where the spatial part $\phi(x)$, is uncoupled with the spin part $\chi(u)$, and $E - eV(x) - \mathcal{E}$ represents the kinetic energy of the system. The spatial part satisfies the one-dimensional Schroedinger equation, and the spin part is independent of the interaction, so that the probability of quantum tunneling is contained in the spatial part and does not depend on the particular value of the spin. If the particle is initially on the left-hand side of the barrier, with an initial kinetic energy $E_0 = E - \mathcal{E}$, then we can determine the quantum probability for crossing for $a = 1$ and different values of the potential width $b$.

The one-dimensional quantum mechanical problem of the spatial part for the same one-dimensional potential depicted in Fig. 6.17 is: [20]

$$\phi(x) = \begin{cases} e^{ikx} + Re^{-ikx}, & x \leq -a, \\ C_1\mathrm{Ai}(D(1-G+\frac{x}{a})) + C_2\mathrm{Bi}(D(1-G+\frac{x}{a})), & -a \leq x \leq 0, \\ C_3\mathrm{Ai}(L(1-G-\frac{x}{b})) + C_4\mathrm{Bi}(L(1-G-\frac{x}{b})), & 0 \leq x \leq b, \\ Te^{ikx}, & x \geq b, \end{cases} \tag{6.27}$$

where $x$ is the same dimensionless position variable as before, and the constants

$$k = \sqrt{\frac{E}{2mc^2}}, \; D = \sqrt[3]{\frac{eV_0a^2}{2mc^2}}, \; L = \sqrt[3]{\frac{eV_0b^2}{2mc^2}}, \; G = \frac{E}{eV_0}. \tag{6.28}$$

Functions $\mathrm{Ai}(x)$ and $\mathrm{Bi}(x)$ are the Airy functions of $x$. The six integration constants $R$, $T$, and $C_i$, $i = 1, 2, 3, 4$, can be obtained by assuming continuity of the functions and their first order derivatives at the separation points of the different regions. The coefficient $|R|^2$ represents the probability of the particle to be reflected by the potential and $|T|^2$ its probability of crossing.

Computing the $T$ amplitude for $a = 1$ and different values of the potential width $b$, and for energies below the top of the barrier $eV_0$, we show in Fig. 6.22, the average probability for quantum tunneling for four different potentials $V_0$ of $10^2$, $10^3$, $10^4$ and $10^5$ Volts. This average probability has been computed by assuming that on the left of the barrier there is a uniform distribution of particles of energies below $eV_0$.

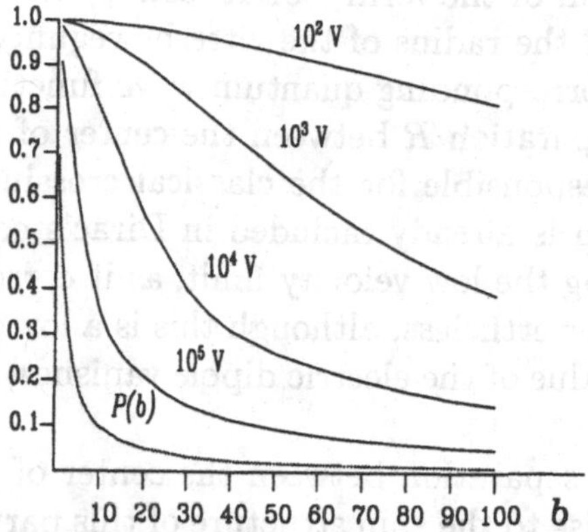


*Figure 6.22.* **Classical and Quantum Probability of crossing for different potentials.**

If we consider for the classical spinning particle the same uniform distribution of particles, then, the function $P(b) = 1 - K_c(b)$, where $K_c(b)$ is the minimum dimensionless kinetic energy for crossing computed before, represents the ratio of the particles that with kinetic energy below the top of the potential cross the barrier because of the spin contribution.

This function $P(b)$, is also depicted in Fig. 6.22. We see that for the different potentials shown in that figure the classical average probability of crossing is smaller than the quantum one, but for stronger potentials this classical probability, coming from the spin contribution, becomes relatively important.

Because the tunnel effect is a function of $\hbar$ and the spin of elementary particles is also of order of $\hbar$ it is very difficult to separate from the outcome of a real experiment involving elementary particles, which part is due to a pure quantum effect and which is the contribution to crossing coming from the spin structure. From (6.25) and (6.26) it is clear that the quantum probability of tunneling is independent of the spin.

To test experimentally this contribution, it will be necessary to perform separate experiments with particles of the same mass and charge but with different values of the spin. Thus, the difference in the outcome will be related to the spin contribution. This can be accomplished for instance, by using ions of the type $A^{++}$ that could be either in a singlet, $(s = 0)$ state or in a triplet $(s = 1)$ state.

But if there exists a contribution to crossing not included in the usual quantum mechanical analysis we have to modify the quantum mechanical equations. To be consistent with the above analysis the Schroedinger-

Pauli equation should be modified to include the additional electric dipole term. A term of the form $-eER\cos\omega t$, where $E$ is the external electric field and $R$ the radius of the zitterbewegung, should be considered to solve the corresponding quantum wave function. This term is of the order of the separation $R$ between the center of mass and center of charge, which is responsible for the classical crossing. This additional electric dipole term is already included in Dirac's equation but is suppressed when taking the low velocity limit, as it corresponds to this low energy example. Nevertheless, although this is a low energy process and the time average value of the electric dipole vanishes, there are very high field gradients.

We see that the separation between the center of mass and center of charge that gives rise to the spin structure of this particle model justifies that this system can cross a potential barrier even if its kinetic energy is below the top of the potential.

## 2.1 SPIN POLARIZED TUNNELING

I like to point out the following ideas to discuss whether they can be useful in connection with the interpretation of the magnetoresistance of polycrystaline films. This is known in the literature as the **spin polarized tunneling.** [21]

The main feature of the "classical" spin polarized tunneling we have seen in the previous section is not a matter of whether tunneling is classical or not, because this is a nonsense question. Matter at this scale is interpreted under quantum mechanical rules. But if we use a model of a classical spinning particle that, when polarized orthogonal to the direction of motion, produces a crossing that is not predicted by the Schroedinger-Pauli equation, it means that this quantum mechanical equation is lacking some term. The coupling term $-\boldsymbol{\mu}\cdot\boldsymbol{B}$, between the magnetic moment and magnetic field that gives rise to the Pauli equation, is inherited from Dirac's electron theory. But Dirac's equation also predicts another term $-\boldsymbol{d}\cdot\boldsymbol{E}$, of the coupling of an instantaneous electric dipole with the electric field. It is this oscillating electric dipole term that we believe is lacking in quantum mechanical wave equations. In general, the average value of this term in an electric field of smooth variation is zero. But in high intensity fields or in intergranular areas in which the effective potentials are low, but their gradients could be very high, this average value should not be negligible.

The conduction of electrons in synterized materials is completely different than the conduction on normal conductors. The material is not a continuous crystal. It is formed by small grains that are bound together by the action of some external pressure. If we can depict roughly the electric current flow, this is done by the jumping of electrons from grain to grain, through a tunneling process in which there is some estimated effective potential barrier confined in the gap between grains. Therefore these materials show in general a huge resistivity when compared with true conductors.

The form of this potential is unknown. The simplest one is to assume a wall of thickness $d$, the average separation between grains, and height $h$. But it can also be estimated as one of the potentials of the former example. What we have shown previously is that for every potential barrier, there is always a minimum energy, below the top of the potential, that electrons above that energy cross with probability 1 when polarized orthogonal to the motion, even within a classical interpretation. But this effect is not predicted by "normal" quantum mechanics because tunneling is spin independent.

Now, let us assume that we are able to estimate some average effective potential barrier in the intergranular zone of this polycristaline material. If the corresponding minimum crossing energy of this barrier for polarized electrons is below the Fermi level, then, when we introduce a magnetic field in the direction of the film and the magnetic domains in the grains become polarized, all electrons above that minimum energy of crossing will flow from grain to grain as in a good conductor, with a classical probability 1. That's all. Here the difficulty is to estimate properly this potential barrier and therefore the corresponding classical crossing energy.

It can be argued that the presence of the magnetic field to polarize electrons produces a change in the energy of particles. Nevertheless, even for a magnetic field of the order of 1 Tesla and in a potential barrier of 1 Volt, the magnetic term $-\boldsymbol{\mu} \cdot \boldsymbol{B}$ contributes with an energy of order of $\pm 5.7 \times 10^{-5}$eV, which does not modify the quantum probability of crossing.

## 3. QUANTUM MECHANICAL POSITION OPERATOR

One of the earlier controversies with Dirac's equation for a free particle was the disagreement between the direction of the linear momentum $\boldsymbol{P}$ and the velocity operator $\boldsymbol{u} \equiv c\boldsymbol{\alpha}$. Dirac himself [22] found that $\boldsymbol{P}$ was related to some time-average value of $\boldsymbol{u}$ with the usual relativistic expected relation $\boldsymbol{P} = H < \boldsymbol{u} > /c^2$. The search was then oriented to find the position of a point $\boldsymbol{q}$ such that its velocity was in the direction of $\boldsymbol{P}$ and $\boldsymbol{P} = (H/c^2)d\boldsymbol{q}/dt$, and that their components $q_i$ could be used as canonical variables.

This point $\boldsymbol{q}$ will be interpreted as a generalization of the center of mass of the particle and its proper definition is important to separate from the total angular momentum $\boldsymbol{J}$ the orbital angular momentum $\boldsymbol{L} = \boldsymbol{q} \times \boldsymbol{P}$ associated to this point. The remaining angular momentum $\boldsymbol{S} = \boldsymbol{J} - \boldsymbol{L}$ will be interpreted as the spin of the system. Therefore an accurate definition of $\boldsymbol{q}$ has associated a definition of spin.

For systems which have a well defined symmetric and conserved energy-momentum tensor $T^{\mu\nu} = T^{\nu\mu}$, $\partial_\nu T^{\mu\nu} = 0$, Pryce [23] was able to define three possibilities which we discuss in connection with the definition of the center of mass of our models.

The conservation of $T^{\mu\nu}$ of a classical system implies that the magnitudes

$$P^\mu = \int_t T^{\mu 0} d^3x, \quad J^{\mu\nu} = \int_t (x^\mu T^{\nu 0} - x^\nu T^{\mu 0}) d^3x = -J^{\nu\mu},$$

are constants of the motion. The above integrals are extended to the whole three-dimensional space taken at a constant time $t$. They define respectively the total four-momentum and total generalized angular momentum at any time $t$.

From the classical viewpoint, $P^\mu$ and $J^{\mu\nu}$ are the generating functions of the infinitesimal Poincaré transformations so that they satisfy the Poisson bracket relations of the Poincaré group. Because $T^{00}$ is considered as the energy density of the system, the first definition by Pryce takes the idea of defining a **center of energy** at a constant time by

$$P^0 q^\mu = \int_t x^\mu T^{00} d^3x,$$

where $P^0 = H/c$. This amounts in terms of the above magnitudes to

$$q^\mu = \frac{ctP^\mu + J^{\mu 0}}{P^0}. \tag{6.29}$$

This leads to $q^0 \equiv ct$ and, because $P^\mu$ and $J^{\mu\nu}$ are conserved, the time derivative of the spatial part of $q^\mu$ satisfies $\boldsymbol{P} = (H/c^2) d\boldsymbol{q}/dt$.

The spatial part of (6.29) matches with our definition of centre of mass of a spinning particle, because $J^{i0} \equiv cK_i$ is our kinematical momentum and the space-space part $J^{ij}$ is the angular momentum $\boldsymbol{J}$, and this expression is equivalent to obtaining $\boldsymbol{q}$ from the general expression of the kinematical momentum

$$\boldsymbol{K} = \frac{H}{c^2}\boldsymbol{q} - t\boldsymbol{P}. \tag{6.30}$$

This expression in the case of spinning Bradyons is written in terms of the charge position $\boldsymbol{r}$ as (see (3.88))

$$\boldsymbol{K} = \frac{H}{c^2}\boldsymbol{r} - t\boldsymbol{P} - \frac{1}{c}\boldsymbol{D},$$

and in the case of Luxons as

$$\boldsymbol{K} = \frac{H}{c^2}\boldsymbol{r} - t\boldsymbol{P} - \frac{1}{c^2}\boldsymbol{Z}\times\boldsymbol{u},$$

as in (3.159), where in both cases the total angular momentum appears as

$$\boldsymbol{J} = \boldsymbol{r}\times\boldsymbol{P} + \boldsymbol{Z}.$$

Function $\boldsymbol{Z} = \boldsymbol{u}\times\boldsymbol{U} + \boldsymbol{W}$ satisfies in the free case the dynamical equation $d\boldsymbol{Z}/dt = \boldsymbol{P}\times\boldsymbol{u}$ and when quantizing the system is the classical equivalent to Dirac's spin operator $\hbar\boldsymbol{\Sigma}/2$. If we consider that the electron is a Luxon, the center of mass position $\boldsymbol{q}$ is related to the position of the charge by

$$\boldsymbol{q} = \boldsymbol{r} - \frac{1}{H}\boldsymbol{Z}\times\boldsymbol{u}. \tag{6.31}$$

The geometric properties of the definition (6.29) are that $q^\mu$ does not transform like a four-vector and, as we saw in Sec. 3.3 of Chapter 3, the spatial components satisfy the Poisson brackets

$$\{\boldsymbol{q}, \boldsymbol{q}\} = -\frac{c^2}{H^2}\boldsymbol{S}, \tag{6.32}$$

and

$$\{\boldsymbol{q}, \boldsymbol{P}\} = 1.$$

Therefore they cannot be used as canonical variables.

The spin observable $\boldsymbol{S} = \boldsymbol{J} - \boldsymbol{q}\times\boldsymbol{P}$ is a constant of the motion for the free particle and satisfies the Poisson brackets

$$\{\boldsymbol{S}, \boldsymbol{S}\} = \boldsymbol{S} - \frac{c^2}{H^2}(\boldsymbol{S}\cdot\boldsymbol{P})\boldsymbol{P}, \quad \{\boldsymbol{S}, \boldsymbol{P}\} = 0, \quad \{\boldsymbol{S}, H\} = 0.$$

It is related with $\boldsymbol{Z}$ by

$$\boldsymbol{S} = \boldsymbol{Z} + \frac{1}{H}(\boldsymbol{Z}\times\boldsymbol{u})\times\boldsymbol{P} = \frac{1}{H}(H - \boldsymbol{u}\cdot\boldsymbol{P})\boldsymbol{Z} + \frac{1}{H}(\boldsymbol{P}\cdot\boldsymbol{Z})\boldsymbol{u}. \tag{6.33}$$

The second Pryce attempt was oriented to obtain a covariant definition. The result is

$$X^\mu = \frac{ctP^\mu}{P^0} + \frac{J^{\mu\nu}P_\nu}{m^2c^2} + \frac{J^{\nu 0}P^\mu P_\nu}{m^2c^2P^0}. \tag{6.34}$$

It also gives $X^0 \equiv ct$ and $\boldsymbol{P} = (H/c^2)d\boldsymbol{X}/dt$ for the free particle, but for the Poisson bracket of the space components

$$\{\boldsymbol{X}, \boldsymbol{X}\} = \frac{1}{m^2c^2}\,\boldsymbol{S}, \tag{6.35}$$

so that they cannot also be used as canonical variables. For the spatial part it leads to the expression

$$\boldsymbol{X} = \boldsymbol{q} + \frac{1}{m^2c^2}\boldsymbol{S} \times \boldsymbol{P}, \tag{6.36}$$

in terms of the previous $\boldsymbol{q}$ and $\boldsymbol{S}$ observables. Both definitions coincide for spinless systems.

To obtain canonical variables we see by inspection of equations (6.32) and (6.35) that the idea to obtain vanishing Poisson's brackets, is to produce a weighted average of the two previous definitions by

$$\widetilde{\boldsymbol{q}} = \frac{H\boldsymbol{q} + mc^2\boldsymbol{X}}{H + mc^2}.$$

Using previous Poisson brackets one gets $\{\widetilde{q}_i, \widetilde{q}_j\} = 0$, and $\{\widetilde{q}_i, P_j\} = \delta_{ij}$. However in this case $\widetilde{q}^\mu$ observable is not covariant. Using the expression of $\boldsymbol{X}$ (6.36) this gives the third Pryce definition of a canonical center of mass

$$\widetilde{\boldsymbol{q}} = \boldsymbol{q} + \frac{1}{m(H + mc^2)}\boldsymbol{S} \times \boldsymbol{P}. \tag{6.37}$$

In all the above expressions, observable $\boldsymbol{q}$ should be replaced by (6.31) whenever we want to express the corresponding position observable in terms of the charge position. The corresponding spin observable $\widetilde{\boldsymbol{S}}$ satisfies $\{\widetilde{\boldsymbol{S}}, \widetilde{\boldsymbol{S}}\} = \widetilde{\boldsymbol{S}}$, and is related to the previous one by

$$\widetilde{\boldsymbol{S}} = \frac{H}{mc^2}\boldsymbol{S} - \frac{\boldsymbol{P} \cdot \boldsymbol{S}}{m(H + mc^2)}\boldsymbol{P}.$$

We see that if $H$, $\boldsymbol{P}$ and $\boldsymbol{S}$ are constants of the motion, $\widetilde{\boldsymbol{S}}$ is also a constant of the motion. For all these spin operators we have

$$\boldsymbol{P} \cdot \boldsymbol{J} = \boldsymbol{P} \cdot \boldsymbol{Z} = \boldsymbol{P} \cdot \boldsymbol{S} = \boldsymbol{P} \cdot \widetilde{\boldsymbol{S}},$$

and the Pauli-Lubanski four-vector is defined in terms of the $\boldsymbol{S}$ observable as

$$W^\mu \equiv (\boldsymbol{P} \cdot \boldsymbol{S}, H\boldsymbol{S}/c).$$

There is however another possibility pointed out by Bacry [24] in which an alternative covariant definition of a position observable can be found which fulfills the requirement of having commuting coordinates, although they cannot be used as canonical variables. In any irreducible representation of the Poincaré group with generators $P^\mu$ and $J^{\mu\nu}$, this position is defined as

$$Y^\mu = \frac{1}{2m^2c^2}\left(J^{\mu\nu}P_\nu + P_\nu J^{\mu\nu}\right). \tag{6.38}$$

It satisfies

$$[Y^\mu, Y^\nu] = 0, \quad [Y^\mu, P^\nu] = i\hbar\left(\frac{P^\mu P^\nu}{m^2c^2} - g^{\mu\nu}\right).$$

For the time and space components and in the classical case in which products of $P^\mu$ and $J^{\mu\nu}$ can be taken in any order it reduces to

$$Y^0 = \frac{1}{m^2c}\boldsymbol{K}\cdot\boldsymbol{P}, \quad \boldsymbol{Y} = \frac{1}{m^2c^2}\left(H\boldsymbol{K} + \boldsymbol{J}\times\boldsymbol{P}\right)$$

that always vanish for an observer for which $\boldsymbol{P} = \boldsymbol{K} = 0$. If we substitute for $\boldsymbol{K}$ its expression (6.30) and take for $\boldsymbol{J} = \boldsymbol{q}\times\boldsymbol{P} + \boldsymbol{S}$, it is transformed into

$$\boldsymbol{Y} = \boldsymbol{q} + \frac{1}{m^2c^2}\left((\boldsymbol{q}\cdot\boldsymbol{P})\boldsymbol{P} + \boldsymbol{S}\times\boldsymbol{P}\right).$$

It differs from Pryce's covariant definition (6.36) in a term along $\boldsymbol{P}$. This implies that the orbital angular momentum associated to both $\boldsymbol{X}$ and $\boldsymbol{Y}$ are the same and therefore it leads to the same spin observable.

This center of mass definition has the nice property that for a two-particle relativistic system where $P^\mu = P_1^\mu + P_2^\mu$ and $J^{\mu\nu} = J_1^{\mu\nu} + J_2^{\mu\nu}$ are the total linear and angular momentum of the system, the center of mass of the system defined according to (6.38) satisfies

$$Y^\mu = \frac{m_1}{m_1 + m_2}Y_1^\mu + \frac{m_2}{m_1 + m_2}Y_2^\mu,$$

in the particular case when the $P_i^\mu$, $i = 1, 2$, are parallel four-vectors.

The final conclusion seems to be that there is no possibility for finding a relativistically covariant definition that at the same time yields canonical variables.

The need of having canonical coordinates is to produce a canonical quantization of the model. But canonical quantization is unnecessary if we have at hand a Lagrangian description as shown in Chapter 4 where quantization can be directly obtained on the kinematical space. The first definition of center of mass has the advantage that even in the case of an interacting particle, in which the interaction does not depend on the acceleration and angular velocity, as in the usual minimal coupling of electrodynamics, the expression of the mechanical linear momentum $\boldsymbol{P}_m$ remains the same as in the free case as

$$\boldsymbol{P}_m = \frac{H_m}{c^2}\frac{d\boldsymbol{q}}{dt}.$$

Here $H_m$ is the mechanical energy, which is also expressed in terms of the degrees of freedom and their derivatives as in the free case. Now dynamical equations for the center of mass motion become

$$\frac{d\boldsymbol{P}_m}{dt} = \boldsymbol{F},$$

where $\boldsymbol{F}$ is the external Lorentz force, defined at the charge position. This does not happen for the other definitions when the particle is no longer free and that is why we consider the first definition, although not covariant, as the center of mass or center of energy of the particle.

In the quantum case if $H = c\boldsymbol{\alpha}\cdot\boldsymbol{P} + mc^2\beta$ is Dirac's Hamiltonian, when acting on one-particle states it satisfies $H\psi = \pm E\psi$, where $E = +c(m^2c^2 + \boldsymbol{P}^2)^{1/2}$.

Total angular momentum in the quantum case is

$$\boldsymbol{J} = \boldsymbol{r}\times\boldsymbol{P} + \boldsymbol{Z} \equiv \boldsymbol{r}\times\boldsymbol{P} + \frac{\hbar}{2}\begin{pmatrix}\boldsymbol{\sigma} & 0\\ 0 & \boldsymbol{\sigma}\end{pmatrix}$$

where $d\boldsymbol{Z}/dt = \boldsymbol{P}\times c\boldsymbol{\alpha}$, for the free particle.

When restricted to positive and negative energy eigenstates the observables $\widetilde{\boldsymbol{q}}$, $\boldsymbol{P}$, $H/E$ and $\widetilde{\boldsymbol{S}}$ satisfy the same commutation relations as the observables $\boldsymbol{r}$, $\boldsymbol{P}$, $\beta$ and $\hbar\boldsymbol{\Sigma}/2$. The canonical transformation that links both sets of variables, obtained first by Pryce, is known in the literature as the Foldy-Wouthuysen transformation.[25] The basic feature of this transformation is that it takes positive (negative) energy states into positive (negative) energy states, respectively.

This classical position operator $\widetilde{\boldsymbol{q}}$ can be written as

$$\widetilde{\boldsymbol{q}} = \boldsymbol{r} - \frac{1}{H}\boldsymbol{Z}\times\boldsymbol{u} + \frac{1}{m(H+mc^2)}\boldsymbol{S}\times\boldsymbol{P}, \tag{6.39}$$

where we have to replace $\boldsymbol{S}$ by its expression (6.33) and finally in the above formula the following observables by their quantum mechanical equivalents

$$\boldsymbol{Z} = \frac{\hbar}{2}\begin{pmatrix}\boldsymbol{\sigma} & 0\\ 0 & \boldsymbol{\sigma}\end{pmatrix}, \quad \text{and} \quad \boldsymbol{u} = c\boldsymbol{\alpha} = c\begin{pmatrix}0 & \boldsymbol{\sigma}\\ \boldsymbol{\sigma} & 0\end{pmatrix}.$$

This position becomes in the Pryce-Foldy-Wouthuysen representation (PFW for short)

$$\widetilde{\boldsymbol{q}} = e^{-iS}\boldsymbol{r}e^{iS} = \boldsymbol{r} + \frac{i\hbar c}{2E}\beta\boldsymbol{\alpha} + \frac{i\hbar c^2}{2E(E+mc^2)}(\boldsymbol{\alpha}\cdot\boldsymbol{P})\boldsymbol{\alpha}.$$

This representation is generated by the unitary operator $e^{iS}$ which is given by

$$e^{\pm iS} = \frac{mc^2 + E \pm c\boldsymbol{\gamma} \cdot \boldsymbol{P}}{\sqrt{2E(E + mc^2)}}.$$

If we transform the angular momentum with the PFW transformation we get

$$e^{-iS}\boldsymbol{J}e^{iS} = \boldsymbol{J} = \widetilde{\boldsymbol{q}} \times \boldsymbol{P} + \widetilde{\boldsymbol{S}}.$$

Both, $\widetilde{\boldsymbol{q}} \times \boldsymbol{P}$ and $\widetilde{\boldsymbol{S}}$, are separately constants of the motion for the free particle. Foldy and Wouthuysen call $\widetilde{\boldsymbol{S}}$ the **mean spin operator**. Position operator $\widetilde{\boldsymbol{q}}$ also fits with the Newton-Wigner position operator [26] for spin 1/2 particles. Therefore the PFW transformation is that one which changes the spatial argument $\boldsymbol{r}$ of Dirac's spinor $\psi(t, \boldsymbol{r})$, and thus the position of the charge, for the variables $\widetilde{\boldsymbol{q}}$.

Bunge [27] also proposed another position operator that looks like the first two terms of $\widetilde{\boldsymbol{q}}$,

$$\boldsymbol{q}_B = \boldsymbol{r} + \frac{\lambda}{2} i\boldsymbol{\gamma},$$

with $\lambda = \hbar/mc$ is Compton's wave-length, and $\boldsymbol{\gamma} \equiv \beta\boldsymbol{\alpha}$, so that in the center of mass frame $\boldsymbol{P} = 0$ and $E = mc^2$, and $\widetilde{\boldsymbol{q}} = \boldsymbol{q}_B = \boldsymbol{q}$.

We can see that in both Pauli-Dirac and Weyl representations, Dirac's matrices become

$$\boldsymbol{\gamma} = \begin{pmatrix} 0 & \boldsymbol{\sigma} \\ -\boldsymbol{\sigma} & 0 \end{pmatrix},$$

and according to the interpretation we gave for the body frame in Section 4.3 of Chapter 4, this amounts to $3\boldsymbol{e}_2 = i\boldsymbol{\gamma}$ in the sense that vector operator $3\boldsymbol{e}_2$ plays the role of a unit vector directed from the charge to the center of mass and $\lambda = 2R$, where $R$ is the radius of the trajectory of the charge. Then, the relative position vector $\boldsymbol{k} = -Ri\boldsymbol{\gamma}$, so that $\boldsymbol{q}_B = \boldsymbol{r} - \boldsymbol{k}$.

However the velocity $d\boldsymbol{q}_B/dt$ is not a constant of the motion even for the free particle and the additional terms in either $\boldsymbol{q}$ or $\widetilde{\boldsymbol{q}}$ are needed to properly define the position of a point whose motion is along $\boldsymbol{P}$.

## 4. FINSLER STRUCTURE OF KINEMATICAL SPACE

Let $X$ be a given manifold. The tangent bundle of $X$, represented by $TX$, consists of the manifold $X$ together with all its possible tangent vectors at every point. Let us define on this manifold a real function $L$ which, to the pair $(x, y) \in TX$, associates the real number $L(x, y)$. Variables $y^i$ are the components of a tangent vector to the manifold at point $x$.

Among the possible functions $L$ that can be defined on $TX$, we shall restrict ourselves to positive homogeneous functions of first degree in the variables $y$, *i.e.,*:

$$L(x, ky) = kL(x, y), \qquad \text{for } k > 0. \tag{6.40}$$

In this case, taking the derivative of both sides with respect to $k$,

$$\frac{\partial L(x, ky)}{\partial k} = L = \frac{\partial L(x, z)}{\partial z^i} y^i,$$

with $z = ky$, and taking $k = 1$,

$$L = \frac{\partial L(x, y)}{\partial y^i} y^i = F_i(x, y) y^i. \tag{6.41}$$

Similarly $L^2(x, ky) = k^2 L^2(x, y)$, and if we differentiate twice with respect to $k$ and make $k = 1$ again, we get:

$$\frac{\partial^2 L^2(x, y)}{\partial y^i \partial y^j} y^i y^j = 2L^2, \tag{6.42}$$

and therefore $L^2$ has the form of a quadratic form in the variables $y^i$, where the coefficients are functions of $x$ and $y$, such that it can be written as:

$$L^2 = \frac{1}{2} \frac{\partial^2 L^2(x, y)}{\partial y^i \partial y^j} y^i y^j \equiv g_{ij}(x, y)\, y^i y^j. \tag{6.43}$$

If we consider in the manifold $X$, a trajectory $x(\tau)$ and the two close points $x(\tau)$ and $x(\tau + d\tau)$, and multiply (6.43) by $(d\tau)^2$,

$$L^2 (d\tau)^2 = (L d\tau)^2 = g_{ij}(x, y) y^i y^j (d\tau)^2 = g_{ij}(x, y) dx^i dx^j, \tag{6.44}$$

so that the function $g_{ij}(x, y)$ defined on $TX$, can be interpreted as a generalized metric of the manifold $X$, provided some invariant properties are required for $L$. This metric depends not only on the point $x$ but also on the direction $y$ along the path. Then $L\, d\tau$ is just the distance between the two points.

**Definition:** A manifold $X$ with a metric $g_{ij}(x, y)$ that comes from a real positive homogeneous function defined on $TX$, $L(x, y)$, of first degree in the $y$ variables, is called a **Finsler Space.** [28]

This is the case of the invariant Lagrangian functions that define classical elementary particles, when considered written in terms of the kinematical variables and their derivatives, and thus, the kinematical space of a generalized Lagrangian system can be transformed into a Finsler space, and the path integral of $Ld\tau$ is the length of the trajectory in $X$-space. As we shall see, the classical trajectories in $X$-space, are geodesics of the Finsler space $X$.

## 4.1 PROPERTIES OF THE METRIC

The components of the tangent vector $y$ transform like the contravariant components of a vector on the manifold $X$. If the function $L(x,y)$ is invariant under transformations of $X$ space that can be extended to $TX$, then the functions $F_i = \partial L(x,y)/\partial y^i$ transform like the covariant components of a vector on $X$,

$$y^{i'} = \dot{x}^{i'} = \frac{\partial x^{i'}}{\partial x^j}\dot{x}^j = \frac{\partial x^{i'}}{\partial x^j}y^j,$$

$$F_i(x,y) = \frac{\partial L(x,y)}{\partial y^i} = \frac{\partial L(x',y')}{\partial y^{j'}}\frac{\partial y^{j'}}{\partial y^i} = \frac{\partial y^{j'}}{\partial y^i}F_{j'}(x,y) = \frac{\partial x^{j'}}{\partial x^i}F_{j'}(x,y).$$

The metric $g_{ij}(x,y)$ is thus a symmetric covariant second-rank tensor on the manifold $X$, such that under arbitrary changes of coordinates on $X$, $x' = x'(x)$ transforms according to

$$g_{i'j'}(x',y') = \frac{\partial x^k}{\partial x^{i'}}\frac{\partial x^l}{\partial x^{j'}}\,g_{kl}(x,y). \tag{6.45}$$

We can define the covariant components of the tangent vector $y$ with the use of the metric tensor by $y_j = g_{ji}(x,y)y^i$, and since $L = \partial L/\partial y^j\, y^j$, and $L^2 - L(\partial L/\partial y^j)\,y^j = g_{ij}y^iy^j$ it implies that $y_j = g_{ji}y^i = L(\partial L/\partial y^j)$. Because $L^2 = y_iy^i$, $L$ represents the absolute value of the tangent vector, and $F_j = \partial L/\partial y^j = y_j/L$ are the covariant components of the unit tangent vector.

If we differentiate $y_j$ with respect to $y^i$ we get:

$$\frac{\partial y_j}{\partial y^i} = \frac{\partial}{\partial y^i}\left(L\frac{\partial L}{\partial y^j}\right) = \frac{\partial}{\partial y^i}\left(\frac{1}{2}\frac{\partial L^2}{\partial y^j}\right) = \frac{1}{2}\frac{\partial^2 L^2}{\partial y^i\partial y^j} = g_{ij}(x,y).$$

The magnitudes $g^{ij}(x,y)$ defined by $g^{ij}(x,y)g_{jk}(x,y) = \delta^i_k$ transform like the contravariant components of a second-rank tensor on $X$, and therefore we can use $g^{ij}(x,y)$ and $g_{ij}(x,y)$ for rising and lowering indexes of tensor magnitudes on $X$, as in Riemannian metric spaces.

If the metric tensor $g_{ij}(x,y)$ is independent of $y$, then the Finsler space $X$ becomes a Riemann space, and in that case the symmetric covariant tensor of third rank $C_{ijk}(x,y)$, defined by

$$C_{ijk}(x,y) \equiv \frac{1}{2}\frac{\partial g_{ij}(x,y)}{\partial y^k} = \frac{1}{4}\frac{\partial^3 L^2(x,y)}{\partial y^i\partial y^j\partial y^k}, \tag{6.46}$$

which is called Cartan's torsion tensor, vanishes identically. The Riemann spaces are manifolds with a connection that comes from a metric that is independent of the tangent vector $y$, and in consequence, they are torsion-free spaces.

Since the metric tensor is a homogeneous function of zero degree in the variables $y$, then Euler's theorem on homogeneous functions, implies:

$$\frac{\partial g_{ij}(x,y)}{\partial y^k}y^k = 0 \quad \Rightarrow \quad C_{ijk}(x,y)y^k = 0,$$

and the symmetric tensor $C_{ijkl}(x,y) = \partial C_{ijk}(x,y)/\partial y^l$ satisfies:

$$y^i C_{ijkl} = -C_{jkl}.$$

Another interesting tensor is the symmetric tensor $h_{ij}(x,y)$:

$$h_{ij}(x,y) = g_{ij}(x,y) - F_i(x,y)F_j(x,y), \tag{6.47}$$

which is a homogeneous function of zero degree in the variables $y^j$, and is denoted as the angular metric tensor, in the sense that it comes from $g_{ij}$ by the subtraction of the tensor constructed with the components of the unit vector $F_i$ and $F_j$.

Since $F_i = y_i/L$, taking the derivative with respect to $y^j$, it gives:

$$\frac{\partial F_i}{\partial y^j} = \frac{g_{ij}(x,y)}{L} - \frac{y_i}{L^2}\frac{\partial L}{\partial y^j} = \frac{1}{L}\left(g_{ij}(x,y) - F_iF_j\right)$$

i.e.,

$$h_{ij}(x,y) = L\,\frac{\partial F_i(x,y)}{\partial y^j}.$$

If we use the condition that $F_i$ is a homogeneous function of zero degree of $y^j$, then

$$h_{ij}(x,y)\,y^j = 0,$$

and

$$g^{ij}(x,y)h_{ij}(x,y) = g^{ij}(x,y)g_{ij}(x,y) - g_{ij}(x,y)F_i(x,y)F_j(x,y) = N - 1,$$

where $N$ is the dimension of the manifold $X$.

## 4.2 GEODESICS ON FINSLER SPACE

Given the manifold $X$, the distance between two points $x_1$ and $x_2$ is defined as the minimum value of the functional

$$d[C] = \int_{x_1}^{x_2} L(x, dx),$$

along a curve $C$ joining the two fixed points $x_1$ and $x_2$, since $L(x, dx) = ds$, is the arc length between the points $x$ and $x + dx$. If the curve $C$ is parametrized in terms of a parameter $\tau$, then this integral is independent of the path parameter. It can be written as

$$\int_{\tau_1}^{\tau_2} L(x(\tau), \dot{x}(\tau)d\tau) = \int_{\tau_1}^{\tau_2} L(x(\tau), \dot{x}(\tau))d\tau.$$

The geodesics joining $x_1$ with $x_2$ are curves on $X$ that satisfy Euler-Lagrange's equations:

$$\frac{d}{d\tau}\left(\frac{\partial L(x,\dot{x})}{\partial \dot{x}^i}\right) - \frac{\partial L(x,\dot{x})}{\partial x^i} = 0. \tag{6.48}$$

Because

$$\frac{\partial L(x,\dot{x})}{\partial \dot{x}^i} = l_i = \frac{\dot{x}_i}{L} = g_{ij}(x,\dot{x})\frac{\dot{x}^j}{L},$$

and also $L^2(x,\dot{x}) = g_{ij}(x,\dot{x})\dot{x}^i\dot{x}^j$, it gives

$$\frac{\partial L(x,\dot{x})}{\partial x^i} = \frac{1}{2}L(x,\dot{x})^{-1}\dot{x}^j\dot{x}^k\frac{\partial g_{jk}(x,\dot{x})}{\partial x^i}, \tag{6.49}$$

and substituted in (6.48) and after multiplying the result by $L(x,\dot{x})$ it leads to:

$$g_{ij}(x,\dot{x})\left[\frac{d^2x^j}{d\tau^2} - \dot{x}^j\frac{d}{d\tau}(\ln L(x,\dot{x}))\right] + \dot{x}^j\dot{x}^k\left[\frac{\partial g_{ij}}{\partial x^k} - \frac{1}{2}\frac{\partial g_{jk}}{\partial x^i}\right] = 0.$$

After contraction with $g^{li}(x,\dot{x})$, it gives:

$$\frac{d^2x^l}{d\tau^2} - \dot{x}^l\frac{d}{d\tau}(\ln L(x,\dot{x})) + \dot{x}^j\dot{x}^k g^{li}\Gamma_{jik} = 0,$$

where ${\Gamma_j}^l{}_k = g^{li}\Gamma_{jik}$ are the Finslerian Christoffel symbols, that define the connection on $X$-space. They are expressed in terms of the metric in the same form as in the case of a Riemann space:

$${\Gamma_j}^l{}_k = \frac{1}{2}g^{li}\left[\frac{\partial g_{ij}}{\partial x^k} + \frac{\partial g_{ik}}{\partial x^j} - \frac{\partial g_{jk}}{\partial x^i}\right]. \tag{6.50}$$

If we choose as a path parameter the arc length, then the tangent vector $\dot{x}$ is of unit length, i.e., $L(x,\dot{x}) = 1$, and $\ln L = 0$, and the equations of geodesics become

$$\frac{d^2x^l}{d\tau^2} + {\Gamma_j}^i{}_k\dot{x}^j\dot{x}^k = 0. \tag{6.51}$$

The Christoffel symbols ${\Gamma_j}^l{}_k$, are not tensor magnitudes and transform in a different way than in the case of a Riemann space under a change of coordinates of the manifold $X$. However, equations (6.51) keep the same form under these transformations.

## 4.3 EXAMPLES

The pointlike nonrelativistic particle has a kinematical space spanned by the variables $x^\mu \equiv (t, \boldsymbol{r})$, $\mu = 0,1,2,3$, and $L = m\dot{\boldsymbol{r}}^2/2\dot{t}$. If considered as a Finsler space, the metric should be:

$$g_{00} = \frac{3}{4}m^2\frac{(\dot{\boldsymbol{r}}^2)^2}{\dot{t}^4}, \quad g_{0i} = g_{i0} = -m^2\frac{\dot{\boldsymbol{r}}^2\dot{x}_i}{\dot{t}^3},$$

$$g_{ij} = g_{ji} = \frac{1}{2}m^2\frac{\dot{\boldsymbol{r}}^2}{\dot{t}^2}\delta_{ij} + m^2\frac{\dot{x}_i\dot{x}_j}{\dot{t}^2}.$$

Nevertheless, in the Galilei case the Lagrangian is not invariant under Galilei transformations and therefore is not invariant under arbitrary transformations. It turns out that the above quantities $g_{\mu\nu}$ do not transform like the components of a tensor. We have no Finsler structure in this case.

For the pointlike relativistic particle the kinematical variables are also $x^\mu \equiv (ct, \boldsymbol{r})$ and $L = -mc\sqrt{c^2\dot{t}^2 - \dot{\boldsymbol{r}}^2}$, the kinematical space is a Finsler space thus giving rise to a metric

$$g_{\mu\nu} = m^2c^2\eta_{\mu\nu},$$

where $\eta_{\mu\nu} \equiv \mathrm{diag}(1,-1,-1,-1)$ is Minkowski's metric tensor. In this particular case, the metric is independent of the tangent vector $\dot{x}$. The Finsler metric in this case is in fact a Riemannian metric. For spinless particles the metric is not direction dependent and thus the kinematical space is a flat Riemann space.

The metric transforms as a second rank covariant tensor under the changes of coordinates of the kinematical space $X$ when restricted to Poincaré transformations, and because its components are independent of the coordinates and their derivatives it transforms like a tensor under arbitrary transformations.

General Relativity is the extension of a free point particle with a flat kinematical space $X$ to a general kinematical space endowed with a Riemann metric $g_{\mu\nu}(x)$. These metric coefficients are in general functions of the space-time point $x$ and are interpreted as the potentials that produce the gravitational effects on the particle, such that a freely falling particle in this gravitational background follows a geodesic in this space. The source of the gravitational field is the energy-momentum tensor $T^{\mu\nu}$, such that the metric satisfies Einstein's equations

$$G^{\mu\nu} = -8\pi G T^{\mu\nu}. \tag{6.52}$$

Einstein's tensor $G^{\mu\nu}$ is written in terms of Ricci tensor $R^{\mu\nu}$ and the curvature scalar $R = R^{\mu}{}_{\mu}$ as $G^{\mu\nu} = R^{\mu\nu} - g^{\mu\nu}\, R/2$, and $G$ is Newton's gravitational constant. The Ricci tensor is a contraction of the Riemann-Christoffel curvature tensor $R^{\lambda}{}_{\mu\nu\sigma}$ by $R_{\mu\nu} \equiv R^{\lambda}{}_{\mu\lambda\nu}$ which is written in terms of the first and second derivatives of $g^{\mu\nu}$ in the form:

$$R^{\lambda}{}_{\mu\nu\sigma} = \frac{\partial \Gamma^{\lambda}_{\mu\nu}}{\partial x^{\sigma}} - \frac{\partial \Gamma^{\lambda}_{\mu\sigma}}{\partial x^{\nu}} + \Gamma^{\rho}_{\mu\nu}\Gamma^{\lambda}_{\sigma\rho} - +\Gamma^{\rho}_{\mu\sigma}\Gamma^{\lambda}_{\nu\rho},$$

where the Riemann-Christoffel symbols $\Gamma^{\lambda}_{\mu\nu}$ are expressed in terms of the metric by the same expression as in the Finsler space (6.50). Einstein's equations (6.52) when solved in terms of the unknowns $g_{\mu\nu}$ will give us the metric of the kinematical space of our point particle system in the presence of gravitation. In $T^{\mu\nu}$ are included all contributions of the external matter and fields and also the contribution of the energy-momentum of our point particle.

In the relativistic case, Lagrangians for elementary spinning particles are invariant under the Poincaré group $\mathcal{P}$. But their kinematical space is larger than the space-time manifold of our point particle. Even the Finsler metric coefficients for a free spinning particle also depend on the derivatives of the kinematical variables. Therefore the natural generalization to introduce gravitation for spinning particles is to consider that the kinematical space of our material system $X$ is in fact a Finsler space endowed with a general metric $g_{\mu\nu}(x, \dot{x})$ that will differ from the one of a free particle to take into account the gravitational effects, and that depends on both $x$ and $\dot{x}$. We thus see that the presence of spin establishes a direction dependent force thus confirming the idea that if we have gravitational effects related to the spin structure, then, the gravitational force will be direction dependent. [29]

This is consistent with the usual treatment of spin in General Relativity produced by Papapetrou *et. al.* in which they show that the orbits of spinning particles will differ from the geodesics on space-time. [30] As we have seen they are in fact geodesics on a larger manifold where the metric is direction dependent.

## 5. EXTENDING THE KINEMATICAL GROUP

We started in Sec. 11. of the first chapter by considering first the space-time translation group $\{\mathbb{R}^4, +\}$ as the kinematical group to implement the Special Relativity Principle. Later we increased the complexity of the group by considering new transformations like rotations and boosts, and arrived at the Galilei and Poincaré groups, where nonrelativistic and relativistic models of elementary particles, respectively, were worked out.

Is this the end of the story? Clearly not. Once we increased the number of parameters of the kinematical group, more and more classical variables were available to describe new degrees of freedom and consequently elementary objects of a more complex structure arose. Even more, when the number of classical variables grows, we have the chance to try new transformations involving these new variables, thus increasing the number of parameters of the symmetry group.

Particle physicists consider that to describe hadronic matter, besides the Poincaré group, some kind of *internal* kinematical group of the $SU(n)$ type is necessary to define new *internal* observables, like isospin, hypercharge and many others, that arise in high energy interactions.

We shall consider next the possibility of enlarging the Galilei and Poincaré groups by defining new space-time transformations like dilations, local rotations and local Lorentz transformations. In the next section we shall enlarge the Poincaré group to the Conformal group of Minkowski space-time.

## 5.1 SPACE-TIME DILATIONS

Space-time dilations act on the kinematical variables and their derivatives in the form

$$t' = e^{\lambda}t, \quad \boldsymbol{r}' = e^{\lambda}\boldsymbol{r}, \quad \boldsymbol{u}' = \boldsymbol{u}, \quad \boldsymbol{\rho}' = \boldsymbol{\rho},$$
$$\dot{t}' = e^{\lambda}\dot{t}, \quad \dot{\boldsymbol{r}}' = e^{\lambda}\dot{\boldsymbol{r}}, \quad \dot{\boldsymbol{u}}' = \dot{\boldsymbol{u}}, \quad \boldsymbol{\omega}' = \boldsymbol{\omega}.$$

Any group element of this Abelian one-parameter group is characterized by the normal parameter $\lambda$. The neutral element is given by $\lambda = 0$, and the composition law is $\lambda'' = \lambda' + \lambda$.

Translation invariant Lagrangians whose dependence on $\dot{t}$ and $\dot{\boldsymbol{r}}$ is only through the dependence on the velocity $\boldsymbol{u} = \dot{\boldsymbol{r}}/\dot{t}$ are also invariant. Terms of either form $\boldsymbol{\omega}^2 - \dot{\boldsymbol{u}}^2$ or $\boldsymbol{\omega} \cdot \dot{\boldsymbol{u}}$ are invariant. In the relativistic case, the Poincaré invariant terms $\boldsymbol{\alpha}^2 - \boldsymbol{\omega}^2$ and $\boldsymbol{\alpha} \cdot \boldsymbol{\omega}$, given in (3.95) and (3.96), respectively, are also dilation invariant. However terms of the form $c^2\dot{t}^2 - \dot{\boldsymbol{r}}^2$ transform with a constant factor so that for instance the point particle relativistic Lagrangian is not invariant but transforms with a global factor and therefore the dynamical equations, for the free point particle, remain invariant.

The Bradyons described by Lagrangians (3.97) and (3.98) are not invariant under dilations. For Luxons we shall analyze their invariance under the more general conformal transformations in the next section, but the photon Lagrangian (3.151) is invariant under dilations and the possible Lagrangian terms in (3.176) are invariant, while those in (3.177) transform with a constant global factor because of their explicit dependence on the variable $\dot{t}$.

If a Lagrangian is invariant under this group, an infinitesimal transformation of dimensionless parameter $\delta\lambda$ produces the variations $\delta t = t\delta\lambda$ and $\delta \boldsymbol{r} = \boldsymbol{r}\delta\lambda$ and thus Noether's theorem defines the following constant of the motion

$$D = tH - \boldsymbol{r} \cdot \boldsymbol{P}, \tag{6.53}$$

with dimensions of action. Taking the $\tau$-derivative of this expression this yields $H = \boldsymbol{P} \cdot \boldsymbol{u}$, as it happens in the case of massless Luxons, as for the photon, but not for the electron. This suggests that if we assume dilation invariance for massive Luxons, then necessarily the Lagrangian must depend on some additional kinematical variables that also transform under this group, to obtain on the right-hand side of (6.53) more additional terms.

## 5.2 LOCAL ROTATIONS

Let us consider, for instance, what happens if we think about the possibility of rotating the local frame attached to the particle without modifying the remaining kinematical variables. This means that we have in addition to $\mathcal{G}$ or $\mathcal{P}$ a new three-parameter group of transformations of the kinematical variables, isomorphic to the rotation group but restricted in its action to only the orientation variables. Let us represent by $R(\boldsymbol{\nu})$ one of these transformations. Then the kinematical variables and their derivatives transform under this **local** $SO(3)$, or $SO(3)_L$ for short, of parameter $\boldsymbol{\nu}$ in the form:

$$t' = t, \quad \boldsymbol{r}' = \boldsymbol{r}, \quad \boldsymbol{u}' = \boldsymbol{u}, \quad R(\boldsymbol{\rho}') = R(\boldsymbol{\nu})R(\boldsymbol{\rho}),$$

and

$$\dot{t}' = \dot{t}, \quad \dot{\boldsymbol{r}}' = \dot{\boldsymbol{r}}, \quad \dot{\boldsymbol{u}}' = \dot{\boldsymbol{u}}, \quad \boldsymbol{\omega}' = R(\boldsymbol{\nu})\boldsymbol{\omega}.$$

Nonrelativistic Lagrangians constructed from terms of the form $\boldsymbol{\omega} \cdot \dot{\boldsymbol{u}}$ which are invariant under rotations are no longer invariant under these local rotations, because $\boldsymbol{\omega}$ rotates but $\dot{\boldsymbol{u}}$ does not. Therefore, requirement of invariance of dynamical equations under a local $SO(3)_L$ rejects terms of the above form thus restricting the free Lagrangians to depend only on terms of the form $\boldsymbol{\omega}^2 - \dot{\boldsymbol{u}}^2$.

Let us assume that our Lagrangian is of the above invariant form. Once we have a new invariance group we can define the corresponding Noether's constants of the motion. In this case the three new constants of the motion define a vector magnitude $\boldsymbol{I}$ we shall call the I-spin, which for both nonrelativistic and relativistic systems reduces to the magnitude $\boldsymbol{W} \equiv \partial L/\partial \boldsymbol{\omega}$. In fact it is a constant of the motion, because if the dependence of the Lagrangian on the orientation variables is only through $\boldsymbol{\omega}^2$ terms, then $\boldsymbol{W} \sim \boldsymbol{\omega}$ and its dynamical equations are $d\boldsymbol{W}/dt = \boldsymbol{\omega} \times \boldsymbol{W} = 0$, as we have seen in previous chapters.

The new generators of the enlarged kinematical group, $\boldsymbol{I}$, satisfy in addition to the commutation relations of $SO(3)_L$,

$$[\boldsymbol{I}, \boldsymbol{I}] = \boldsymbol{I},$$

the commutation relations with the old generators of the Galilei group

$$[\boldsymbol{I}, H] = [\boldsymbol{I}, \boldsymbol{P}] = [\boldsymbol{I}, \boldsymbol{K}] = 0, \quad [\boldsymbol{I}, \boldsymbol{J}] = \boldsymbol{I},$$

so that they are invariant under translations and Galilei boosts and transform like vectors under rotations.

For the relativistic case they no longer commute with the $\boldsymbol{K}$ but they satisfy

$$[\boldsymbol{I}, \boldsymbol{K}] = -\boldsymbol{K}, \quad [\boldsymbol{I}, \boldsymbol{J}] = \boldsymbol{I}.$$

## 5.3 LOCAL LORENTZ TRANSFORMATIONS

As seen above, local rotations amount to rotating the local frame, or triad, associated to the particle. We can extend this, in the relativistic case, to the rotation of the local tetrad associated to the particle, without modifying the space-time variables. Thus, in addition to the Poincaré group, we have a local Lorentz group $\mathcal{L}_L$ of transformations acting only on the local tetrad $\boldsymbol{e}_\alpha$. But this looks like a gauge group in field theory, where to describe interactions it is assumed that the total Lagrangian density must be invariant under a local transformation group. This gauge group is a set of transformations that transform the phase of the fields without changing the space-time coordinates.

However, this gauge group cannot be considered as a kinematical group because in general the free Lagrangian is not invariant under its transformations. It is the whole Lagrangian, with the inclusion of the additional gauge fields that mediate in between the interaction, which has to be invariant. Therefore, in field theory, what we have is a **dynamical symmetry** that restricts the possible interaction terms.

Whether or not the above discussion about local groups can be interpreted as a dynamical or kinematical symmetry is an important subject that deserves more work. If it is a kinematical symmetry we can explore the additional group parameters to use them as additional kinematical variables. These will give rise to new degrees of freedom to be used to explore new physical properties. But if it is a dynamical symmetry, then no additional degrees of freedom are necessary, but it produces a constraint on the possible interaction terms to be considered in the Lagrangian.

We shall not explore these possibilities any longer in this book, but it can be taken as a plausible conjecture for future work.

# 6. CONFORMAL INVARIANCE

In this section we shall consider the Conformal group as a plausible extension of the Poincaré group, because it has played a historical role in connection with electromagnetism. Conformal invariance of Maxwell's equations has been known since the early days of this century, [31] so that we shall briefly review next the Conformal group structure and analyze the conformal invariance of the proposed models for the photon and electron, when considered as a kinematical group.

## 6.1 CONFORMAL GROUP

In a metric space the angle $\theta$ between two curves, intersecting at a point, is given by

$$\cos\theta = \frac{g_{\mu\nu} dx^\mu dy^\nu}{\sqrt{g_{\mu\nu} dx^\mu dx^\nu}\,\sqrt{g_{\mu\nu} dy^\mu dy^\nu}}. \tag{6.54}$$

The differential elements $dx^\mu$ and $dy^\mu$, respectively, are the corresponding coordinates of the arc elements of the two curves. A transformation of the space that preserves the angles between intersecting curves is named a **conformal transformation**.

Let us consider the mapping from the Euclidean plane into the unit sphere given by the stereographic projection. This application preserves the angles between two intersecting curves in the plane and their corresponding images on the unit sphere. If we now rotate this unit sphere, this rotation induces back a transformation of the points of the plane, in general non-linear, that also preserves the angles between the above curves. It turns out that the whole rotation group $SO(3)$ when acting on this sphere is also a group of conformal transformations of our initial Euclidean plane. Therefore the $SO(3)$ group is a subgroup of the Conformal group of flat two-dimensional space.

Let us consider those transformations of Minkowski space-time such that

$$ds'^2 = \eta_{\mu\nu} dx^{\mu'} dx^{\nu'} = \Omega(x)^2\, \eta_{\mu\nu} dx^\mu dx^\nu = \Omega(x)^2\, ds^2, \tag{6.55}$$

where $\Omega(x)^2$ is a global factor. Then, according to (6.54), angles between curves in Minkowski space are conserved, and the transformation is a conformal transformation. These are examples of conformal transformations.

If we have a pseudometric manifold $\mathbb{R}^{p+q}$, with metric of signature $(p,q)$, with $p$ space-like variables and $q$ time-like, *i.e.*, the differential arc element can be written in the form

$$ds^2 = -dx_1^2 - dx_2^2 - \cdots - dx_p^2 + dx_{p+1}^2 + \cdots + dx_{p+q}^2, \tag{6.56}$$

then, the group of inhomogeneous linear transformations $ISO(p,q)$, including translations and pseudo-orthogonal transformations, preserves the metric and therefore is a group of conformal transformations of the manifold. But, what is the largest connected group of conformal transformations of a metric space? It can be shown [32] that the *Conformal Group* of this manifold is the pseudo-orthogonal group of a manifold of larger dimension, one more dimension in each kind of variables, *i.e.*, the group $SO(p+1,q+1)$. According to this the conformal group of the plane, considered as the Euclidean plane of signature $(2,0)$, will be $SO(3,1)$, and the conformal group of Minkowski space-time, with metric of signature $(3,1)$, is the group $SO(4,2)$.

The proof is based on the generalized stereographic projection on a larger pseudo-metric space. We start with a manifold $\mathbb{R}^{p+q}$ and we go into a new space $\mathbb{R}^{(p+1)+(q+1)}$. We map every point of the initial space of coordinates $x^\mu$ onto a point on the null cone of the second manifold of coordinates $z^a \equiv (u,(u-v)x^\mu, v)$, with one new space-like variable $u$ and another $v$, time-like, and therefore the norm of the image point is $-u^2 + (u-v)^2 x_\mu x^\mu + v^2 = 0$. If we choose

$$u^2 + (u-v)^2 x^2 = v^2 + (u-v)^2 y^2 = 1,$$

where $x^2 \equiv x_1^2 + \cdots + x_p^2$ and $y^2 \equiv x_{p+1}^2 + \cdots + x_{p+q}^2$, respectively, and therefore $x_\mu x^\mu = -x^2 + y^2$, these two conditions define in the new space of dimension $p+q+2$, a submanifold of dimension $p+q$ homeomorphic to the initial space. The group $SO(p+1,q+1)$ maps this submanifold on itself and therefore preserves the angles between curves.

Among the transformations of this group we find those of the form

$$A = \begin{pmatrix} 1 & 0 & 0 \\ 0 & R & 0 \\ 0 & 0 & 1 \end{pmatrix}, \tag{6.57}$$

with $R \in SO(p,q)$, that corresponds to the application $(u,(u-v)x^\mu, v) \to (u,(u-v)x'^\mu, v)$, with $x' = Rx$, and therefore $x'^\mu x'_\mu = x^\mu x_\mu$. But also of the form

$$D(\lambda) = \begin{pmatrix} \cosh\lambda & 0 & \sinh\lambda \\ 0 & \mathbb{I} & 0 \\ \sinh\lambda & 0 & \cosh\lambda \end{pmatrix}, \tag{6.58}$$

where $\mathbb{I}$ is the $(p+q) \times (p+q)$ unit matrix, and corresponds to

$$\begin{aligned} u' &= \cosh\lambda\, u + \sinh\lambda\, v, &(6.59)\\ (u'-v')x'^{\mu} &= (u-v)x^{\mu}, &(6.60)\\ v' &= \cosh\lambda\, v + \sinh\lambda\, u, &(6.61)\end{aligned}$$

that map the point $x^{\mu}$ on the point $x'^{\mu}$, i.e.,

$$x'^{\mu} = \frac{(u-v)x^{\mu}}{(\cosh\lambda - \sinh\lambda)(u-v)} = e^{\lambda}x^{\mu}. \qquad (6.62)$$

It is the group of **dilations** of the original space, with normal (or canonical) parameter $\lambda$. The transformations of the form

$$P(a^{\mu}) = \begin{pmatrix} 1 + a_{\mu}a^{\mu}/2 & a_{\nu} & -a_{\mu}a^{\mu}/2 \\ a^{\mu} & \mathbb{I} & -a^{\mu} \\ a_{\mu}a^{\mu}/2 & a_{\nu} & 1 - a_{\mu}a^{\mu}/2 \end{pmatrix}, \qquad (6.63)$$

correspond to

$$\begin{aligned} u' &= u + \frac{1}{2}a_{\mu}a^{\mu}(u-v) + a_{\nu}x^{\nu}(u-v), &(6.64)\\ (u'-v')x'^{\mu} &= (u-v)x^{\mu} + a^{\mu}(u-v), &(6.65)\\ v' &= v + \frac{1}{2}a_{\mu}a^{\mu}(u-v) + a_{\nu}x^{\nu}(u-v), &(6.66)\end{aligned}$$

and therefore the image of point $x^{\mu}$ is the point

$$x'^{\mu} = \frac{(x^{\mu} + a^{\mu})(u-v)}{u-v} = x^{\mu} + a^{\mu}, \qquad (6.67)$$

*i.e.*, is the group of translations of normal parameters $a^{\mu}$. Finally, the transformations

$$R(b^{\mu}) = \begin{pmatrix} 1 + b_{\mu}b^{\mu}/2 & -b_{\nu} & b_{\mu}b^{\mu}/2 \\ -b^{\mu} & \mathbb{I} & -b^{\mu} \\ -b_{\mu}b^{\mu}/2 & b_{\nu} & 1 - b_{\mu}b^{\mu}/2 \end{pmatrix}, \qquad (6.68)$$

correspond to

$$\begin{aligned} u' &= u + \frac{1}{2}b_{\mu}b^{\mu}(u+v) - b_{\nu}x^{\nu}(u-v), &(6.69)\\ (u'-v')x'^{\mu} &= (u-v)x^{\mu} - b^{\mu}(u+v), &(6.70)\\ v' &= v - \frac{1}{2}b_{\mu}b^{\mu}(u+v) + b_{\nu}x^{\nu}(u-v). &(6.71)\end{aligned}$$

The transformed point of $x^\mu$ is

$$x'^\mu = \frac{(u-v)x^\mu - b^\mu(u+v)}{(u-v) - 2b_\nu x^\nu(u-v) + b_\mu b^\mu(u+v)} = \frac{x^\mu - b^\mu(u+v)/(u-v)}{1 - 2b_\nu x^\nu + b_\mu b^\mu(u+v)/(u-v)},$$

but, since

$$-u^2 + (u-v)^2 x_\mu x^\mu + v^2 = 0,$$

it gives $u^2 - v^2 = (u-v)^2 x_\mu x^\mu$, i.e., $(u+v)/(u-v) = x_\mu x^\mu$, and the image is the point

$$x'^\mu = \frac{x^\mu - b^\mu x_\mu x^\mu}{1 - 2b_\mu x^\mu + b_\mu b^\mu x_\mu x^\mu}. \tag{6.72}$$

These nonlinear transformations, are called **pure conformal transformations**. Parameters $b^\mu$, are normal or canonical parameters because as we can easily check $R(a)R(b) = R(a+b)$.

## 6.2 CONFORMAL GROUP OF MINKOWSKI SPACE

The Conformal group is the group of transformations that leave invariant Maxwell's equations in empty space. Then they leave invariant the null path of a photon, $ds^2 \equiv (dx^0)^2 - d\boldsymbol{r}^2 = 0$, and also the differential path of the charge of the electron (see Sec. 4.2).

Since the Poincaré group leaves invariant the arc $(ds)^2$, it is a subgroup of the Conformal group. It contains the space-time translations, of infinitesimal generators $P_\mu$, and Lorentz transformations of generators $J_{\mu\nu} = x_\mu\partial_\nu - x_\nu\partial_\mu$.

The linear transformation of dimensionless parameter $\lambda$,

$$t' = e^\lambda t, \quad \boldsymbol{r}' = e^\lambda \boldsymbol{r}, \quad \forall\lambda \in \mathbb{R},$$

gives rise to

$$c^2dt'^2 - d\boldsymbol{r}'^2 = e^{2\lambda}(c^2dt^2 - d\boldsymbol{r}^2) = 0,$$

and therefore these transformations generate a one-parameter subgroup, the subgroup of space-time dilations. The infinitesimal generator is

$$D = t\frac{\partial}{\partial t} + \boldsymbol{r}\cdot\nabla = x^\mu\partial_\mu. \tag{6.73}$$

The non-linear transformation of normal parameter $a^\mu$,

$$x'^\mu = \frac{x^\mu - a^\mu x^2}{1 - 2a\cdot x + a^2x^2}, \tag{6.74}$$

where we have written $a \cdot x \equiv a^\mu x_\mu$ and $a^2 \equiv a^\mu a_\mu$, is a pure Conformal transformation. If we call $N = 1 - 2a \cdot x + a^2 x^2$, then $(x')^2 = x^2/N$. If the transformation is infinitesimal, it can be expanded to first order in the group parameters

$$x'^\mu = \frac{x^\mu - a^\mu x^2}{1 - 2a \cdot x + O(a^2)} \simeq (x^\mu - a^\mu x^2)(1+2a \cdot x) \simeq x^\mu - a^\mu x^2 + 2(a \cdot x)x^\mu,$$

and for the infinitesimal arc element

$$(dx')^2 = \frac{(dx)^2}{N^2},$$

and if $(dx)^2 = 0$, then also $(dx')^2 = 0$. The neutral element corresponds to $a = 0$, and the infinitesimal generators $R_\mu$, are given by

$$R_\mu = 2x_\mu x^\nu \partial_\nu - x^2 \partial_\mu, \tag{6.75}$$

that commute among themselves and together with $D$ and the generators of the Poincaré group, lead to the commutation relations of the **Conformal group $\mathcal{C}$** as

$$[J_{\mu\nu}, J_{\rho\sigma}] = \eta_{\nu\rho} J_{\mu\sigma} + \eta_{\mu\sigma} J_{\nu\rho} - \eta_{\mu\rho} J_{\nu\sigma} - \eta_{\nu\sigma} J_{\mu\rho}, \tag{6.76}$$

$$[J_{\mu\nu}, P_\sigma] = -\eta_{\mu\sigma} P_\nu + \eta_{\nu\sigma} P_\mu, \quad [J_{\mu\nu}, R_\sigma] = -\eta_{\mu\sigma} R_\nu + \eta_{\nu\sigma} R_\mu, \tag{6.77}$$

$$[P_\mu, P_\nu] = 0, \quad [R_\mu, R_\nu] = 0, \quad [P_\mu, R_\nu] = 2\eta_{\mu\nu} D - 2J_{\mu\nu}, \tag{6.78}$$

$$[J_{\mu\nu}, D] = 0, \quad [P_\mu, D] = P_\mu, \quad [R_\mu, D] = -R_\mu. \tag{6.79}$$

It is a 15 parameter group that is locally isomorphic to the groups $SU(2,2)$ and $SO(4,2)$.

From the point of view of $SO(4,2)$, if we consider Minkowski spacetime with coordinates $x^0 = ct$, $x^i$, $i = 1,2,3$, and we enlarge this space by considering two new variables, one space-like coordinate $x^4$ and another time-like $x^5$, the metric on this manifold is $G = \mathrm{diag}(1,-1,-1,-1,-1,1)$, *i.e.*, $g_{00} = g_{55} = 1$, $g_{ii} = -1$, $i = 1,\ldots,4$, and the remaining $g_{ij}$ vanishing, then the 15 infinitesimal generators take the form

$$J_{ab} = -J_{ba} = x_a \partial_b - x_b \partial_a, \quad a,b = 0,1,2,3,4,5, \tag{6.80}$$

and satisfy the commutation relations of $SO(4,2)$,

$$[J_{ab}, J_{cd}] = g_{bc} J_{ad} + g_{ad} J_{bc} - g_{ac} J_{bd} - g_{bd} J_{ac}. \tag{6.81}$$

A matrix representation of these relations is

$$[J_{ab}]^c{}_d = \delta^c_a g_{bd} - \delta^c_b g_{ad}, \tag{6.82}$$

where $c$ is the row index and $d$ the column index.

If as usual we reserve Greek indices $\mu = 0, 1, 2, 3$ for the Minkowski space-time part, then we identify the Lorentz group generators with the corresponding $J_{\mu\nu}$ with $\mu, \nu = 0, 1, 2, 3$. Let us set $J_i = \frac{1}{2}\epsilon_{ijk}J_{jk}$, $K_i = J_{0i}$, $D = J_{45}$ and $A_\mu = J_{\mu 4}$ and $B_\mu = J_{\mu 5}$. If we define $P_\mu = A_\mu + B_\mu$ and $R_\mu = A_\mu - B_\mu$, we obtain the above commutation relations and we thus check the local isomorphism between the Conformal group and $SO(4,2)$.

From the matrix point of view we can write:

$$D(\lambda) = \exp(\lambda D) = \begin{pmatrix} \mathbb{I} & \mathbf{0} & \mathbf{0} \\ \mathbf{0} & \cosh\lambda & \sinh\lambda \\ \mathbf{0} & \sinh\lambda & \cosh\lambda \end{pmatrix}, \tag{6.83}$$

$$P(a^\mu) = \begin{pmatrix} \mathbb{I} & -a^\mu & a^\mu \\ a_\nu & 1 - a_\mu a^\mu/2 & a_\mu a^\mu/2 \\ a_\nu & -a_\mu a^\mu/2 & 1 + a_\mu a^\mu/2 \end{pmatrix} = \exp(P_\mu a^\mu), \tag{6.84}$$

$$R(b^\mu) = \begin{pmatrix} \mathbb{I} & b^\mu & b^\mu \\ -b_\nu & 1 - b_\mu b^\mu/2 & -b_\mu b^\mu/2 \\ b_\nu & b_\mu b^\mu/2 & 1 + b_\mu b^\mu/2 \end{pmatrix} = \exp(R_\mu b^\mu), \tag{6.85}$$

where $\mathbb{I}$ is a $4\times 4$ unit matrix. The matrix representation of the different generators takes the form:

$$J_1 = \begin{pmatrix} \cdot & \cdot & \cdot & \cdot & \cdot & \cdot \\ \cdot & \cdot & \cdot & \cdot & \cdot & \cdot \\ \cdot & \cdot & \cdot & 1 & \cdot & \cdot \\ \cdot & \cdot & -1 & \cdot & \cdot & \cdot \\ \cdot & \cdot & \cdot & \cdot & \cdot & \cdot \\ \cdot & \cdot & \cdot & \cdot & \cdot & \cdot \end{pmatrix}, \quad J_2 = \begin{pmatrix} \cdot & \cdot & \cdot & \cdot & \cdot & \cdot \\ \cdot & \cdot & \cdot & -1 & \cdot & \cdot \\ \cdot & \cdot & \cdot & \cdot & \cdot & \cdot \\ \cdot & 1 & \cdot & \cdot & \cdot & \cdot \\ \cdot & \cdot & \cdot & \cdot & \cdot & \cdot \\ \cdot & \cdot & \cdot & \cdot & \cdot & \cdot \end{pmatrix},$$

$$J_3 = \begin{pmatrix} \cdot & \cdot & \cdot & \cdot & \cdot & \cdot \\ \cdot & \cdot & 1 & \cdot & \cdot & \cdot \\ \cdot & -1 & \cdot & \cdot & \cdot & \cdot \\ \cdot & \cdot & \cdot & \cdot & \cdot & \cdot \\ \cdot & \cdot & \cdot & \cdot & \cdot & \cdot \\ \cdot & \cdot & \cdot & \cdot & \cdot & \cdot \end{pmatrix}, \quad K_1 = \begin{pmatrix} \cdot & 1 & \cdot & \cdot & \cdot & \cdot \\ 1 & \cdot & \cdot & \cdot & \cdot & \cdot \\ \cdot & \cdot & \cdot & \cdot & \cdot & \cdot \\ \cdot & \cdot & \cdot & \cdot & \cdot & \cdot \\ \cdot & \cdot & \cdot & \cdot & \cdot & \cdot \\ \cdot & \cdot & \cdot & \cdot & \cdot & \cdot \end{pmatrix},$$

$$K_2 = \begin{pmatrix} \cdot & \cdot & 1 & \cdot & \cdot & \cdot \\ \cdot & \cdot & \cdot & \cdot & \cdot & \cdot \\ 1 & \cdot & \cdot & \cdot & \cdot & \cdot \\ \cdot & \cdot & \cdot & \cdot & \cdot & \cdot \\ \cdot & \cdot & \cdot & \cdot & \cdot & \cdot \\ \cdot & \cdot & \cdot & \cdot & \cdot & \cdot \end{pmatrix}, \quad K_3 = \begin{pmatrix} \cdot & \cdot & \cdot & 1 & \cdot & \cdot \\ \cdot & \cdot & \cdot & \cdot & \cdot & \cdot \\ \cdot & \cdot & \cdot & \cdot & \cdot & \cdot \\ 1 & \cdot & \cdot & \cdot & \cdot & \cdot \\ \cdot & \cdot & \cdot & \cdot & \cdot & \cdot \\ \cdot & \cdot & \cdot & \cdot & \cdot & \cdot \end{pmatrix},$$

where the dots should be replaced by zeroes, and where we have chosen the first four rows and columns as the corresponding ones of the usual Minkowski space-time. The remaining generators are

$$P_0 = \begin{pmatrix} \cdot & \cdot & \cdot & \cdot & 1 & -1 \\ \cdot & \cdot & \cdot & \cdot & \cdot & \cdot \\ \cdot & \cdot & \cdot & \cdot & \cdot & \cdot \\ \cdot & \cdot & \cdot & \cdot & \cdot & \cdot \\ 1 & \cdot & \cdot & \cdot & \cdot & \cdot \\ 1 & \cdot & \cdot & \cdot & \cdot & \cdot \end{pmatrix}, \quad P_1 = \begin{pmatrix} \cdot & \cdot & \cdot & \cdot & \cdot & \cdot \\ \cdot & \cdot & \cdot & \cdot & 1 & -1 \\ \cdot & \cdot & \cdot & \cdot & \cdot & \cdot \\ \cdot & \cdot & \cdot & \cdot & \cdot & \cdot \\ \cdot & -1 & \cdot & \cdot & \cdot & \cdot \\ \cdot & -1 & \cdot & \cdot & \cdot & \cdot \end{pmatrix},$$

$$P_2 = \begin{pmatrix} \cdot & \cdot & \cdot & \cdot & \cdot & \cdot \\ \cdot & \cdot & \cdot & \cdot & \cdot & \cdot \\ \cdot & \cdot & \cdot & \cdot & 1 & -1 \\ \cdot & \cdot & \cdot & \cdot & \cdot & \cdot \\ \cdot & \cdot & -1 & \cdot & \cdot & \cdot \\ \cdot & \cdot & -1 & \cdot & \cdot & \cdot \end{pmatrix}, \quad P_3 = \begin{pmatrix} \cdot & \cdot & \cdot & \cdot & \cdot & \cdot \\ \cdot & \cdot & \cdot & \cdot & \cdot & \cdot \\ \cdot & \cdot & \cdot & \cdot & \cdot & \cdot \\ \cdot & \cdot & \cdot & \cdot & 1 & -1 \\ \cdot & \cdot & \cdot & -1 & \cdot & \cdot \\ \cdot & \cdot & \cdot & -1 & \cdot & \cdot \end{pmatrix},$$

$$R_0 = \begin{pmatrix} \cdot & \cdot & \cdot & \cdot & 1 & 1 \\ \cdot & \cdot & \cdot & \cdot & \cdot & \cdot \\ \cdot & \cdot & \cdot & \cdot & \cdot & \cdot \\ \cdot & \cdot & \cdot & \cdot & \cdot & \cdot \\ 1 & \cdot & \cdot & \cdot & \cdot & \cdot \\ -1 & \cdot & \cdot & \cdot & \cdot & \cdot \end{pmatrix}, \quad R_1 = \begin{pmatrix} \cdot & \cdot & \cdot & \cdot & \cdot & \cdot \\ \cdot & \cdot & \cdot & \cdot & 1 & 1 \\ \cdot & \cdot & \cdot & \cdot & \cdot & \cdot \\ \cdot & \cdot & \cdot & \cdot & \cdot & \cdot \\ \cdot & -1 & \cdot & \cdot & \cdot & \cdot \\ \cdot & 1 & \cdot & \cdot & \cdot & \cdot \end{pmatrix},$$

$$R_2 = \begin{pmatrix} \cdot & \cdot & \cdot & \cdot & \cdot & \cdot \\ \cdot & \cdot & \cdot & \cdot & \cdot & \cdot \\ \cdot & \cdot & \cdot & \cdot & 1 & 1 \\ \cdot & \cdot & \cdot & \cdot & \cdot & \cdot \\ \cdot & \cdot & -1 & \cdot & \cdot & \cdot \\ \cdot & \cdot & 1 & \cdot & \cdot & \cdot \end{pmatrix}, \quad R_3 = \begin{pmatrix} \cdot & \cdot & \cdot & \cdot & \cdot & \cdot \\ \cdot & \cdot & \cdot & \cdot & \cdot & \cdot \\ \cdot & \cdot & \cdot & \cdot & \cdot & \cdot \\ \cdot & \cdot & \cdot & \cdot & 1 & 1 \\ \cdot & \cdot & \cdot & -1 & \cdot & \cdot \\ \cdot & \cdot & \cdot & 1 & \cdot & \cdot \end{pmatrix},$$

$$D = \begin{pmatrix} \cdot & \cdot & \cdot & \cdot & \cdot & \cdot \\ \cdot & \cdot & \cdot & \cdot & \cdot & \cdot \\ \cdot & \cdot & \cdot & \cdot & \cdot & \cdot \\ \cdot & \cdot & \cdot & \cdot & \cdot & \cdot \\ \cdot & \cdot & \cdot & \cdot & \cdot & -1 \\ \cdot & \cdot & \cdot & \cdot & -1 & \cdot \end{pmatrix},$$

where their infinitesimal transformations affect the other two additional dimensions.

We can easily check that they satisfy the commutation relations:

$$[\boldsymbol{J},\boldsymbol{J}] = -\boldsymbol{J},\ [\boldsymbol{J},\boldsymbol{K}] = -\boldsymbol{K},\ [\boldsymbol{J},\boldsymbol{P}] = -\boldsymbol{P},\ [\boldsymbol{J},\boldsymbol{R}] = -\boldsymbol{R},\ [\boldsymbol{K},\boldsymbol{K}] = \boldsymbol{J},$$

$$[D,\boldsymbol{J}] = [D,\boldsymbol{K}] = 0,\ [D,P_\mu] = -P_\mu,\ [D,R_\mu] = R_\mu,\ [P_0,R_0] = -2D,$$

$$[P_\nu, P_\mu] = [R_\nu, R_\mu] = 0,\ [P_0, \boldsymbol{R}] = -2\boldsymbol{K},\ [R_0, \boldsymbol{P}] = -2\boldsymbol{K},$$
$$[P_0, \boldsymbol{J}] = 0, \quad [P_0, \boldsymbol{K}] = -\boldsymbol{P}, \quad [R_0, \boldsymbol{J}] = 0, \quad [R_0, \boldsymbol{K}] = -\boldsymbol{R},$$
$$[\boldsymbol{P}, \boldsymbol{K}] = -P_0, \quad [\boldsymbol{R}, \boldsymbol{K}] = -R_0, \quad [P_i, R_j] = 2\delta_{ij}\, D - 2\epsilon_{ijk} J_k.$$

In the case of the group $SU(2,2)$, it leaves invariant on the space $\mathbb{C}^4$ a quadratic form defined by means of the metric $G = \text{diag}(1,1,-1,-1)$. If an arbitrary infinitesimal transformation is written in the form $A = \mathbb{I} + \epsilon M$, with $\epsilon$ the infinitesimal group parameter, the invariance leads to $A^\dagger G A = G$, and in terms of the generators $M^\dagger G + GM = 0$, and since $G^2 = \mathbb{I}$, it implies $M^\dagger = -GMG$, and if we assume that $M$ matrices are written in $2 \times 2$ blocks of the form

$$M = \begin{pmatrix} P & Q \\ R & S \end{pmatrix},$$

it gives rise to

$$P^\dagger = -P, \quad S^\dagger = -S, \quad Q^\dagger = R, \quad \text{Trace}(P + S) = 0. \tag{6.86}$$

The four-dimensional representation of this Lie algebra is spanned by 15 traceless $4 \times 4$ matrices. We can choose as a basis of the Lie algebra the following: Seven skewhermitian matrices

$$\boldsymbol{J} = \frac{i}{2}\begin{pmatrix} \boldsymbol{\sigma} & 0 \\ 0 & \boldsymbol{\sigma} \end{pmatrix}, \quad B_0 = \frac{i}{2}\begin{pmatrix} \mathbb{I} & 0 \\ 0 & -\mathbb{I} \end{pmatrix}, \quad \boldsymbol{A} = \frac{i}{2}\begin{pmatrix} \boldsymbol{\sigma} & 0 \\ 0 & -\boldsymbol{\sigma} \end{pmatrix}, \tag{6.87}$$

and eight hermitian

$$A_0 = \frac{i}{2}\begin{pmatrix} 0 & \mathbb{I} \\ -\mathbb{I} & 0 \end{pmatrix}, \quad \boldsymbol{B} = \frac{i}{2}\begin{pmatrix} 0 & \boldsymbol{\sigma} \\ -\boldsymbol{\sigma} & 0 \end{pmatrix}, \tag{6.88}$$

$$\boldsymbol{K} = -\frac{1}{2}\begin{pmatrix} 0 & \boldsymbol{\sigma} \\ \boldsymbol{\sigma} & 0 \end{pmatrix}, \quad D = \frac{1}{2}\begin{pmatrix} 0 & \mathbb{I} \\ \mathbb{I} & 0 \end{pmatrix}. \tag{6.89}$$

If we define as in the previous case of $SO(4,2)$, the matrices $P_\mu = A_\mu + B_\mu$ and $R_\mu = A_\mu - B_\mu$, the 15 matrices $\boldsymbol{J}$, $\boldsymbol{K}$, $D$, $P_\mu$ and $R_\mu$ satisfy the commutation relations of the Lie algebra of $SU(2,2)$ which is isomorphic to that of $SO(4,2)$.

If in the case of Dirac's algebra (see Sec. 4.4 of Chapter 4) we take $\hbar = 1$, then we can relate the 15 traceless matrices with the 15 generators of the Conformal group in the form:

$$\boldsymbol{J} = i\boldsymbol{S}, \quad B_0 = iZ_3, \quad \boldsymbol{A} = \frac{3i}{2}\boldsymbol{e}_3, \tag{6.90}$$

$$A_0 = Z_2, \quad \boldsymbol{B} = \frac{3}{2}\boldsymbol{e}_2, \quad \boldsymbol{K} = -\frac{3}{2}\boldsymbol{e}_1, \quad D = Z_1. \tag{6.91}$$

In this way

$$P_\mu = \frac{i}{2}\begin{pmatrix} \sigma^\mu & \sigma^\mu \\ -\sigma^\mu & -\sigma^\mu \end{pmatrix}, \quad R_\mu = \frac{i}{2}\begin{pmatrix} \sigma_\mu & -\sigma_\mu \\ \sigma_\mu & -\sigma_\mu \end{pmatrix}. \tag{6.92}$$

Dirac,[33] obtains the following representation of the Lie algebra of the Conformal group. He denotes by $\beta_a = (\gamma_0, \gamma_1, \gamma_2, \gamma_3, \gamma_5, i\mathbb{I})$ this set of six matrices and by $\lambda_a = (\gamma_0, \gamma_1, \gamma_2, \gamma_3, \gamma_5, -i\mathbb{I})$, $a = 0, 1, 2, 3, 4, 5$, the same set except the last matrix that is taken of opposite sign, and since $\gamma$ matrices anticommute, then the 15 magnitudes $J_{ab} = -\frac{1}{2}\beta_a\lambda_b = -J_{ba}$ satisfy the commutation rules of $SU(2,2)$. If we work in the Pauli-Dirac representation, and as before we define $D = J_{45}$, $A_\mu = J_{\mu 4}$, $B_\mu = J_{\mu 5}$ and $J_i = \frac{1}{2}\epsilon_{ijk}J_{jk}$, $K_i = J_{0i}$, then

$$\boldsymbol{J} = \frac{i}{2}\begin{pmatrix} \boldsymbol{\sigma} & 0 \\ 0 & \boldsymbol{\sigma} \end{pmatrix}, \quad B_0 = \frac{i}{2}\begin{pmatrix} \mathbb{I} & 0 \\ 0 & -\mathbb{I} \end{pmatrix}, \quad \boldsymbol{A} = \frac{i}{2}\begin{pmatrix} \boldsymbol{\sigma} & 0 \\ 0 & -\boldsymbol{\sigma} \end{pmatrix}, \tag{6.93}$$

$$A_0 = \frac{i}{2}\begin{pmatrix} 0 & \mathbb{I} \\ -\mathbb{I} & 0 \end{pmatrix}, \quad \boldsymbol{B} = \frac{i}{2}\begin{pmatrix} 0 & \boldsymbol{\sigma} \\ -\boldsymbol{\sigma} & 0 \end{pmatrix}, \tag{6.94}$$

$$\boldsymbol{K} = -\frac{1}{2}\begin{pmatrix} 0 & \boldsymbol{\sigma} \\ \boldsymbol{\sigma} & 0 \end{pmatrix}, \quad D = \frac{1}{2}\begin{pmatrix} 0 & \mathbb{I} \\ \mathbb{I} & 0 \end{pmatrix}, \tag{6.95}$$

similar to the previous one.

In the Weyl representation we get

$$\boldsymbol{J} = \frac{i}{2}\begin{pmatrix} \boldsymbol{\sigma} & 0 \\ 0 & \boldsymbol{\sigma} \end{pmatrix}, \quad B_0 = \frac{i}{2}\begin{pmatrix} 0 & \mathbb{I} \\ \mathbb{I} & 0 \end{pmatrix}, \quad \boldsymbol{A} = \frac{i}{2}\begin{pmatrix} 0 & \boldsymbol{\sigma} \\ \boldsymbol{\sigma} & 0 \end{pmatrix}, \tag{6.96}$$

$$A_0 = \frac{i}{2}\begin{pmatrix} 0 & \mathbb{I} \\ -\mathbb{I} & 0 \end{pmatrix}, \quad \boldsymbol{B} = \frac{i}{2}\begin{pmatrix} 0 & \boldsymbol{\sigma} \\ -\boldsymbol{\sigma} & 0 \end{pmatrix}, \tag{6.97}$$

$$\boldsymbol{K} = \frac{1}{2}\begin{pmatrix} \boldsymbol{\sigma} & 0 \\ 0 & -\boldsymbol{\sigma} \end{pmatrix}, \quad D = -\frac{1}{2}\begin{pmatrix} \mathbb{I} & 0 \\ 0 & -\mathbb{I} \end{pmatrix}, \tag{6.98}$$

such that

$$A_\mu = \frac{i}{2}\begin{pmatrix} 0 & \sigma^\mu \\ \sigma_\mu & 0 \end{pmatrix}, \quad B_\mu = \frac{i}{2}\begin{pmatrix} 0 & \sigma^\mu \\ -\sigma_\mu & 0 \end{pmatrix}, \tag{6.99}$$

$$P_\mu = i\begin{pmatrix} 0 & \sigma^\mu \\ 0 & 0 \end{pmatrix}, \quad R_\mu = i\begin{pmatrix} 0 & 0 \\ \sigma_\mu & 0 \end{pmatrix}. \tag{6.100}$$

In any of these representations, the generator $D = \frac{i}{2}\gamma_5$, and therefore the infinitesimal generator of dilations is related to the **chirality operator** $\gamma^5 = \gamma^0\gamma^1\gamma^2\gamma^3$.

## 6.3 CONFORMAL OBSERVABLES OF THE PHOTON

Since the Poincaré group $\mathcal{P}$ is a subgroup of the Conformal group $\mathcal{C}$ and the manifold spanned by the variables $(t, \boldsymbol{r}, \boldsymbol{u}, \boldsymbol{\rho})$ with $u = c$, which describes the kinematical variables of Luxons, is a homogeneous space of $\mathcal{P}$, it is also a homogeneous space of $\mathcal{C}$. Let us consider first the case of photons which are described by the same kinematical space manifold, but now considered as a homogeneous space of $\mathcal{C}$.

Under space-time dilations orientation variables are not affected and thus, the transformation of the kinematical variables and their derivatives is

$$t' = e^{\lambda}t, \quad \boldsymbol{r}' = e^{\lambda}\boldsymbol{r}, \quad \boldsymbol{u}' = \boldsymbol{u}, \quad \dot{\boldsymbol{u}}' = \dot{\boldsymbol{u}}, \quad \boldsymbol{\rho}' = \boldsymbol{\rho}, \quad \boldsymbol{\omega}' = \boldsymbol{\omega}.$$

Therefore the photon Lagrangian (3.151) is invariant under this one-parameter group of transformations. Then, since parameter $\lambda$ is dimensionless, Noether's theorem defines a new constant of the motion $D$, with dimensions of action.

$$D = x^{\mu}\partial L/\partial \dot{x}^{\mu} = x^{\mu}P_{\mu} = tH - \boldsymbol{r}\cdot\boldsymbol{P}, \tag{6.101}$$

and if we use the expressions of these observables for the photon we get

$$D = \hbar(\omega t - \frac{\epsilon}{c}\boldsymbol{r}\cdot\boldsymbol{\omega}) = \hbar(\omega t - \boldsymbol{r}\cdot\boldsymbol{k}), \tag{6.102}$$

where $\boldsymbol{\omega}$ is the angular velocity of the photon and $\boldsymbol{k}$ the wave number. We can identify the observable $D$ with the Lorentz invariant internal action. It is equal to $\hbar$ times the Lorentz invariant internal phase of the photon.

From $\dot{D} = 0$ we get again $H = \boldsymbol{P}\cdot\boldsymbol{u}$, the relation between the energy and linear momentum.

Under pure conformal transformations the velocity of the photon is conserved, and the angular variables, and thus the angular velocity $\boldsymbol{\omega}$, remain invariant. Invariance of the Lagrangian under these transformations produce the following constants of the motion

$$R_{\mu} = 2x_{\mu}x^{\nu}P_{\nu} - x^2P_{\mu} = 2x_{\mu}D - x^2P_{\mu}. \tag{6.103}$$

If we set $Q \equiv R_0$, it takes the form

$$Q = 2ctD - (c^2t^2 - \boldsymbol{r}^2)H/c, \tag{6.104}$$

and $\dot{Q} = 0$, since $\dot{D} = 0$, it gives

$$2c\dot{t}D - 2(c^2t\dot{t} - \boldsymbol{r}\cdot\dot{\boldsymbol{r}})\frac{H}{c} = 0, \tag{6.105}$$

i.e., $D = tH - \boldsymbol{r}\cdot\boldsymbol{P}$, the expression of observable $D$. The remaining three are written as

$$\boldsymbol{R} = 2\boldsymbol{r}D - (c^2t^2 - \boldsymbol{r}^2)\boldsymbol{P}, \tag{6.106}$$

and $\dot{\boldsymbol{R}} = 0$ gives rise to

$$D\boldsymbol{u} - (c^2t - \boldsymbol{r}\cdot\boldsymbol{u})\boldsymbol{P} = 0, \tag{6.107}$$

which shows the linear relationship between the linear momentum and the velocity of the photon. Substitution of $\boldsymbol{P} = H\boldsymbol{u}/c^2$ brings again the definition of $D$.

If at time $t = 0$ the photon bursts from the origin of the reference frame $\boldsymbol{r}(0) = 0$, then all these constant observables take the value $D = R_\mu = 0$.

## 6.4 CONFORMAL OBSERVABLES OF THE ELECTRON

Let us consider only the Poincaré group enlarged with space-time dilations. An element of this group is parametrized by $(g, \lambda)$, where by $g$ we mean the usual parametrization of $\mathcal{P}$, and $\lambda$ the new dimensionless normal group parameter of the new transformations. Let us consider then the homogeneous space of this group spanned by the variables $(t, \boldsymbol{r}, \boldsymbol{u}, \boldsymbol{\rho}, \alpha)$, with $u = c$, and therefore $(t, \boldsymbol{r}, \boldsymbol{u}, \boldsymbol{\rho})$ is a point of the previously considered kinematical space of the electron and $\alpha$ the new dimensionless variable that describes a new degree of freedom of this system, as suggested in Sec. 5.1. Therefore, the Lagrangian will also show, in general, dependence on $\alpha$ and $\dot{\alpha}$. Under a transformation of parameter $\lambda$, the kinematical variables and their derivatives transform as:

$$t' = e^\lambda t, \quad \boldsymbol{r}' = e^\lambda \boldsymbol{r}, \quad \boldsymbol{u}' = \boldsymbol{u}, \quad \boldsymbol{\rho}' = \boldsymbol{\rho}, \quad \alpha' = \alpha + \lambda,$$

$$\dot{t}' = e^\lambda \dot{t}, \quad \dot{\boldsymbol{r}}' = e^\lambda \dot{\boldsymbol{r}}, \quad \dot{\boldsymbol{u}}' = \dot{\boldsymbol{u}}, \quad \boldsymbol{\omega}' = \boldsymbol{\omega}, \quad \dot{\alpha}' = \dot{\alpha}.$$

If the free Lagrangian is independent of $t$, $\boldsymbol{r}$ and $\alpha$, and the dependence on $\dot{t}$ and $\dot{\boldsymbol{r}}$ is only through variable $\boldsymbol{u} = \dot{\boldsymbol{r}}/\dot{t}$, then, it is invariant under this transformation and the corresponding conserved observable takes the form:

$$D = tH - \boldsymbol{r}\cdot\boldsymbol{P} - \mathcal{D}, \tag{6.108}$$

where the term $\mathcal{D} = \partial L/\partial\dot{\alpha}$, comes from the dependence of the Lagrangian on the derivative of the new degree of freedom $\dot{\alpha}$. This term is necessarily nonvanishing, because $\dot{D} = 0$ leads to

$$H - \boldsymbol{P}\cdot\boldsymbol{u} - \frac{d\mathcal{D}}{dt} = 0, \tag{6.109}$$

and compared with Dirac's Hamiltonian, the last term is identified with

$$\frac{d\mathcal{D}}{dt} = \frac{1}{c^2}\left(\frac{d\boldsymbol{u}}{dt} \times \boldsymbol{u}\right) \cdot \boldsymbol{Z}. \tag{6.110}$$

When analyzed in the center of mass frame and in the Pauli-Dirac representation, $d\boldsymbol{u}/dt \times \boldsymbol{u} \sim \boldsymbol{e}_3$ and this observable becomes

$$\frac{d\mathcal{D}}{dt} = \frac{c}{R_0} T_3 = \frac{c\hbar}{2R_0}\begin{pmatrix} \mathbb{I} & 0 \\ 0 & -\mathbb{I} \end{pmatrix} = \frac{\hbar}{2}\omega\beta = \beta mc^2, \tag{6.111}$$

where $\omega = c/R_0$ is the zitterbewegung frequency and in terms of $\beta$ Dirac's matrix.

The important feature is that if we consider that our system has as kinematical space a homogeneous space of the enlarged group, then necessarily the Lagrangian has to depend on this new degree of freedom $\alpha$ and its derivative $\dot{\alpha}$. Otherwise, if the variable $\alpha$ is not considered as a kinematical variable, the conserved quantity $D$ will be of the form $D = tH - \boldsymbol{r} \cdot \boldsymbol{P}$ and this leads to contradictions when compared with Dirac's Hamiltonian.

Integration of (6.111) in the center of mass frame yields

$$\mathcal{D}(t) = \mathcal{D}(0) + \beta mc^2 t.$$

Then observable $\mathcal{D}$ looks like the action, or some internal phase in units of $\hbar$, of the internal motion of the charge, in the positive or negative sense according to the particle or antiparticle state we consider.

Under more general conformal transformations, we need the knowledge of the explicit Lagrangian to see how it transforms in order to apply Noether's theorem properly.

In this way if we consider that the electron symmetry group is the complete group $\mathcal{C}$, then necessarily the kinematical space must contain, at least, the additional variable $\alpha$ that describes some internal phase.

# 7. CLASSICAL LIMIT OF QUANTUM MECHANICS

This is a controversial subject in the sense that to what extent two different formalisms can be related in some peculiar manner. Because we have at hand a more detailed classical description of elementary particles we produce an alternative interpretation of this subject.

One often reads in textbooks on quantum mechanics that *Classical Mechanics* (CM) *is the limit $\hbar \to 0$ of Quantum Mechanics* (QM). [34, 35, 36, 37] Of course there are exceptions. For instance in Feynman's Lectures on Physics [38] we read: '*In the classical limit, the quantum mechanics* *will agree with Newtonian Mechanics'*. Both expressions are equivalent if by CM we understand the classical mechanics of spinless (or point) particles, *i.e.*, Newtonian Mechanics. We think this is the meaning the mentioned authors try to express. But the first statement as it stands, might lead to wrong interpretations if considered literally, as some of the quoted references might suggest. In order to clarify this idea of the classical limit of QM, let us consider the following simplified diagram of Fig. 6.23.

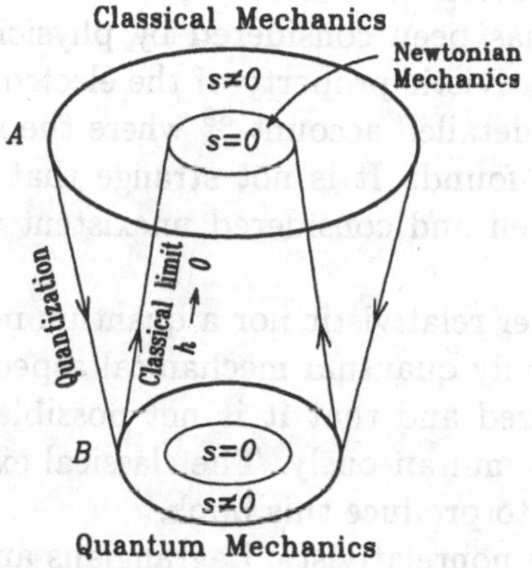


*Figure 6.23.* Classical Limit of Quantum Mechanics.

The set $A$ represents the whole body of knowledge of CM which includes two subsets, the subset ($s = 0$) or domain of spinless particles or Newtonian Mechanics and the subset ($s \neq 0$) of spinning particles. If we restrict CM to satisfy the additional requirement of the Uncertainty Principle we enter into the more restricted body of knowledge of QM, represented by the smaller set $B$ in which we also have the two subdomains of spinless and spinning particles. If in this more constrained domain we perform now the additional restriction of taking the limit $\hbar \to 0$ it is doubtful that after these two restrictions we shall reach the wider and less restrictive domain of the whole CM.

In QM we have that the measurement of any two observables $C$ and $D$ is not in general compatible, and the uncertainty in their simultaneous measurement is related to its commutator $[C, D] \simeq \hbar$ which is of order of Planck's constant $\hbar$. But we also have in QM that the particle states satisfy eigenvalue equations for the spin of the form $S^2|\psi>= s(s+1)\hbar^2|\psi>$, where the right-hand side is also a function of $\hbar$. It turns out that when performing the limit $\hbar \to 0$ we get $[C, D] = 0$ and also $S^2|\psi> = 0$, i.e., the physics of compatible observables of spinless systems.

With this analysis we see that the limit $\hbar \to 0$ of QM is in fact closer to Newtonian Mechanics and not to the whole domain of CM. Perhaps one of the reasons for the identification of CM with Newtonian Mechanics in the mentioned references lies in the fact that since the early days of the quantum theory we are used to work in QM with spinning systems, while the CM of spinning particles is still waiting for a complete development and improvement at least equivalent to the one we have achieved in the quantum domain. Even we can remember here that for many years, spin has been considered by physicists a strict quantum mechanical and relativistic property of the electron, as was pointed out by Levy-Leblond's detailed account [39] where the relevant references on this matter can be found. It is not strange that the recent history of physics had forgotten and considered unexistent the $(s \neq 0)$ region of CM.

The spin is neither relativistic nor a quantum mechanical property of the electron. The only quantum mechanical aspect of the electron spin is that it is quantized and that it is not possible to measure any two of its components simultaneously. The classical explanation of spin has been the challenge to produce this book.

For example, the nonrelativistic Lagrangians analyzed in Chapter 2

$$L = \frac{m}{2}\left(\frac{dr}{dt}\right)^2 - \frac{m}{2\omega^2}\left(\frac{du}{dt}\right)^2,$$

and

$$L = \frac{m}{2}\left(\frac{dr}{dt}\right)^2 + \frac{I}{2}\omega^2,$$

the magnitudes $\omega = 2mc^2/\hbar$ and $I = mR_0^2$, $R_0 = \hbar/2mc$, are related to Planck's constant $\hbar$. In the limit $\hbar \to 0$, both Lagrangians have as a limit the point-like Newtonian particle. Similarly we get the same limit for the most general non-relativistic spinning particle (2.135).

For the relativistic Lagrangians (3.97) and (3.98) the same thing happens, that in the limit $S \sim \hbar \to 0$ both systems transform in the point-like relativistic particle. Parameter $b$ in Lagrangians (3.107) and (3.108) is $b \sim \hbar^2$, and thus vanishes in the proposed limit. The photonic Lagrangian

$$L = \epsilon\frac{S}{c}\boldsymbol{u} \cdot \boldsymbol{\omega},$$

vanishes in the limit $S \equiv \hbar \to 0$. In all cases, taking the limit $\hbar \to 0$ is equivalent to suppressing the spin content of the system.

## Notes

1 A.O. Barut, *Electrodynamics and Classical Theory of Fields and Particles*, Dover, N.Y. (1980).

2 J.D. Jackson, *Classical Electrodynamics*, John Wiley and Sons, NY 3rd. ed. (1998).

3 P.A.M. Dirac, *Proc. Roy. Soc. London* **A117**, 610 (1928).

4 J.M. Levy-Leblond, *Comm. Math. Phys.* **6**, 286 (1967).

5 A. Proca, *Compt. Rend.* **202**, 1420 (1936); *Journ. Phys. Radium*, **49**, 245 (1988).

6 F.J. Belinfante, *Phys. Rev.* **92**, 997 (1953).

7 P.A. Moldauer and K.M. Case, *Phys. Rev.* **102**, 279 (1956).

8 V.S. Tumanov, *Sov. Phys. JETP*, **19**, 1182 (1964).

9 S. Weinberg, in *Lectures on Elementary Particles and Quantum Field Theory*, edited by S. Deser, M. Grisaru and H. Pendleton, MIT press, Cambridge, MA (1970), p. 283.

10 R. Jackiw, *Phys. Rev.* D **57**, 2635 (1998).

11 S. Ferrara, M. Porrati and V.L. Telegdi, *Phys. Rev.* D **46**, 3529 (1992).

12 M. Rivas, J.M.Aguirregabiria and A. Hernández, *Phys. Lett.* **A 257**, 21 (1999).

13 P.A.M. Dirac, *Proc. Roy. Soc. London*, **A117**, 610 (1928).

14 D. Hestenes, *Space-Time algebra*, Gordon and Breach, NY (1966); D. Hestenes and G. Sobczyk, *Clifford Algebra to Geometric Calculus*, D. Reidel Pub. Co. Dordrecht, (1984).

15 C.G. Darwin, *Proc. Roy. Soc. (London)*, **A118**, 654 (1928).

16 J.J. Sakurai, *Advanced Quantum Mechanics*, Addison-Wesley, Reading Mass. (1967), p.119.

17 V.I. Fushchich, A. Nikitin and V.A. Sagolub, *Rep. Math. Phys.* **13**, 175 (1978).

18 J.M. Aguirregabiria, *Dynamics Solver*, computer program for solving different kinds of dynamical systems, which is available from his author through the web site <http://tp.lc.ehu.es/jma.html> at the server of the Theoretical Physics dept. of The University of the Basque Country, Bilbao (Spain).

19 J.M. Levy-Leblond, *Galilei Group and Galilean Invariance*, in E.M. Loebl, *Group Theory and its applications*, Acad. Press, NY (1971), vol. 2, p. 221.

20 L. Landau and E. Lifchitz, *Mécanique quantique*, Mir Moscow (1988), 3rd. edition.

21 V.N. Dobrovolsky, D.I. Sheka and B.V. Chernyachuk, Surface Science **397**, 333 (1998); P. Raychaudhuri, T.K. Nath, A.K. Nigam and R. Pinto, cond-mat/9805258, preprint.

22 P.A.M. Dirac, *The Principles of Quantum mechanics*, Oxford Univ. Press, 4th ed. (1967).

23 M.H.L. Pryce, *Proc. Roy. Soc.* **A195**, 62 (1948).

24 H. Bacry, *Phys. Lett.* **5**, 37 (1963).

25 L.L. Foldy and S.A. Wouthuysen, *Phys. Rev.* **78**, 29 (1950).

26 T.D. Newton and E.P. Wigner, *Rev. Mod. Phys.* **21**, 3 (1949).

27 M. Bunge, *Nuovo Cimento* **1**, 977 (1955).

28 H. Rund, *The Hamilton-Jacobi theory in the calculus of variations*, Krieger Pub. Co., N.Y. (1973), ch. 3; G.S. Asanov, *Finsler geometry, Relativity and Gauge theories*, Reidel Pub. Co, Dordrecht (1985).

29 G.S. Asanov, *Finsler geometry, Relativity and Gauge theories*, Reidel Pub. Co, Dordrecht, Holland (1985).

30 A. Papapetrou, *Proc. Roy. Soc.* **A 209**, 248 (1951). A. Papapetrou and E. Corinaldesi, *Proc. Roy. Soc.* **A 209**, 259 (1951).

31 H. Bateman, *Proc. Lond. Math. Soc.* **8**, 228 (1910); **8**, 469 (1910); E. Cunningham, **10**, 77 (1910).

32 B. Felsager, *Geometry, Particles and Fields*, Odense Univ. Press (1981).

33 P.A.M. Dirac, *Ann. of Math.* **37**, 429 (1936).

34 J. Glimm and A. Jaffe, *Quantum Physics*, Springer-Verlag, Berlin (1987), p. 3: "Classical mechanics is the limit $\hbar \to 0$ of quantum mechanics. Nonrelativistic (Newtonian) mechanics is the limit $c \to \infty$ of special Relativity."

35 L. Landau and E. M. Lifshitz, *Quantum Mechanics*, Pergamon, Oxford (1974), p. 25: "The transition from quantum mechanics to classical mechanics, corresponding to large phase, can be formally described as a passage to the limit $\hbar \to 0$...".

36 E. Merzbacher, *Quantum Mechanics*, John Wiley, NY (1970), p. 3: "We thus expect that classical mechanics is contained in quantum mechanics as a limiting form ($\hbar \to 0$)".

37 A. Messiah, *Mécanique Quantique*, Dunod, Paris (1959) ch. VI p. 180: "à la limite où $\hbar \to 0$, les lois de la Mécanique Quantique doivent se réduire à celles de la Mécanique Classique".

38 R.P. Feynman, R.B. Leighton and M. Sands, *The Feynman Lectures on Physics*, vol. III, Addison-Wesley, Reading, Mass. (1965).

39 J. M. Levy-Leblond, *Riv. Nuovo Cim.*, **4**, 99, (1974).

# Epilogue

This is a book about spin. In these pages we have explored a new formalism for elementary particle physics that produces, among other things, a classical description of spin.

This new look at the subject has depicted a clear picture of the zitterbewegung, thus justifying on classical grounds the dipole properties of charged spinning particles. The price paid for it is that classical elementary particles look like extended objects. But, nevertheless, it is not necessary to talk about macroscopic properties like shape or size, and the analytical description is done in terms of positions and velocities.

To describe the evolution of a particle we have to distinguish between two points, the center of mass and the center of charge. The center of charge is that point where the values of the external fields are computed to produce Newton-like dynamical equations for the center of mass. The above field values are used to determine the total external force which gives rise to the variation of the mechanical linear momentum. The motion of the center of charge around the center of mass is a generalization, in the relativistic case, of an isotropic harmonic motion. In addition to this we also need to describe the evolution of the orientation of the system which seems to play no role in the electromagnetic dipole structure but contributes to the total spin. This partially gives an answer to Professor Barut's quotation at the beginning of the book.

The definition of spin is related to a proper definition of the center of mass, which can be defined accurately for a single particle, although its definition for a compound system is not free from difficulties. The center of mass position of a particle, very well defined for every inertial observer, cannot be described in terms of a canonical covariant four-vector.

The structure of the spin observable is twofold. One part is related to the zitterbewegung, and thus to the separation between the center of mass and center of charge, and another that comes from the rotation of the particle. The first part gives rise to a spin of an (anti)orbital type in the sense that it has the direction opposite to the expected orbital angular momentum of a moving point. It is directly related to the magnetic moment of the particle produced by the motion of the charge and always quantizes with integer values. The spin 1/2 part of a fermionic system comes from the second contribution due to the rotation of the particle as a whole, as in the case of a rigid body. Mechanical experiments devised to measure the spin of the particle are not able to separate the two parts and, therefore, when expressing the magnetic moment of the particle in terms of the total spin, this introduces the concept of gyromagnetic ratio.

A charged elementary spinning particle can be interpreted as a first order approximation, as a point (the center of mass) in which we locate the scalar properties mass $m$ and charge $e$. We also locate at this point a magnetic moment $\boldsymbol{\mu}$, as an intrinsic property, of the same value as the classical magnetic moment associated to the motion of the center of charge around the center of mass. In addition to this, the particle has an instantaneous electric dipole $\boldsymbol{d}$ which is oscillating with the zitterbewegung frequency in a plane orthogonal to the spin. Its magnitude is the product of the charge times the radius of the zitterbewegung, which for the electron is just half Compton's wave length. The presence of this instantaneous electric dipole has proven to explain the Darwin term of Dirac's equation and produces also a measurable classical contribution to the crossing of a potential barrier. This approximate model is sufficient for a low energy analysis, but in high energy physics or for instance a very close electron-electron interaction, the exact positions of the center of charge and mass of both particles are necessary to properly state and solve dynamical equations.

The quantum mechanical formalism shows that it is not necessary to base quantization upon a previous classical canonical formalism. A Lagrangian formalism, the special relativity principle associated to a specific kinematical group, together with Feynman's quantization method, are sufficient to produce the quantum scenario in which spin, and many other observables, are obtained by group theoretical methods. But in order to obtain a classical dynamical description in terms of end points, and therefore formally closer to quantum dynamics, we have worked out a Lagrangian approach where the important variables are the end-point variables, called kinematical variables here. The only restriction we proposed for the kinematical variables is that they must lie on a homogeneous space of the kinematical group of symmetries if they are going to describe an elementary particle. To achieve this goal, one of the usual assumptions of Lagrangian mechanics has been withdrawn: Lagrangians are allowed to depend, in general, on higher order derivatives.

This has been the main reason for the revisited generalized Lagrangian formalism developed in the first chapter, where the definition of classical elementary particle is stated. The dependence of the Lagrangian on higher order derivatives and the explicit construction of the different observables as functions of the kinematical variables, are the features that distinguish this approach from previous ones. This becomes more evident when we compare the formalism with the other models of spinning particles, discussed in Chapter 5.

The whole formalism is a group theoretical one and it goes very close to the standard quantum mechanics of one-particle systems. We have been able to define the quantum mechanical operators equivalent to each one of the different terms that constitute the classical spin and other fundamental observables of the particle. Even a kind of 'correspondence principle' has been announced as a result of their formal expressions.

To be consistent with the quantum mechanical framework, the classical description of a Dirac particle or, basically, of an electron or a quark, suggests that the charge of the particle is moving in circles at the speed of light around the center of mass. No other classical model produced by this formalism fulfils Dirac's equation when quantized.

This feature of a point charge moving at the speed of light is not contradictory with Special Relativity if the center of mass of the system can never reach this velocity. Nevertheless this raises new questions. One important question is the problem of radiation. The charge of a free spinning particle has an accelerated but radiationless motion, because mechanical observables like energy, linear momentum and spin are conserved. Radiation must be related to the acceleration of the center of mass, and perhaps to the corresponding motion of the center of charge, *i.e.*, to the change of the above mechanical attributes. Therefore, the analysis of radiation reaction has to be revisited.

There are no great differences between the non-relativistic and relativistic approach, as far as the pictorial description of elementary particles is concerned. Because the basic variables are the same the spin observables have the same basic form, although Lorentz invariance gives rise to more complicated analytical expressions in the relativistic framework. There is no non-relativistic limit of the model that satisfies Dirac's equation. Nevertheless, for specific purposes, simple non-relativistic models with zitterbewegung and rotation can be designed.

But at the same time the proposed formalism gives some hints for its extension to include other kinematical groups. If matter, at the elementary level, has more intrinsic properties than just mass and spin, this justifies that we have to look for larger symmetry groups than the Galilei or Poincaré groups. But not only as symmetry groups. We have to obtain from them the new kinematical variables necessary to describe the additional information we are searching for. This is another open task and deserves more work.

Isospin conservation, in a quantum field theory context, is considered as a dynamical symmetry instead of a kinematical one because it is related to an $SU(2)$ local gauge invariance of the interaction Lagrangian. A glance at the invariance under a local rotation group of a free Lagrangian has given the possibility of using local angular variables to describe the internal orientation, but considered as new variables that remain invariant under the spatial rotation group. We are not sure whether the new observables can be related with isospin, although they satisfy the same commutation relations as the isospin operators and commute with translations. But invariance under this group restricts a little bit more the kind of suitable Lagrangians for elementary particles.

The whole approach is independent of the kinematical group which we use to define the basic symmetries of the physical system. The more complex the kinematical group the more restricted are the kind of models that can be depicted, because of the greater number of constraints imposed by the symmetry principles. But at the same time the model will be described by more variables thus showing a more complex or richer structure.

Some of the perspectives the formalism suggests have been presented as conjectures. In some cases, one of the reasons has been the computational difficulties for obtaining analytical solutions. In others, the possibility of finding even a convergent numerical routine. There are still many open questions. Our hope is that the application of the formalism to old and new problems will necessarily produce corrections to improve our understanding of some physical phenomena.

# References

Science is a collective job and it is difficult to determine where the different ideas come from, either from published works or from formal or informal discussions. Some of the basic references related to classical and quantum spinning particles, used while writing this book with the spin as a basic subject matter, are listed in alphabetical order. I apologize for the references missed and not included in this list.


[1] J.M. Aguirregabiria, *Dynamics Solver*, computer program for solving different kinds of dynamical systems, which is available from his author through the web site `<http://tp.lc.ehu.es/jma.html>` at the server of the Theoretical Physics Dept. of The University of the Basque Country, Bilbao, Spain.

[2] R. Arens, *Classical relativistic particles*, Comm. Math. Phys. **21**, 139 (1971).

[3] G.S. Asanov, *Finsler geometry, Relativity and Gauge theories*, Reidel Pub. Co, Dordrecht (1985).

[4] M.V. Atre and N. Mukunda, *Classical Particles with Internal structure: General formalism and application to first-order Internal Spaces*, J. Math. Phys. **27**, 2908 (1986).

[5] M.V. Atre and N. Mukunda, *Classical Particles with Internal Structure. II. Second-order Internal Spaces*, J. Math. Phys. **28**, 792 (1987).

[6] H. Bacry, *La précession du spin des particules dans in champ quelconque*, Compt. Rend. Acad. Sci., **253**, 389 (1961).

[7] H. Bacry, *Thomas's classical theory of Spin*, Nuovo Cimento, **26**, 1164 (1962).

[8] H. Bacry, *A covariant position operator. Application to a system of two particles*, Phys. Lett. **5**, 37 (1963).

[9] H. Bacry, *Position and polarization operators in relativistic and nonrelativistic mechanics*, J. Math. Phys. **5**, 109 (1964).

[10] H. Bacry, *Space-Time and Degrees of Freedom of the Elementary Particle*, Commun. Math. Phys. **5**, 97, (1967).

[11] H. Bacry, *A generalized interpretation of the Lorentz and Thomas-Bargmann-Michel-Telegdi equations*, Ann. Phys. **236**, 286 (1994).

[12] H. Bacry and A. Kihlberg, *Wavefunctions on Homogeneous spaces*, J. Math. Phys. **10**, 2132 (1969).

[13] H. Bacry and J.M. Levy-Leblond, *Possible Kinematics*, J. Math. Phys. **9**, 1605 (1968).

[14] V. Bargmann, *Irreducible unitary representations of the Lorentz group*, Ann. Math. **48**, 568 (1947).

[15] V. Bargmann, *On unitary ray representations of continuous groups*, Ann. Math. **59**, 1 (1954).

[16] V. Bargmann, *Note on Wigner's Theorem on Symmetry Operations*, J. Math. Phys. **5**, 862 (1964).

[17] V. Bargmann, L. Michel and V.L. Telegdi, *Precession of the Polarization of particles moving in a homogeneous electromagnetic field*, Phys. Rev. Lett. **2**, 435, (1959).

[18] A.O. Barut, *Electrodynamics and Classical Theory of Fields and Particles*, Dover, N.Y. (1980).

[19] A.O. Barut, *Excited states of zitterbewegung*, Phys. Lett. **B 237**, 436 (1990).

[20] A.O. Barut and A.J. Bracken, *Zitterbewegung and the internal geometry of the electron*, Phys. Rev. **D 23**, 2454 (1981).

[21] A.O. Barut and M.G. Cruz, *Classical Relativistic Spinning Particle with Anomalous Magnetic Moment: the Precession of Spin*, J. Phys. **A 26**, 6499 (1993).

[22] A.O. Barut and M.G. Cruz, *On the zitterbewegung of the relativistic electron*, Eur. J. Phys. **1**, 119 (1994).

[23] A.O. Barut and R. Raczka, *Theory of group representations and applications*, PWN, Warszawa (1980).

[24] A.O. Barut and N. Zanghi, *Classical model of the Dirac Electron*, Phys. Rev. Lett. **52**, 2009 (1984).

[25] H. Bateman, *The transformation of the electrodynamical equations*, Proc. Lond. Math. Soc. **8**, 223 (1910).

[26] W. Beiglbock, *The center-of-mass in Einsteins Theory of gravitation*, Commum. math. Phys. **5**, 106 (1967).

[27] R.G. Beil, *Moving frame transport and gauge transformations*, Found. Phys. **25**, 717 (1995);

[28] R.G. Beil, *Poincaré transport of frames*, Found. Phys. **25**, 1577 (1995).

[29] L. Bel and E. Ruiz, *Relativistic Schroedinger equations*, J. Math. Phys. **29**, 1840 (1988).

[30] F.J. Belinfante, *Intrinsic Magnetic Moment of Elementary Particles of Spin 3/2*, Phys. Rev. **92**, 997 (1953).

[31] R.A. Berg, *Position and intrinsic spin operators in quantum theory*, J. Math. Phys. **6**, 34 (1965).

[32] W. Bernreuther and M. Suzuki, *The electric dipole moment of the electron*, Rev. Mod. Phys. **63**, 313 (1991).

[33] R. A. Beth, *Mechanical Detection and Measurement of the Angular Momentum of Light*, Phys. Rev. **50**, 115 (1936).

[34] A. Bette, *On a pointlike relativistic massive and spinning particle*, J. Math. Phys. **25**, 2456 (1984).

[35] H.J. Bhabha, *Relativistic wave equations for the elementary particles*, Rev. Mod. Phys. **17**, 200 (1945).

[36] H.J. Bhabha and H.C. Corben, *General classical theory of spinning particles in a Maxwell field*, Proc. Roy. Soc. London **A 278**, 273 (1941).

[37] F. Bopp and R. Haag, *Über die Möglichkeit von Spinmodellen*, Z. Naturforschg. **5a**, 644 (1950).

[38] L.C. Biedenharn, M.Y. Han and H. van Dam, *Two-component Poincaré-invariant equations for massive charged leptons*, Phys. Rev. Lett. **27**, 1167 (1971).

[39] L.C. Biedenharn, M.Y. Han and H. van Dam, *Two-component alternative to Dirac's equation*, Phys. Rev. **D 6**, 500 (1972).

[40] L.C. Biedenharn, M.Y. Han and H. van Dam, *Composite systems viewed as relativistic quantal rotators: Vectorial and spinorial models*, Phys. Rev. **D 22**, 1938 (1980).

[41] L.C. Biedenharn and J.D. Louck, *Angular Momentum in Quantum Physics. Theory and Application*, Cambridge U. P., Cambridge, England, (1989).

[42] M. Borneas, *A quantum equation of motion with higher derivatives*, Am. J. Phys. **40**, 248 (1972).

[43] M. Bunge, *A picture of the electron*, Nuovo Cimento **1**, 977 (1955).

[44] M. Bunge, *Survey of the interpretations of quantum mechanics*, Am. J. Phys. **24**, 272 (1956).

[45] P. Caldirola, *A new model of classical electron*, Supl. Nuovo Cimento **3**, 297(1956).

[46] J.F.Cariñena, M.A. del Olmo and M. Santander, *Kinematic groups and dimensional analysis*, J. Phys. **A 14**, 1 (1981).

[47] H. Casimir, *Über die konstruktion einer zu den irreduzibelen darstellungen halbeinfacher kontinuierlicher gruppen gehörigen differentialgleichung*, Proc. Roy. Acad. Amsterd. **34**, 844 (1931).

[48] G. Cavalleri, *Schroedinger's equation as a consequence of Zitterbewegung*, Lett. Nuovo Cimento, **43**, 285 (1985).

[49] M. Chaichian, R. González Felipe and D. Louis Martínez, *Spinning relativistic particle in an external electromagnetic field*, Phys. Lett. **A 236**, 188 (1997).

[50] G.C. Constantelos, *Integrals of motion for Lagrangians Including Higher Order Derivatives*, Nuovo Cimento, **B 21**, 279 (1974).

[51] R. Courant and D. Hilbert, *Methods of Mathematical Physics*, Vol. 1, Interscience, N.Y. (1970).

[52] H.C. Corben, *Spin in classical and quantum theory*, Phys. Rev. **121**, 1833 (1961).

[53] H.C. Corben, *Spin precession in classical relativistic mechanics*, Nuovo Cimento, **20**, 529 (1961).

[54] H.C. Corben, *Classical and quantum theories of spinning particles*, Holden-Day, San Francisco (1968).

[55] H.C. Corben, *Quantized relativistic rotator*, Phys. Rev. **D 30**, 2683 (1984).

[56] H.C. Corben, *Factors of 2 in magnetic moments, spin-orbit coupling, and Thomas precession*, Am. J. Phys. **61**, 551 (1993).

[57] H.C. Corben, *Structure of a spinning point particle at rest*, Int. J. Theor. Phys. **34**, 19 (1995).

[58] E. Cunningham, *The Principle of Relativity in Electrodynamics and an Extension Thereof*, Proc. Lond. Math. Soc. **10**, 77 (1910).

[59] J.P.Dahl, *The Spinning Electron*, Mat. Fys. Medd. Dan. Vid. Selsk. **39** No 12, 1 (1977).

[60] R.H. Dalitz, *The collected works of P.A.M. Dirac, 1924-1948*, Cambridge U.P., Cambridge (1995).

[61] C.G. Darwin, *The electron as a vector wave*, Proc. Roy. Soc. Lon. **A116**, 227 (1927).

[62] C.G. Darwin, *The wave equations of the electron*, Proc. Roy. Soc. Lon. **A118**, 654 (1928).

[63] C. Dewdney, P.R. Holland, A. Kyprianidis and J.P. Vigier, *Spin and non-locality in quantum mechanics*, Nature, **336**, 536 (1988).

[64] P.A.M. Dirac, *The Quantum Theory of the Electron*, Proc. Roy. Soc. Lon. **A117**, 610 (1928).

[65] P.A.M. Dirac, *The quantum theory of the electron. Part II.*, Proc. Roy. Soc. Lon. **A118**, 351 (1928).

[66] P.A.M. Dirac, *The Lagrangian in quantum mechanics*, Phys. Z. der Sowjetunion **3**, 64 (1933).

[67] P.A.M. Dirac, *Homogeneous variables in classical dynamics*, Proc. Cambridge Phil. Soc. **29**, 389 (1933).

[68] P.A.M. Dirac, *Wave equations in conformal space*, Ann. of Math. **37**, 429 (1936).

[69] P.A.M. Dirac, *The Principles of Quantum mechanics*, Oxford Univ. Press, 4th ed. Oxford (1967).

[70] A.R. Edmonds, *Angular Momentum in Quantum Mechanics*, Princeton U. P., Princeton N.J. (1957).

[71] J.R. Ellis, *A canonical formalism for an acceleration dependent Lagrangian*, J. Phys. **A 8**, 496 (1975).

[72] E.E. Entralgo and V.V. Kuryshkin, *Classical theory of a point spin-particle and structural point objects*, Nuovo Cimento **A 103**, 561 (1990).

[73] E.E. Entralgo, B. Cabrera and J. Portieles, *Towards the problem of two point particles with spin*, Nuovo Cimento **A 111**, 1185 (1998).

[74] B. Felsager, *Geometry, Particles and Fields*, Odense Univ. Press, (1981).

[75] S. Ferrara, M. Porrati and V.L. Telegdi, *$g = 2$ as the natural value of the tree-level gyromagnetic ratio of elementary particles*, Phys. Rev. **D 46**, 3529 (1992).

[76] H. Feshbach and F. Villars, *Elementary relativistic wave mechanics of spin 0 and spin 1/2 particles*, Rev. Mod. Phys. **30**, 24 (1958).

[77] R.P. Feynman and A.R. Hibbs, *Quantum Mechanics and Path Integrals*, MacGraw Hill, N.Y., (1965).

[78] R.P. Feynman, R.B. Leighton and M. Sands, *The Feynman Lectures on Physics*, vol. III, Addison-Wesley, Reading, Mass. (1965).

[79] D. Finkelstein, *Internal Structure of Spinning Particles*, Phys. Rev. **100**, 924 (1955).

[80] G.N. Fleming, *Covariant position operators, spin and locality*, Phys. Rev. **137**, 188 (1965).

[81] L.L. Foldy, *The electromagnetic properties of Dirac particles*, Phys. Rev. **78**, 688 (1952).

[82] L.L. Foldy and S.A. Wouthuysen, *On the Dirac theory of spin 1/2 particles and its non-relativistic limit*, Phys. Rev. **78**, 29 (1950).

[83] J. Frenkel, *Die elektrodynamik des rotierenden elektrons*, Zeits. f. Phys. **37**, 243 (1926).

[84] V.I. Fushchich, A. Nikitin and V.A. Sagolub, *On the non-relativistic motion equations in the Hamiltonian form*, Rep. Math. Phys. **13**, 175 (1978).

[85] I.M. Gelfand and S.V. Fomin, *Calculus of Variations*, Prentice Hall, Englewood Cliffs, NJ (1963).

[86] J. Glimm and A. Jaffe, *Quantum Physics*, Springer-Verlag, Berlin, (1987).

[87] R.J. Gould, *The intrinsic magnetic moment of elementary particles*, Am. J. Phys. **64**, 597 (1996).

[88] Z. Grossmann and A. Peres, *Classical theory of the Dirac electron*, Phys. Rev. **132**, 2346 (1963).

[89] F. Gürsey, *Relativistic kinematics of a classical point particle in spinor form*, Nuovo Cimento **5**, 784 (1957).

[90] R. Gurtler and D. Hestenes, *Consistency in the formulation of the Dirac, Pauli and Schroedinger theories*, J. Math. Phys. **16**, 573 (1975).

[91] F. Halbwachs, *Lagrangian formalism for a classical relativistic particle endowed with internal structure*, Prog. Theor. Phys. **24**, 291 (1960).

[92] A.J. Hanson and T. Regge, *The Relativistic Spherical Top*, Ann. Phys. **87**, 498, (1974).

[93] A. Hanson, T. Regge and C. Teitelboim, *Constrained Hamiltonian systems*, Academia Nazionale dei Lincei, **22** (1976).

[94] C.F. Hayes and J.M. Jankowski, *Quantization of generalized mechanics*, Nuovo Cimento **B 58**, 494 (1968).

[95] D. Hestenes, *Space-Time algebra*, Gordon and Breach, NY (1966).

[96] D. Hestenes, *Vectors, spinors, and complex numbers in classical and quantum physics*, Am. J. Phys. **39**, 1013 (1971).

[97] D. Hestenes, *Local observables in the Dirac theory*, J. Math. Phys. **14**, 893 (1973).

[98] D. Hestenes and G. Sobczyk, *Clifford Algebra to Geometric Calculus*, D. Reidel Pub. Co., Dordrecht, (1984).

[99] R.P. Holland, *Causal interpretation of a system of two spin-1/2 particles*, Phys. Rep. **169**, 293 (1988).

[100] H. Hönl, *Feldmechanik des elektrons und der elementarteilchen*, Ergeb. Exacten Naturwiss. **26**, 291 (1952).

[101] H. Hönl and A. Papapetrou, *Über die innere Bewegung des elektrons. I*, Zeit. f. Phys. **112**, 512 (1939).

[102] K. Huang, *On the zitterbewegung of the Dirac electron*, Am. J. Phys. **20**, 479 (1952).

[103] V.W. Hughes and T. Kinoshita, *Anomalous g values of the electron and muon*, Rev. Mod. Phys. **71**, 133 (1999).

[104] E. Inönü and E.P. Wigner, *Representations of the Galilei group*, Nuovo Cimento **9**, 705 (1952).

[105] C. Itzykson and A. Voros, *Classical electrodynamics of point particles*, Phys. Rev. **D 5**, 2939 (1972).

[106] R. Jackiw, *$g = 2$ as a gauge condition*, Phys. Rev. **D 57**, 2635 (1998).

[107] R. Jackiw and S. Weinberg, *Weak-interaction corrections to the muon magnetic moment and to the muonic-atom energy levels*, Phys. Rev. **D 5**, 2396 (1972).

[108] R.C. Jennison, *A new classical relativistic model of the electron*, Phys. Lett. **A 141**, 377 (1989).

[109] T.F. Jordan and N. Mukunda, *Lorentz-covariant position operators for spinning particles*, Phys. Rev. **132**, 1842 (1963).

[110] M. K.-H. Kiessling, *Classical electron theory and conservation laws*, Phys. Lett. **A 258**, 197 (1999).

[111] A.A. Kirillov, *Cours sur la thèorie des reprèsentations des groupes*, Ed. Univ. Moscow, Moscow, (1971).

[112] A.A. Kirillov, *Élements de la theorie des représentations*, Mir, Moscow (1974).

[113] J.G. Koestler and J.A. Smith, *Some developments in generalized classical mechanics*, Am. J. Phys. **33**, 140 (1965).

[114] B. Kostant, *Lecture Notes in Mathematics* Vol. **170**, Springer-Verlag, Berlin, (1970).

[115] J.G. Krüger and D.K. Callebaut, *Comments on generalized mechanics*, Am. J. Phys. **36**, 557 (1968).

[116] L. Landau and E. M. Lifshitz, *Quantum Mechanics*, Pergamon, Oxford, (1974).

[117] J. M. Levy-Leblond, *Non-relativistic particles and wave equations*, Comm. Math. Phys. **6**, 286, (1967).

[118] J.M. Levy-Leblond, *Group theoretical foundations of Classical Mechanics: The Lagrangian gauge problem*, Comm. Math. Phys. **12**, 64 (1969).

[119] J.M. Levy-Leblond, *Minimal electromagnetic coupling as a consequence of Lorentz invariance*, Ann. Phys. **57**, 481 (1970).

[120] J.M. Levy-Leblond, *Conservation laws for gauge-variant Lagrangians in classical mechanics*, Am. J. Phys. **39**, 502 (1971).

[121] J.M. Levy-Leblond, *Galilei Group and Galilean Invariance*, in E.M. Loebl, *Group Theory and its applications*, Acad. Press, N.Y. (1971), vol. 2, p. 221.

[122] J. M. Levy-Leblond, *The Pedagogical Role and Epistemological Significance of Group Theory in Quantum Mechanics*, Riv. Nuovo Cim. **4**, 99, (1974).

[123] J.M. Levy-Leblond, *Quantum heuristics of angular momentum*, Am. J. Phys. **44**, 719 (1976).

[124] J.M. Levy-Leblond, *Classical charged particles revisited: renormalising mass and spin*, Eur. J. Phys. **10**, 265 (1989).

[125] J.M. Levy-Leblond, *A geometrical quantum phase effect*, Phys. Lett. **A 125**, 441 (1987).

[126] J.M. Levy-Leblond, *Enigmas of the SpħinX*, Found. Mod. Phys. **26**, 226 (1993).

[127] J.M. Levy-Leblond, *Position-dependent effective mass and Galilean invariance*, Phys. Rev. **A 52**, 1845 (1995).

[128] B. Liebowitz, *A model of the electron*, Nuovo Cimento **A 63**, 1235 (1969).

[129] C.A. Lopez, *Stability of an extended model of the electron*, Phys. Rev. **D 33**, 2489 (1986).

[130] C.A. Lopez, *Internal structure of a classical spinning electron*, Gen. Rel. Grav. **24**, 285 (1992).

[131] L. Lubanski, *Neue Bewegungsgleichungen materieller systeme in Minkowskischer welt*, Acta Phys. Pol. **6**, 356 (1937).

[132] J. Maddox, *Where zitterbewegungen may lead*, Nature **325**, 306 (1987).

[133] M. Mathisson, *Neue mechanik materieller systeme*, Acta Phys. Pol. **6**, 163 (1937).

[134] M. Mathisson, *Das zitternde elektron und seine dynamik*, Acta Phys. Pol. **6**, 218 (1937).

[135] E. Merzbacher, *Quantum Mechanics*, John Wiley, N.Y. (1970).

[136] A.Messiah, *Mécanique Quantique*, Dunod, Paris, (1959).

[137] P.A. Moldauer and K.M. Case, *Properties of Half-integral spin Dirac-Fierz-Pauli particles*, Phys. Rev. **102**, 279 (1956).

[138] N. Mukunda, H. van Dam and L.C. Biedenharn, *Composite systems viewed as relativistic quantal rotators: Vectorial and spinorial models*, Phys. Rev. **D 22**, 1938 (1980).

[139] P.L. Nash, *A Lagrangian theory of the classical spinning electron*, J. Math. Phys. **25**, 2104 (1984).

[140] K. Nagy, *Relativistic equation of motion for spinning particles*, Acta Phys. Hung. **7**, 325 (1957).

[141] T.D: Newton and E.P. Wigner, *Localized States for Elementary Systems*, Rev. Mod. Phys. **21**, 400 (1949).

[142] A.G. Nikitin, *On exact Foldy-Wouthuysen transformation*, J. Phys. **A 31**, 3297 (1998).

[143] P. Nyborg, *On classical theories of spinning particles*, Nuovo Cimento, **23**, 47 (1962).

[144] H.C. Ohanian, *What is spin?*, Am. J. Phys. **54**, 500 (1986).

[145] P.J. Olver, *Applications of Lie Groups to Differential Equations*, Springer-Verlag, N.Y. (1986).

[146] M. Ostrogradsky, Mem. Acad. St. Petersburg, **6**, 385 (1850).

[147] A. Papapetrou, *Spinning test-particles in general relativity. I*, Proc. Roy. Soc. **A 209**, 248 (1951).

[148] A. Papapetrou and E. Corinaldesi, *Spinning test-particles in general relativity. II*, Proc. Roy. Soc. **A 209**, 259 (1951).

[149] P. Pearle, *Classical Electron Models*, in *Electromagnetism: Paths to Research*, D. Teplitz (editor), p. 211, Plenum Press, NY (1982).

[150] L. de la Peña, J.L. Jimenez and R. Montemayor, *The classical motion of an extended charged particle revisited*, Nuovo Cimento, **B 69**, 71 (1982).

[151] G.A. Perkins, *The direction of the zitterbewegung: A hidden variable*, Found. Phys. **6**, 237 (1976).

[152] A. Proca, *Mécanique du point*, J. Phys. Radium, **15**, 5 (1954).

[153] A. Proca, *Particules des trés grandes vitesses en mécanique spinorielle*, Nuovo Cimento **2**, 962 (1955).

[154] M.H.L. Pryce, *The mass-centre in the restricted theory of relativity and its connexion with the quantum theory of elementary particles*, Proc. Roy. Soc. Lond. **A 195**, 62 (1948).

[155] K. Rafanelli and R. Schiller, *Classical motions of spin-1/2 particles*, Phys. Rev. **135**, 279 (1964).

[156] E. Recami and G. Salesi, *Kinematics and Hydrodynamics of spinning particles*, Phys. Rev. **A 57**, 98 (1998).

[157] F. Riahi, *On Lagrangians with higher order derivatives*, Am. J. Phys. **40**, 386 (1972).

[158] O.W. Richardson, *A mechanical effect accompanying magnetization*, Phys. Rev. **26**, 248 (1908).

[159] F. Riewe, *Generalized mechanics of a spinning particle*, Lett. Nuovo Cimento, **1**, 807 (1971).

[160] F. Riewe, *Relativistic classical spinning-particle mechanics*, Nuovo Cimento, **B 8**, 271 (1972).

[161] M. Rivas, *Classical Particle Systems: I. Galilei free particles*, J. Phys. **A 18**, 1971 (1985).

[162] M. Rivas, *Classical Relativistic Spinning Particles*, J. Math. Phys. **30**, 318 (1989).

[163] M. Rivas, *Quantization of generalized spinning particles. New derivation of Dirac's equation*, J. Math. Phys. **35**, 3380 (1994).

[164] M. Rivas, *Is there a classical spin contribution to the tunnel effect?*, Phys. Lett. **A 248**, 279 (1998).

[165] M. Rivas, J.M. Aguirregabiria and A. Hernández, *A pure kinematical explanation of the gyromagnetic ratio $g = 2$ of leptons and charged bosons*, Phys. Lett. **A 257**, 21 (1999).

[166] M. Rivas, M. Valle and J.M. Aguirregabiria, *Composition law and contractions of the Poincaré group*, Eur. J. Phys. **6**, 128 (1986).

[167] T.R. Robinson, *Mass and charge distribution of the classical electron*, Phys. Lett. **A 200**, 335 (1989).

[168] L.M.C.S. Rodrigues and P.R. Rodrigues, *Further developments in generalized classical mechanics*, Am. J. Phys. **38**, 557 (1970).

[169] W.A. Rodrigues, J. Vaz and E. Recami, *A generalization of Dirac non-linear electrodynamics, and spinning charged particles*, Found. Phys. **23**, 469 (1993).

[170] F. Rohrlich, *Classical charged particles*, Addison-Wesley, Reading, Mass. (1965).

[171] F. Rohrlich, *The dynamics of a charged sphere and the electron*, Am. J. Phys. **65**, 1051 (1997).

[172] N. Rosen, *Particle spin and rotation*, Phys. Rev. **82**, 621 (1951).

[173] E.G.P. Rowe, *Rest frames for a point particle in special relativity*, Am. J. Phys. **64**, 1184 (1996).

[174] E.G.P. Rowe and G.T. Rowe, *The classical equations of motion for a spinning point particle with charge and magnetic moment*, Phys. Rep. **149**, 287 (1987).

[175] H. Rund, *The Hamilton-Jacobi theory in the calculus of variations*, Krieger Pub. Co., N.Y. (1973).

[176] L. Ryder, *Relativistic treatment of inertial spin effects*, J. Phys. **A 31**, 2465 (1998).

[177] J.J. Sakurai, *Advanced Quantum Mechanics*, Addison-Wesley, Reading, Mass. (1967).

[178] G. Salesi and E. Recami, *Effects of spin on the cyclotron frequency for a Dirac electron*, Phys. Lett. **A 267**, 219 (2000).

[179] A. Schild and J.A. Schlosser, *Fokker action principle for particles with charge, spin and magnetic moment*, J. Math. Phys. **6**, 1299 (1965).

[180] S.S. Schweber, *On Feynman quantization*, J. Math. Phys. **3**, 831 (1962).

[181] W. Shockley, *"Hidden linear momentum" related to the $\boldsymbol{\alpha} \cdot \boldsymbol{E}$ term for a Dirac-electron wave packet in an electric field*, Phys. Rev. Lett. **20**, 343 (1968).

[182] J. Souriau, *Structure des systèmes dynamiques*, Dunod, Paris, (1970).

[183] R.F. Streater and A.S. Wightman, *PCT, Spin & Statistics, and all that*, Benjamin, New York, (1964).

[184] V. Tapia, *Constrained generalized mechanics. The second order case*, Nuovo Cimento **B 90**, 15 (1985).

[185] L.H. Thomas, *The motion of the spinning electron*, Nature, **117**, 514 (1926).

[186] L.H. Thomas, *The kinematics of an electron with an axis*, Phil. Mag. **3**, 1 (1927).

[187] S. Tomonaga, *The Story of Spin*, University of Chicago Press, Chicago, (1997).

[188] G.E. Uhlenbeck and S.A. Goudsmit, *Spinning electrons and the structure of spectra*, Nature (London), **117**, 264 (1926).

[189] J. Vaz, *A spinning particle model including radiation reaction*, Phys. Lett. **B 345**, 448 (1995).

[190] J. Vaz and W.A. Rodrigues, *Zitterbewegung and the electromagnetic field of the electron*, Phys. Lett. **B 319**, 203 (1993).

[191] N. Ja. Vilenkin, *Fonctions spéciales et Théorie de la représentation des groups*, Dunod, Paris (1969).

[192] M. Visser, *A classical model for the electron*, Phys. Lett. **A 139**, 99 (1989).

[193] S. Weinberg, in *Lectures on Elementary Particles and Quantum Field Theory*, edited by S. Deser, M. Grisaru and H. Pendleton, MIT press, Cambridge, MA, (1970), p. 283.

[194] M.H.L. Weyssenhof, *Further contributions to the dynamics of spin-particles moving with a velocity smaller than that of light*, Acta Phys. Pol. **9**, 26 (1947).

[195] M.H.L. Weyssenhof, *Further contributions to the dynamics of spin-particles moving with the velocity of light*, Acta Phys. Pol. **9**, 34 (1947).

[196] M.H.L. Weyssenhof, *On two relativistic models of Dirac's electron*, Acta Phys. Pol. **9**, 46 (1947).

[197] M.H.L. Weyssenhof and A. Raabe, *Relativistic dynamics of spin-fluids and spin-particles*, Acta Phys. Pol. **9**, 7 (1947).

[198] M.H.L. Weyssenhof and A. Raabe, *Relativistic dynamics of spin-particles moving with the velocity of light*, Acta Phys. Pol. **9**, 19 (1947).

[199] A.S. Wightman, *On the localizability of quantum mechanical systems*, Rev. Mod. Phys. **34**, 845 (1962).

[200] E.T. Whittaker, *Analytical Dynamics*, Cambridge Univ. Press, Cambridge, (1927).

[201] E.P. Wigner, *On Unitary Representations of the Inhomogeneous Lorentz Group*, Ann. Math. **40**, 149, (1939).

[202] E.P. Wigner, *Group theory and its application to the quantum mechanics of atomic spectra*, Acad. Press, N.Y. (1959).

[203] K. Yee and M. Bander, *Equations of motion for spinning particles in an external electromagnetic and gravitational fields*, Phys. Rev. **D 48**, 2797 (1993).

# Index

## E



## F



## G



## H



## I



## K



## L

# M



# N



# O



# P



# Q



# R

# Fundamental Theories of Physics

*Series Editor: Alwyn van der Merwe, University of Denver, USA*

1. M. Sachs: *General Relativity and Matter*. A Spinor Field Theory from Fermis to Light-Years. With a Foreword by C. Kilmister. 1982 ISBN 90-277-1381-2
2. G.H. Duffey: *A Development of Quantum Mechanics*. Based on Symmetry Considerations. 1985 ISBN 90-277-1587-4
3. S. Diner, D. Fargue, G. Lochak and F. Selleri (eds.): *The Wave-Particle Dualism*. A Tribute to Louis de Broglie on his 90th Birthday. 1984 ISBN 90-277-1664-1
4. E. Prugovečki: *Stochastic Quantum Mechanics and Quantum Spacetime*. A Consistent Unification of Relativity and Quantum Theory based on Stochastic Spaces. 1984; 2nd printing 1986 ISBN 90-277-1617-X
5. D. Hestenes and G. Sobczyk: *Clifford Algebra to Geometric Calculus*. A Unified Language for Mathematics and Physics. 1984 ISBN 90-277-1673-0; Pb (1987) 90-277-2561-6
6. P. Exner: *Open Quantum Systems and Feynman Integrals*. 1985 ISBN 90-277-1678-1
7. L. Mayants: *The Enigma of Probability and Physics*. 1984 ISBN 90-277-1674-9
8. E. Tocaci: *Relativistic Mechanics, Time and Inertia*. Translated from Romanian. Edited and with a Foreword by C.W. Kilmister. 1985 ISBN 90-277-1769-9
9. B. Bertotti, F. de Felice and A. Pascolini (eds.): *General Relativity and Gravitation*. Proceedings of the 10th International Conference (Padova, Italy, 1983). 1984 ISBN 90-277-1819-9
10. G. Tarozzi and A. van der Merwe (eds.): *Open Questions in Quantum Physics*. 1985 ISBN 90-277-1853-9
11. J.V. Narlikar and T. Padmanabhan: *Gravity, Gauge Theories and Quantum Cosmology*. 1986 ISBN 90-277-1948-9
12. G.S. Asanov: *Finsler Geometry, Relativity and Gauge Theories*. 1985 ISBN 90-277-1960-8
13. K. Namsrai: *Nonlocal Quantum Field Theory and Stochastic Quantum Mechanics*. 1986 ISBN 90-277-2001-0
14. C. Ray Smith and W.T. Grandy, Jr. (eds.): *Maximum-Entropy and Bayesian Methods in Inverse Problems*. Proceedings of the 1st and 2nd International Workshop (Laramie, Wyoming, USA). 1985 ISBN 90-277-2074-6
15. D. Hestenes: *New Foundations for Classical Mechanics*. 1986 ISBN 90-277-2090-8; Pb (1987) 90-277-2526-8
16. S.J. Prokhovnik: *Light in Einstein's Universe*. The Role of Energy in Cosmology and Relativity. 1985 ISBN 90-277-2093-2
17. Y.S. Kim and M.E. Noz: *Theory and Applications of the Poincaré Group*. 1986 ISBN 90-277-2141-6
18. M. Sachs: *Quantum Mechanics from General Relativity*. An Approximation for a Theory of Inertia. 1986 ISBN 90-277-2247-1
19. W.T. Grandy, Jr.: *Foundations of Statistical Mechanics*. Vol. I: *Equilibrium Theory*. 1987 ISBN 90-277-2489-X
20. H.-H von Borzeszkowski and H.-J. Treder: *The Meaning of Quantum Gravity*. 1988 ISBN 90-277-2518-7
21. C. Ray Smith and G.J. Erickson (eds.): *Maximum-Entropy and Bayesian Spectral Analysis and Estimation Problems*. Proceedings of the 3rd International Workshop (Laramie, Wyoming, USA, 1983). 1987 ISBN 90-277-2579-9
22. A.O. Barut and A. van der Merwe (eds.): *Selected Scientific Papers of Alfred Landé*. [*1888-1975*]. 1988 ISBN 90-277-2594-2

# Fundamental Theories of Physics

23. W.T. Grandy, Jr.: *Foundations of Statistical Mechanics.* Vol. II: *Nonequilibrium Phenomena.* 1988 ISBN 90-277-2649-3
24. E.I. Bitsakis and C.A. Nicolaides (eds.): *The Concept of Probability.* Proceedings of the Delphi Conference (Delphi, Greece, 1987). 1989 ISBN 90-277-2679-5
25. A. van der Merwe, F. Selleri and G. Tarozzi (eds.): *Microphysical Reality and Quantum Formalism, Vol. 1.* Proceedings of the International Conference (Urbino, Italy, 1985). 1988 ISBN 90-277-2683-3
26. A. van der Merwe, F. Selleri and G. Tarozzi (eds.): *Microphysical Reality and Quantum Formalism, Vol. 2.* Proceedings of the International Conference (Urbino, Italy, 1985). 1988 ISBN 90-277-2684-1
27. I.D. Novikov and V.P. Frolov: *Physics of Black Holes.* 1989 ISBN 90-277-2685-X
28. G. Tarozzi and A. van der Merwe (eds.): *The Nature of Quantum Paradoxes.* Italian Studies in the Foundations and Philosophy of Modern Physics. 1988 ISBN 90-277-2703-1
29. B.R. Iyer, N. Mukunda and C.V. Vishveshwara (eds.): *Gravitation, Gauge Theories and the Early Universe.* 1989 ISBN 90-277-2710-4
30. H. Mark and L. Wood (eds.): *Energy in Physics, War and Peace.* A Festschrift celebrating Edward Teller's 80th Birthday. 1988 ISBN 90-277-2775-9
31. G.J. Erickson and C.R. Smith (eds.): *Maximum-Entropy and Bayesian Methods in Science and Engineering.* Vol. I: *Foundations.* 1988 ISBN 90-277-2793-7
32. G.J. Erickson and C.R. Smith (eds.): *Maximum-Entropy and Bayesian Methods in Science and Engineering.* Vol. II: *Applications.* 1988 ISBN 90-277-2794-5
33. M.E. Noz and Y.S. Kim (eds.): *Special Relativity and Quantum Theory.* A Collection of Papers on the Poincaré Group. 1988 ISBN 90-277-2799-6
34. I.Yu. Kobzarev and Yu.I. Manin: *Elementary Particles. Mathematics, Physics and Philosophy.* 1989 ISBN 0-7923-0098-X
35. F. Selleri: *Quantum Paradoxes and Physical Reality.* 1990 ISBN 0-7923-0253-2
36. J. Skilling (ed.): *Maximum-Entropy and Bayesian Methods.* Proceedings of the 8th International Workshop (Cambridge, UK, 1988). 1989 ISBN 0-7923-0224-9
37. M. Kafatos (ed.): *Bell's Theorem, Quantum Theory and Conceptions of the Universe.* 1989 ISBN 0-7923-0496-9
38. Yu.A. Izyumov and V.N. Syromyatnikov: *Phase Transitions and Crystal Symmetry.* 1990 ISBN 0-7923-0542-6
39. P.F. Fougère (ed.): *Maximum-Entropy and Bayesian Methods.* Proceedings of the 9th International Workshop (Dartmouth, Massachusetts, USA, 1989). 1990 ISBN 0-7923-0928-6
40. L. de Broglie: *Heisenberg's Uncertainties and the Probabilistic Interpretation of Wave Mechanics.* With Critical Notes of the Author. 1990 ISBN 0-7923-0929-4
41. W.T. Grandy, Jr.: *Relativistic Quantum Mechanics of Leptons and Fields.* 1991 ISBN 0-7923-1049-7
42. Yu.L. Klimontovich: *Turbulent Motion and the Structure of Chaos.* A New Approach to the Statistical Theory of Open Systems. 1991 ISBN 0-7923-1114-0
43. W.T. Grandy, Jr. and L.H. Schick (eds.): *Maximum-Entropy and Bayesian Methods.* Proceedings of the 10th International Workshop (Laramie, Wyoming, USA, 1990). 1991 ISBN 0-7923-1140-X
44. P. Pták and S. Pulmannová: *Orthomodular Structures as Quantum Logics.* Intrinsic Properties, State Space and Probabilistic Topics. 1991 ISBN 0-7923-1207-4
45. D. Hestenes and A. Weingartshofer (eds.): *The Electron.* New Theory and Experiment. 1991 ISBN 0-7923-1356-9

# Fundamental Theories of Physics

46. P.P.J.M. Schram: *Kinetic Theory of Gases and Plasmas.* 1991 ISBN 0-7923-1392-5
47. A. Micali, R. Boudet and J. Helmstetter (eds.): *Clifford Algebras and their Applications in Mathematical Physics.* 1992 ISBN 0-7923-1623-1
48. E. Prugovečki: *Quantum Geometry.* A Framework for Quantum General Relativity. 1992 ISBN 0-7923-1640-1
49. M.H. Mac Gregor: *The Enigmatic Electron.* 1992 ISBN 0-7923-1982-6
50. C.R. Smith, G.J. Erickson and P.O. Neudorfer (eds.): *Maximum Entropy and Bayesian Methods.* Proceedings of the 11th International Workshop (Seattle, 1991). 1993 ISBN 0-7923-2031-X
51. D.J. Hoekzema: *The Quantum Labyrinth.* 1993 ISBN 0-7923-2066-2
52. Z. Oziewicz, B. Jancewicz and A. Borowiec (eds.): *Spinors, Twistors, Clifford Algebras and Quantum Deformations.* Proceedings of the Second Max Born Symposium (Wrocław, Poland, 1992). 1993 ISBN 0-7923-2251-7
53. A. Mohammad-Djafari and G. Demoment (eds.): *Maximum Entropy and Bayesian Methods.* Proceedings of the 12th International Workshop (Paris, France, 1992). 1993 ISBN 0-7923-2280-0
54. M. Riesz: *Clifford Numbers and Spinors* with Riesz' Private Lectures to E. Folke Bolinder and a Historical Review by Pertti Lounesto. E.F. Bolinder and P. Lounesto (eds.). 1993 ISBN 0-7923-2299-1
55. F. Brackx, R. Delanghe and H. Serras (eds.): *Clifford Algebras and their Applications in Mathematical Physics.* Proceedings of the Third Conference (Deinze, 1993) 1993 ISBN 0-7923-2347-5
56. J.R. Fanchi: *Parametrized Relativistic Quantum Theory.* 1993 ISBN 0-7923-2376-9
57. A. Peres: *Quantum Theory: Concepts and Methods.* 1993 ISBN 0-7923-2549-4
58. P.L. Antonelli, R.S. Ingarden and M. Matsumoto: *The Theory of Sprays and Finsler Spaces with Applications in Physics and Biology.* 1993 ISBN 0-7923-2577-X
59. R. Miron and M. Anastasiei: *The Geometry of Lagrange Spaces: Theory and Applications.* 1994 ISBN 0-7923-2591-5
60. G. Adomian: *Solving Frontier Problems of Physics: The Decomposition Method.* 1994 ISBN 0-7923-2644-X
61. B.S. Kerner and V.V. Osipov: *Autosolitons.* A New Approach to Problems of Self-Organization and Turbulence. 1994 ISBN 0-7923-2816-7
62. G.R. Heidbreder (ed.): *Maximum Entropy and Bayesian Methods.* Proceedings of the 13th International Workshop (Santa Barbara, USA, 1993) 1996 ISBN 0-7923-2851-5
63. J. Peřina, Z. Hradil and B. Jurčo: *Quantum Optics and Fundamentals of Physics.* 1994 ISBN 0-7923-3000-5
64. M. Evans and J.-P. Vigier: *The Enigmatic Photon.* Volume 1: The Field $B^{(3)}$. 1994 ISBN 0-7923-3049-8
65. C.K. Raju: *Time: Towards a Constistent Theory.* 1994 ISBN 0-7923-3103-6
66. A.K.T. Assis: *Weber's Electrodynamics.* 1994 ISBN 0-7923-3137-0
67. Yu. L. Klimontovich: *Statistical Theory of Open Systems.* Volume 1: A Unified Approach to Kinetic Description of Processes in Active Systems. 1995 ISBN 0-7923-3199-0; Pb: ISBN 0-7923-3242-3
68. M. Evans and J.-P. Vigier: *The Enigmatic Photon.* Volume 2: Non-Abelian Electrodynamics. 1995 ISBN 0-7923-3288-1
69. G. Esposito: *Complex General Relativity.* 1995 ISBN 0-7923-3340-3

# Fundamental Theories of Physics

70. J. Skilling and S. Sibisi (eds.): *Maximum Entropy and Bayesian Methods.* Proceedings of the Fourteenth International Workshop on Maximum Entropy and Bayesian Methods. 1996 ISBN 0-7923-3452-3
71. C. Garola and A. Rossi (eds.): *The Foundations of Quantum Mechanics Historical Analysis and Open Questions.* 1995 ISBN 0-7923-3480-9
72. A. Peres: *Quantum Theory: Concepts and Methods.* 1995 (see for hardback edition, Vol. 57) ISBN Pb 0-7923-3632-1
73. M. Ferrero and A. van der Merwe (eds.): *Fundamental Problems in Quantum Physics.* 1995 ISBN 0-7923-3670-4
74. F.E. Schroeck, Jr.: *Quantum Mechanics on Phase Space.* 1996 ISBN 0-7923-3794-8
75. L. de la Peña and A.M. Cetto: *The Quantum Dice.* An Introduction to Stochastic Electrodynamics. 1996 ISBN 0-7923-3818-9
76. P.L. Antonelli and R. Miron (eds.): *Lagrange and Finsler Geometry.* Applications to Physics and Biology. 1996 ISBN 0-7923-3873-1
77. M.W. Evans, J.-P. Vigier, S. Roy and S. Jeffers: *The Enigmatic Photon.* Volume 3: Theory and Practice of the $B^{(3)}$ Field. 1996 ISBN 0-7923-4044-2
78. W.G.V. Rosser: *Interpretation of Classical Electromagnetism.* 1996 ISBN 0-7923-4187-2
79. K.M. Hanson and R.N. Silver (eds.): *Maximum Entropy and Bayesian Methods.* 1996 ISBN 0-7923-4311-5
80. S. Jeffers, S. Roy, J.-P. Vigier and G. Hunter (eds.): *The Present Status of the Quantum Theory of Light.* Proceedings of a Symposium in Honour of Jean-Pierre Vigier. 1997 ISBN 0-7923-4337-9
81. M. Ferrero and A. van der Merwe (eds.): *New Developments on Fundamental Problems in Quantum Physics.* 1997 ISBN 0-7923-4374-3
82. R. Miron: *The Geometry of Higher-Order Lagrange Spaces.* Applications to Mechanics and Physics. 1997 ISBN 0-7923-4393-X
83. T. Hakioğlu and A.S. Shumovsky (eds.): *Quantum Optics and the Spectroscopy of Solids.* Concepts and Advances. 1997 ISBN 0-7923-4414-6
84. A. Sitenko and V. Tartakovskii: *Theory of Nucleus.* Nuclear Structure and Nuclear Interaction. 1997 ISBN 0-7923-4423-5
85. G. Esposito, A.Yu. Kamenshchik and G. Pollifrone: *Euclidean Quantum Gravity on Manifolds with Boundary.* 1997 ISBN 0-7923-4472-3
86. R.S. Ingarden, A. Kossakowski and M. Ohya: *Information Dynamics and Open Systems.* Classical and Quantum Approach. 1997 ISBN 0-7923-4473-1
87. K. Nakamura: *Quantum versus Chaos.* Questions Emerging from Mesoscopic Cosmos. 1997 ISBN 0-7923-4557-6
88. B.R. Iyer and C.V. Vishveshwara (eds.): *Geometry, Fields and Cosmology.* Techniques and Applications. 1997 ISBN 0-7923-4725-0
89. G.A. Martynov: *Classical Statistical Mechanics.* 1997 ISBN 0-7923-4774-9
90. M.W. Evans, J.-P. Vigier, S. Roy and G. Hunter (eds.): *The Enigmatic Photon.* Volume 4: New Directions. 1998 ISBN 0-7923-4826-5
91. M. Rédei: *Quantum Logic in Algebraic Approach.* 1998 ISBN 0-7923-4903-2
92. S. Roy: *Statistical Geometry and Applications to Microphysics and Cosmology.* 1998 ISBN 0-7923-4907-5
93. B.C. Eu: *Nonequilibrium Statistical Mechanics.* Ensembled Method. 1998 ISBN 0-7923-4980-6

# Fundamental Theories of Physics

94. V. Dietrich, K. Habetha and G. Jank (eds.): *Clifford Algebras and Their Application in Mathematical Physics*. Aachen 1996. 1998 ISBN 0-7923-5037-5
95. J.P. Blaizot, X. Campi and M. Ploszajczak (eds.): *Nuclear Matter in Different Phases and Transitions*. 1999 ISBN 0-7923-5660-8
96. V.P. Frolov and I.D. Novikov: *Black Hole Physics*. Basic Concepts and New Developments. 1998 ISBN 0-7923-5145-2; PB 0-7923-5146
97. G. Hunter, S. Jeffers and J-P. Vigier (eds.): *Causality and Locality in Modern Physics*. 1998 ISBN 0-7923-5227-0
98. G.J. Erickson, J.T. Rychert and C.R. Smith (eds.): *Maximum Entropy and Bayesian Methods*. 1998 ISBN 0-7923-5047-2
99. D. Hestenes: *New Foundations for Classical Mechanics (Second Edition)*. 1999 ISBN 0-7923-5302-1; PB ISBN 0-7923-5514-8
100. B.R. Iyer and B. Bhawal (eds.): *Black Holes, Gravitational Radiation and the Universe*. Essays in Honor of C. V. Vishveshwara. 1999 ISBN 0-7923-5308-0
101. P.L. Antonelli and T.J. Zastawniak: *Fundamentals of Finslerian Diffusion with Applications*. 1998 ISBN 0-7923-5511-3
102. H. Atmanspacher, A. Amann and U. Müller-Herold: *On Quanta, Mind and Matter Hans Primas in Context*. 1999 ISBN 0-7923-5696-9
103. M.A. Trump and W.C. Schieve: *Classical Relativistic Many-Body Dynamics*. 1999 ISBN 0-7923-5737-X
104. A.I. Maimistov and A.M. Basharov: *Nonlinear Optical Waves*. 1999 ISBN 0-7923-5752-3
105. W. von der Linden, V. Dose, R. Fischer and R. Preuss (eds.): *Maximum Entropy and Bayesian Methods Garching, Germany 1998*. 1999 ISBN 0-7923-5766-3
106. M.W. Evans: *The Enigmatic Photon Volume 5: O(3) Electrodynamics*. 1999 ISBN 0-7923-5792-2
107. G.N. Afanasiev: *Topological Effects in Quantum Mecvhanics*. 1999 ISBN 0-7923-5800-7
108. V. Devanathan: *Angular Momentum Techniques in Quantum Mechanics*. 1999 ISBN 0-7923-5866-X
109. P.L. Antonelli (ed.): *Finslerian Geometries A Meeting of Minds*. 1999 ISBN 0-7923-6115-6
110. M.B. Mensky: *Quantum Measurements and Decoherence Models and Phenomenology*. 2000 ISBN 0-7923-6227-6
111. B. Coecke, D. Moore and A. Wilce (eds.): *Current Research in Operation Quantum Logic*. Algebras, Categories, Languages. 2000 ISBN 0-7923-6258-6
112. G. Jumarie: *Maximum Entropy, Information Without Probability and Complex Fractals*. Classical and Quantum Approach. 2000 ISBN 0-7923-6330-2
113. B. Fain: *Irreversibilities in Quantum Mechanics*. 2000 ISBN 0-7923-6581-X
114. T. Borne, G. Lochak and H. Stumpf: *Nonperturbative Quantum Field Theory and the Structure of Matter*. 2001 ISBN 0-7923-6803-7
115. J. Keller: *Theory of the Electron*. A Theory of Matter from START. 2001 ISBN 0-7923-6819-3
116. M. Rivas: *Kinematical Theory of Spinning Particles*. Classical and Quantum Mechanical Formalism of Elementary Particles. 2001 ISBN 0-7923-6824-X

KLUWER ACADEMIC PUBLISHERS – DORDRECHT / BOSTON / LONDON